STRUCTURAL THEORY
AND ANALYSIS

Other engineering titles from Macmillan

Malcolm Bolton: *A Guide to Soil Mechanics*
J. G. A. Croll and A. C. Walker: *Elements of Structural Stability*
R. T. Fenner: *Computing for Engineers*
J. A. Fox: *An Introduction to Engineering Fluid Mechanics*, Second Edition
N. Jackson: *Civil Engineering Materials*, Second Edition
W. H. Mosley and J. H. Bungey: *Reinforced Concrete Design*
Stuart S. J. Moy: *Plastic Methods for Steel and Concrete Structures*
J. Uren and W. F. Price: *Surveying for Engineers*
E. M. Wilson: *Engineering Hydrology*, Second Edition

Structural Theory and Analysis

J. D. Todd

Fellow of St. Edmund Hall and
University Lecturer in Engineering Science
University of Oxford

Second Edition

MACMILLAN

First edition 1974
Reprinted 1977 (with corrections), 1979
Second edition 1981
Reprinted 1983, 1984

Published by
Higher and Further Education Division
MACMILLAN PUBLISHERS LTD
London and Basingstoke
Companies and representatives
throughout the world

Typeset in Great Britain
by Reproduction Drawings Ltd, Sutton, Surrey

Printed in Hong Kong

ISBN 0 333 32311 4 (hardcover)
ISBN 0 333 32312 2 (paperback)

CONTENTS

PREFACE

This text has been written primarily for the benefit of undergraduates and I suppose that the preface should be directed towards enlightening the reader about the contents and the reasons for writing a book. I have often wondered if students read a preface; they have in all probability been directed to a particular text by their teachers in the subject, and are far more concerned about the number of worked examples or whether the book appears to be too difficult or too easy for their tastes. However there are a few remarks that I should like to make.

I have been concerned for a number of years with lecturing and demonstrating topics in both structures and elasticity. In addition I have been extremely fortunate in teaching undergraduates in pairs using a tutorial system. This has I hope enabled me to discover the areas of structures that appear to be difficult and that are not sufficiently well covered by existing texts. I am also very conscious of the fundamental errors that are made year after year by successive generations of students. I hope that I shall perhaps eliminate some of these difficulties and also point out the common pitfalls.

The book is intended to serve a student as a basic text in structures. It is not possible to cover all the material required for the various degree courses in a single volume and, by the time the final year of a course is reached, books of a more specialised nature will no doubt be required. The first two chapters are concerned with statically determinate structures; it is still essential to have a firm grasp and feel for this subject, as it is one of the foundations of structures. The other main foundation is elasticity; it would therefore not be possible to make a book on structures self contained without including some elements of elasticity. In three chapters I have attempted to cover all the basic elasticity theory, but I have not dwelt on the more mathematical aspects. An attempt has been made to use a consistent sign-convention throughout the text.

The emphasis in structural teaching has been changing over the past few years, particularly as computers have become freely available. It is however still necessary to have a real understanding of basic structural theory. In chapter 6 the fundamental ideas of stiffness and flexibility are introduced. Also included are examples using work and energy methods in the solution of redundant structures and in the determination of displacements. A separate chapter covers the elements of moment distribution and slope deflection—these are still important methods of analysis that can give a rapid solution if a structure is not too complicated. A

knowledge of matrices is required in chapter 8—it is assumed that all engineers
now receive instruction in this subject as part of their mathematics course.
This chapter discusses stiffness and flexibility methods, and although these
are often covered by a book devoted entirely to the subject, I have attempted
to provide sufficient information in one chapter working entirely from the member
approach. It may possibly be a little indigestible at a first reading, but the reader
will have secured sufficient basic knowledge from the previous chapters and with a
little perseverance will win through, particularly as there are several examples
worked out in detail using both stiffness and flexibility.

The final three chapters discuss topics that I believe should be covered
by a fundamental text. Unfortunately it has not been possible when discussing
stability to cover more than the behaviour of a strut in isolation. The last chapter
does I hope provide an adequate introduction to plastic collapse so that more
definitive works can be read without difficulty.

SI units have been used when any numerical values have been required. Problems for solution are provided at the end of each chapter; a number of these have
been taken from Oxford examination papers, by permission of the Clarendon
Press, Oxford, but I am responsible for the answers provided.

There can of course be no original material in a book of this nature and
I am indebted to the many different sources that have been used over the
years in forming my ideas in the presentation of the subject matter.

J. D. TODD

Preface to the second edition

The opportunity has been taken to add extra explanation to improve the clarity
of some subject matter. There are some additions to chapter 3 mainly concerned
with members that are loaded such that the yield stress is exceeded. The slope-
deflection equations in chapter 7 have been written in matrix notation. One of
the main changes is in the first part of chapter 8; it was found that the approach
used in the first edition was a little difficult for a student meeting the applications of stiffness and flexibility for the first time. The new approach is very much
simpler, starting with the derivation of the stiffness matrix for a helical spring.
In chapter 10 stability is now introduced from a potential energy approach, this
should enable the reader to gain a much better understanding and feel for the
subject.

1 PLANE STATICS

1.1 Introduction

In the first place we need to define what is meant by the term *structure*. The reader may well have a mental picture of a bridge, dam or large building; any of these would most certainly be classed as a structure, but they severely limit the use of a term which can be applied in a much wider sense. A car, aeroplane, milk crate or even the chair that you are sitting on are all structures. They are all designed to carry a particular form of applied loading and, what is most important, in the majority of cases only deform by a small amount when the loading is applied. All engineering structures or structural elements are subject to external forces or loads. These will induce other external forces or reactions at the points of support of the structure. Consider the chair: it is designed to carry an external load, provided by your own mass; forces are transmitted through the various members of the chair and the reactions act at the ends of the legs. In the case of an aeroplane in level flight, there must be a vertical component of reaction acting on the wings as a lift force to balance the mass of the plane. The structure of the wing is so designed that it is capable of transmitting this force to the fuselage.

Certain types of structures, for example a dam, rely partly on their large mass to resist the applied forces; we shall not concern ourselves with these types of structure in this text. If a structure is stationary or moving with a constant velocity, the resultant of all the applied loads and reactions will be zero, and the structure is said to be in equilibrium. If however the velocity is not constant it is necessary to consider inertia forces in addition to the applied forces. We shall confine our attention to structures that are in static equilibrium. Very often it will be found that we have to simplify and idealise a structure in order to obtain a theoretical solution that will give us a fairly good idea of the loading in individual members. When the member loads have been deduced we are able to proceed with the design of each member in turn.

The loads that are applied to a structure can be divided into two categories, dead and live loads. The dead loading arises solely from the mass of the structure itself, and it is possible to calculate this to a fairly high degree of accuracy. The live loading may arise from a variety of causes and can often only be estimated. Examples of live loads are moving loads, wind and snow loading, impact and earthquake shocks. In addition to these loadings, members of a structure can have their loading altered by changes in geometry, movements of supports and changes of temperature.

1

1.2 Equations of equilibrium

A system of forces is said to be in equilibrium when the resultant of all the forces
and the resultant of all the moments at one point are equal to zero. For a three-
dimensional system with a set of mutually perpendicular axes, x, y and z, six
conditions must be satisfied. These six conditions can be stated in mathematical
terms as follows:

$$\sum P_x = \sum P_y = \sum P_z = 0 \qquad (1.1)$$

$$\sum M_x = \sum M_y = \sum M_z = 0 \qquad (1.2)$$

where P_x is the component of any force in the x direction and M_x is the moment
of a force P about the x axis.

The equations may be written in vector notation as

$$R = R_x i + R_y j + R_z k = 0 \qquad (1.3)$$

$$M_R = M_{Rx} i + M_{Ry} j + M_{Rz} k = 0 \qquad (1.4)$$

where R is the resultant of the system of forces and M_R is the resultant moment.
It is obvious that $R_x = \sum P_x$ and $M_{Rx} = \sum M_x$, etc.

In fact it is not necessary for the axes to be orthogonal—any three axes can be
chosen for the summation of the forces, and any three axes for the summation of
the moments. The case of parallel axes would of course be excepted.

If the system is coplanar, that is say $P_z = 0$ and $M_x = M_y = 0$, then there will be
only three conditions of equilibrium

$$\sum P_x = \sum P_y = 0 \qquad (1.5)$$

$$\sum M_z = 0 \qquad (1.6)$$

Several special cases arise for a coplanar system

(1) If there are only two forces acting on a body that is in equilibrium,
then the forces must be equal and opposite.

(2) If there are only three forces acting on a body that is in equilibrium,
then the three forces must be concurrent.

(3) A set of coplanar forces not in equilibrium can be reduced to a single
resultant force or resultant moment.

1.3 Supports, reactions and free-body diagrams

The loads applied to a structure must be transmitted via the various members of
the structure to a number of specified points or supports. The resulting forces and
moments at these points or supports are called reactions. The supports can be
classified into three different types—roller, pinned or fixed. It must be emphasised
that these are all ideal support conditions, that is no friction exists in the roller or
pin, and there is no movement or rotation in the fixed support. In practice conditions may well deviate from these ideal situations.

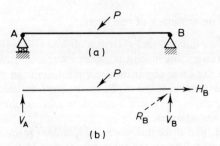

Figure 1.1

The simple beam in figure 1.1a has an inclined force P applied to it. At end A there is a roller support—this will allow rotation or horizontal displacement to take place quite freely—thus there can only be a vertical reaction at A. End B has a pinned support, vertical and horizontal displacement is therefore prevented at this point, but rotation is allowed to take place freely. The reaction at B will be inclined at an angle to the vertical, but may be resolved into two components, one horizontal and the other vertical.

If a line drawing of the beam is now made and the equivalent reactions are applied (figure 1.1b) we have what is termed a *free-body diagram*. V_A and V_B represent the vertical components of reaction at A and B respectively and H_B the horizontal component of the reaction at B. However it would have been possible for V_B and H_B to have been combined and replaced by the single inclined reaction R_B. A horizontal reaction H_A could have been added to the diagram, but this is of course equal to zero.

At this stage it might be as well to state that it is always essential to draw a free-body diagram for a structure. Care must be taken to ensure that all the applied loads and reactions have been shown on the diagram.

In figure 1.2a, the same beam is shown with a fixed support at B, this is some-times called an *encastré* or *built-in support*. In addition to the vertical and hori-zontal components of reaction V_B and H_B, a further reaction or moment M_B is required at B. If the support is rigid this moment will prevent any rotation from taking place. The free-body diagram is shown in figure 1.2b.

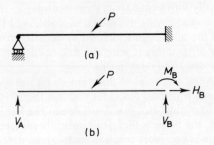

Figure 1.2

1.4 Stability and determinacy of reactions

If we consider a two-dimensional system with known applied loads, the free-body diagram can be drawn showing all the unknown reactions. With the system in static equilibrium the three conditions of equilibrium can be applied, and will result in three equations in terms of the unknown reactions. The equations can then be solved simultaneously. For a complete solution there will in general be a limitation of three unknowns.

If there are less than three unknown independent reactions, there will not be sufficient unknowns to satisfy the three equations and the system will not be in equilibrium. It is then termed *statically unstable* so far as the external supports are concerned.

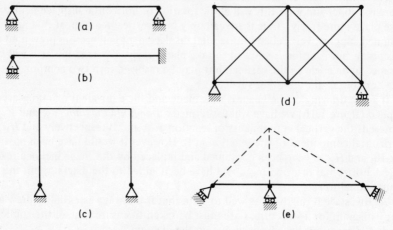

Figure 1.3

If there are more than three unknowns, the equations cannot be completely solved and the system will be statically indeterminate or redundant. For example if there were five unknowns, two of them could be assigned any value, the remaining three could then be found from the equations of equilibrium and would be entirely dependent on the values chosen for the first two reactions. This does not mean that a statically indeterminate system is insoluble, the name implies that the system cannot be solved by the use of statics alone. Additional information will be required about the manner in which the system deforms under the applied loading. This type of structure is generally described as *redundant* and methods of solution will be discussed later in this book.

In figure 1.3 several examples of different structures are shown. The beam, a, has only two unknown vertical reactions and will therefore be an unstable system. The beam, b, has four unknown reactions, one at the left-hand end and three at the right-hand end, the beam is therefore statically indeterminate to the first degree, that is there is one more reaction than can be determined by statics alone. The portal frame, c, is also statically indeterminate to the first degree, as there will be a vertical and horizontal reaction at each support. The frame, d, is statically determinate for reactions, but it will be seen later that it is statically indeterminate so far as the forces in the members are concerned. It should be noted that in all cases the question of statical determinacy is independent of the loading applied to the system. Reverting to system a: if vertical external forces are applied it might be tempting to say that there would only be vertical components of reaction at each support, the support system shown is quite capable of sustaining these, hence the system is stable. However an instability arises if a horizontal force, however, small, is applied. It can be seen that the question of stability and determinacy of the reactions must be independent of the actual loading that is applied to the structure. In effect it would be no use designing a structure to withstand only vertical loads, if it collapsed when a small lateral wind load was applied. In e we have a case where the lines of action of three supports, shown dotted, all pass through a single point. Any loading system that is applied to the beam would have a moment about the point of intersection and this cannot be resisted by the reactions, hence the beam would tend to rotate about the point. Based on this fact a general statement can be made: the reactions must be capable of resisting any small displacement or rotation that is applied to the structure.

In certain cases the number of unknown reactions can be reduced by what is termed an *equation of condition*. This will be introduced in the examples which follow and will be discussed at greater length in section 1.6.

1.5 Calculation of reactions

This can probably best be illustrated by one or two examples, but before dealing with specific cases we ought to consider the different types of loading that may be applied. A concentrated load is assumed to be acting at a point; in actual fact this is an impossibility as there would have to be an infinite stress in the member at the point directly under the load. In practice the material under the load will deform and the load will then be spread over a small area hence reducing the stress

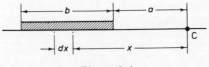

Figure 1.4

concentration. However, for the purpose of calculation it will be sufficiently accurate to assume that the load is acting at a point. Another very common type of load is referred to as a uniformly distributed load (U.D.L.). As the name suggests, the load is distributed along the surface of the member with a constant value per unit length, or per unit area.

Assume that the moment about point C is required for the uniformly distributed load p per unit length, extending over length b, see figure 1.4. Consider an element dx of the load distant x from C. The moment of this 'concentrated' load about C is $px\,dx$. The total moment is found by integrating this expression

$$M_C = \int_a^{a+b} px\,dx = pb\left(a + \frac{b}{2}\right) \qquad (1.7)$$

It is clear from equation 1.7 that the moment can be found by replacing the U.D.L. by a point load, equal in magnitude to its total value, and acting at its centroid.

If the load was not uniformly distributed but was some function of x, the moment could be found in a similar manner. The load would be replaced by a concentrated load equal to the total distributed load and acting at the centroid of the load system.

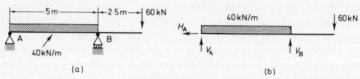

(a)　　　　　　　　　　　　　　　(b)

Figure 1.5

As a numerical example consider the beam in figure 1.5a. The free-body diagram is first drawn as in b. By horizontal resolution it is seen at once that H_A is zero.

Take moments about A for the complete system

$$(40 \times 5 \times 2{\cdot}5) + 60(5 + 2{\cdot}5) = 5V_B$$

$$V_B = 190 \text{ kN}$$

Resolving vertically

$$V_A + V_B = 200 + 60$$

therefore

$$V_A = 70 \text{ kN}$$

Often it is possible to obtain a check on the work by taking moments about some other suitable point, for example ΣM_B

$$(60 \times 2{\cdot}5) = (40 \times 5 \times 2{\cdot}5) - 5V_A$$

therefore

$$V_A = 70 \text{ kN}$$

This last equation is equivalent to saying that the sum of the moments about B is zero. This is true so long as there is no externally applied moment at B.

In figure 1.6a the beam passes over three supports and the portion AD is connected to DC by a pin. From the free-body diagram (figure 1.6b) it would at first appear that the problem is statically indeterminate as there are four unknown reactions. However, as there is a pin at D, the bending moment there will be zero and so we have an additional piece of information. This is known as an equation of condition.

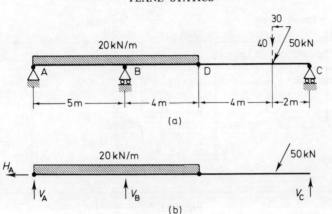

Figure 1.6

Resolving forces horizontally

$$H_A = -30 \text{ kN}$$

The negative sign indicates that H_A is in the opposite direction to that assumed in the diagram. ΣM_D for DC gives

$$6V_C = 40 \times 4$$

therefore

$$V_C = 26.7 \text{ kN}$$

Note when taking moments at the pin, only the moments are included for the forces on the beam DC.

ΣM_A gives

$$(20 \times 9^2)/2 + (40 \times 13) = 5V_B + (26 \cdot 7 \times 15)$$

$$V_B = 186 \text{ kN}$$

Resolving vertically

$$V_A = 220 - 26 \cdot 7 - 186 = 7 \cdot 3 \text{ kN}$$

A check may be obtained by taking moments about B.

A further example involving the use of an equation of condition is the three-pin arch (figure 1.7a). The moment at the pin E will be zero. Then

ΣM_E for portion ECD

$$6H_D = 3V_D$$

ΣM_E for portion ABE

$$3V_A = 4H_A + (30 \times 1)$$

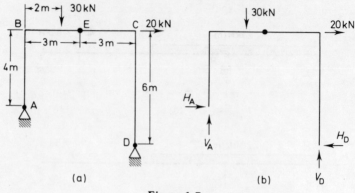

Figure 1.7

Resolving vertically

$$V_A + V_D = 30$$

Resolving horizontally

$$H_A + 20 = H_D$$

Thus there are four equations in terms of the four unknowns. Solving these we obtain

$$V_A = 2 \text{ kN} \quad H_A = -6 \text{ kN} \quad V_D = 28 \text{ kN} \quad H_D = 14 \text{ kN}$$

The negative sign for H_A indicates that the direction of action of H_A is opposite to that shown in figure 1.7b.

1.6 Equation of condition

This has already been introduced in some of the problems solved in the last section. It can be seen that an equation of condition can reduce the degree of indeterminacy of a structure.

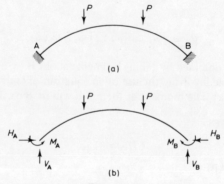

Figure 1.8

The symmetrical arch rib, figure 1.8a, is built-in at both ends and has loads P symmetrically applied. From the free-body diagram, figure 1.8b, it can be seen that there are six unknown reactions. Apparently the arch is indeterminate to the third degree. However if use is made of symmetry, $H_A = H_B$, $M_A = M_B$ and $V_A = V_B$, it can be shown by vertical resolution that the values of V_A and V_B are each equal to P. Thus the arch will only be indeterminate to the second degree. The beam in figure 1.9a has two equal spans and the loading is symmetrical. It can be seen from the free-body diagram figure 1.9b that the problem is indeterminate to the

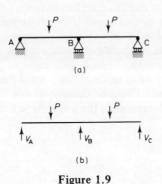

(a)

(b)

Figure 1.9

first degree. There is however an equation of condition which gives $V_A = V_C$ and it is tempting to say that the problem is statically determinate. However the equation of condition will not in this case reduce the degree of indeterminacy. All that the equation states is that the sum of the moments about the centre is zero—this is of course to be expected as there is no external moment applied at the centre support.

It should be emphasised that in general an equation of condition of this type can only be used if there is complete symmetry both for the loading and for the structure. In the example given, both of the beams would have to be of the same length, same cross-section, and made of the same material, for complete symmetry.

1.7 Principle of virtual work

This is an extremely important principle which will have a number of important applications in the field of structural engineering.

If a force is given a displacement in a particular direction, the work done is given by the dot product of the force and the displacement, that is, work is only done when the force and the displacement are in the same direction. Thus it is necessary to take the resolved part of the displacement in the direction of the force. Now consider a small particle acted on by a number of forces P_1, P_2, P_n. The forces are all concurrent and have a resultant P_R. Suppose that the particle is now displaced through a small distance Δ in a particular direction. If the forces do not change magnitude the work done is

$$P_1 . \Delta + P_2 . \Delta + \cdots + P_n . \Delta = P_R . \Delta$$

Had the forces been in equilibrium, the resultant force P_R would have been zero, hence

$$P_1 . \varDelta + P_2 . \varDelta + \cdots + P_n . \varDelta = 0 \qquad (1.8)$$

This is the principle of virtual work which states: *If a particle that is in equilibrium is given a displacement in any direction, then the total work done by the forces acting on the particle is zero.* The only requirement is that the forces must stay constant during the displacement. They can be generalised forces, that is, they need not be linear forces but could be moments or torques. The displacements could also be linear or rotational. It is sometimes referred to as the principle of virtual displacements, since it is not necessary for the displacement to take place—it could be imaginary. We are in fact stating the principle of equilibrium in a slightly different way.

The principle can easily be extended from a small particle to a structure that is in equilibrium. The structure could be thought of as a large number of particles each of which is in equilibrium—thus the principle will still apply.

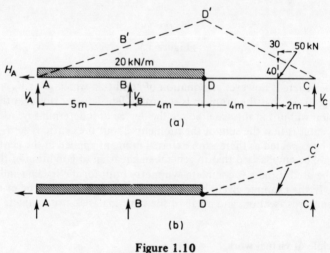

Figure 1.10

We shall now make use of this principle to find the reaction V_B in the problem already solved in figure 1.6. The beam has been redrawn (figure 1.10a). As the reaction at B is required a virtual displacement is applied at point B in the direction of the required reaction that is, vertically. The beam will take up the deformed position as shown in figure 1.10a, all displacements being greatly exaggerated. B has moved to a position B′ while the beam pivots about the end points A and C and hinges about D′.

If the vertical displacment BB′ is v, the centroid of the U.D.L. will move vertically through $9/10v$ and the point of application of the 50 kN load has moved vertically through $3/5v$.

Applying the principle of virtual displacements and noting that only the vertical component of the 50 kN load is required

$$-(180 \times \tfrac{9}{10}v) - (40 \times \tfrac{3}{5}v) + V_B v = 0$$

therefore

$$V_B = 186 \text{ kN}$$

The negative signs are required because the direction of action of the forces is opposite to that of the displacements.

If the reaction V_C is required, a vertical displacement v_1 is applied at C. The beam will now remain on its supports at A and B and pivot about D (figure 1.10b). Thus the work equation becomes

$$-(40 \times \tfrac{2}{3}v_1) + V_C v_1 = 0$$

therefore

$$V_C = 26 \cdot 7 \text{ kN}$$

1.8 Shear force and bending moment

The aim of all structural design is to produce a structure that is capable of carrying the specified forces. Individual members of the structure may have to carry axial and transverse loads, bending moments and torques, thus the design could become extremely complicated.

For the time being the discussion will be limited to a simple straight beam, the applied loads and reactions lie in a single plane which also contains the centroidal axis of the beam.

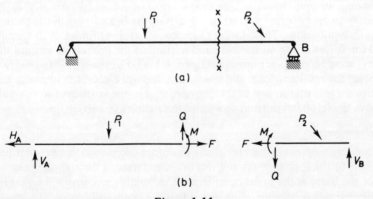

Figure 1.11

Such a beam is shown in figure 1.11a. Imagine the beam to be sectioned at XX. Then the free-body diagram will be as shown in figure 1.11b. To keep each portion of the beam in equilibrium the following forces will be required: an axial force F, a transverse force Q, known as the shear force, and a bending moment M. For equilibrium at the section the corresponding forces in each section must oppose each other; for example if there is a clockwise moment M shown on the R.H.S. this must be opposed by an anti-clockwise moment M on the L.H.S. The values of these forces can be determined from the load applied to the beam and the values of the reactions. A sign convention has to be adopted. In a later chapter we have to use a sign convention for stresses and the convention used here will be similar. A right-handed triad for x, y and z is adopted, the x axis points to the right and the y axis vertically upwards.

Axial force

The axial force is the resultant of all components of applied loads and reactions acting parallel to the axis of the beam at the point considered. Tensile forces will be taken as positive.

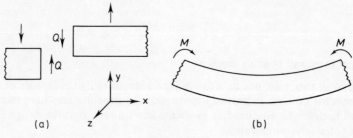

Figure 1.12

Shear force

If a beam is in equilibrium and we imagine it to be sectioned vertically, it will be necessary to apply forces normal to the axis of the beam, lying in the section, to maintain equilibrium. These forces Q are known as shear forces. If the resultant of all the forces acting on the left-hand portion of the beam is downward then a force Q would have to act upwards (figure 1.12a). For the right-hand portion of the beam the resultant force and the shear force will each act in opposite senses. Positive shear is that shown in the diagram, that is, the resultant forces would tend to move the left-hand portion down and the right-hand portion up.

Bending moment

The bending moment is the sum of the moments taken about an axis normal to the plane of the applied loads and reactions, and passing through the centroidal axis of the beam at the point considered. The bending moment will be taken as positive when it is 'sagging'. This means that the beam will be concave upwards—the fibres at the top edge of the beam will be compressed, while those at the bottom edge will be extended. A negative bending moment is termed 'hogging'. Figure 1.12b shows a positive bending moment.

We are now in a position to calculate and sketch both bending-moment and shear-force diagrams for a beam. It is convenient to sketch the beam and then draw the shear-force and bending-moment diagrams underneath. It should be

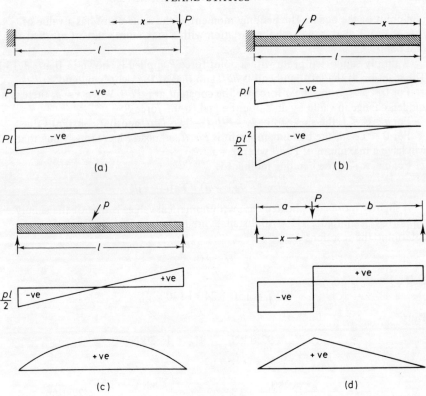

Figure 1.13

emphasised that in all the cases which follow, the beam has been considered as light, that is, making no contribution to either the shear force or bending moment.

Four simple cases are shown in figure 1.13. The first of these, a, is a cantilever with a point load P applied at the free end. Considering a section at a point distant x from the free end, the value of the shear force is $-P$. So that the value of the shear force remains constant along the length of the beam at $-P$. The value of the bending moment is $-Px$ as the beam is hogging. The bending moment will have a maximum value at the fixed end of $-Pl$ and a value of zero at the free end with a linear variation.

Figure 1.13b shows the same cantilever with a U.D.L. of p per unit length. At x the shear force is $-px$ and the bending moment is $-px^2/2$. The shear force has a linear variation from zero at the free end to $-pl$ at the fixed end. The bending moment increases in a parabolic manner from zero at the free end to $-pl^2/2$ at the fixed end.

The shear force and bending moment diagrams for a simply supported beam with a U.D.L. are shown in figure 1.13c. At a point x from the left-hand support the shear force is $-p[(l/2) - x]$, that is, a linear distribution with a zero value at the mid-point of the beam. The bending moment at the same point has a value of $p(lx - x^2)/2$, that is, a parabolic variation with a maximum value of $pl^2/8$ at the mid point.

A simply supported beam with a point force P applied is shown in figure 1.13d. The reaction at the left-hand end is Pb/l and that at the right-hand end Pa/l.

For $0 < x < a$, the shear force will be constant at $-Pb/l$. When $x = a$, there is a sudden change in value by an amount equal to the force P.

For $a < x < l$, the shear force is $-P(b/l - 1) = Pa/l$, another constant value.

For $0 < x < a$, the bending moment is Pbx/l, a linearly increasing value that will have a maximum of Pab/l when $x = a$.

For $a < x < l$, the bending moment is

$$Pbx/l - P(x - a) = Pa(l - x)/l$$

A slightly more complicated case with numerical values is that of the simply supported beam figure 1.14. All distances are in metres. It is first necessary to determine the reactions

$$H_A = 0$$

ΣM_A

$$8V_B = 150 + 240 + (40 \times 9)$$

Thus

$$V_B = 93 \cdot 8 \text{ kN} \qquad V_A = 46 \cdot 2 \text{ kN}$$

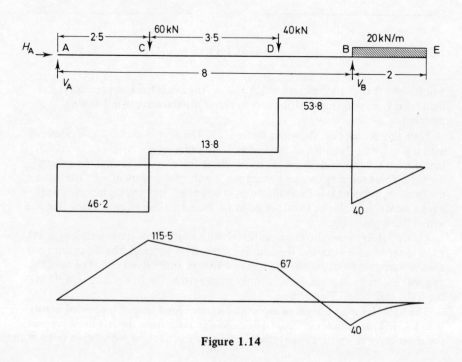

Figure 1.14

The shear force diagram can now be drawn. At A the value is equal to the reaction V_A, that is, $-46\cdot2$ and it stays constant at this value from A to C. At C there is a sudden change of $+60$ and the shear force stays constant at $13\cdot8$ from C to D. At D there is a change of $+40$ and the value stays constant at $53\cdot8$ from D to B. At B the shear force will change from $53\cdot8$ to -40 a total change of $93\cdot8$ equal to the reaction V_B. From B to E there will be a linear decrease with a zero value at E.

The bending moment at A is zero and will increase linearly to $46\cdot2 \times 2\cdot5 =$ $115\cdot5$ kNm at C. There is a linear variation between C and D where the value is $(46\cdot2 \times 6) - (60 \times 3\cdot5) = 67\cdot0$ kNm. Again there is linear variation between D and B where the value is $-20 \times 2^2/2 = -40$ kNm. From B to E there is a parabolic variation with zero bending moment at E.

It can be seen that a point load causes a sudden change in the value of the shear force equal in magnitude to the value of the point load. It also causes a discontinuity in the bending-moment curve. A couple applied to a point on a beam will cause a sudden change in the value of the bending moment equal to the value of the couple, but there will be no change in the value of the shear force at the point of application of the couple.

1.9 Relations between load, shear force and bending moment

Figure 1.15 shows a small element $\mathrm{d}x$ cut from a beam. It is loaded by a force, on the top surface, of average intensity p per unit length, acting upwards. To keep the element in equilibrium it is necessary to add both shear forces and bend-

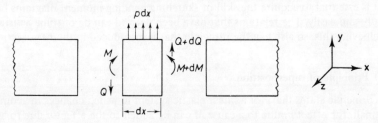

Figure 1.15

ing moments. These will change in value from one side of the element to the other. The sign convention previously adopted has been used. As the load acts in a positive y direction, it is considered positive.

Resolving vertically

$$Q + \mathrm{d}Q = Q - p\,\mathrm{d}x$$

thus

$$\frac{\mathrm{d}Q}{\mathrm{d}x} = -p \qquad\qquad (1.9)$$

Taking moments about the right-hand edge, omitting the term involving the load as a second order quantity

$$M + dM + Q\,dx = M$$

thus

$$\frac{dM}{dx} = -Q \tag{1.10}$$

Making use of equation 1.10, we can express equation 1.9 as

$$\frac{d^2M}{dx^2} = p \tag{1.11}$$

Equation 1.10 shows that if the shear force is negative then the rate of change of bending moment is positive, so that the value of the bending moment is increasing. Also the value of the bending moment will be a maximum or a minimum when the shear force is zero. Exceptions to this would be when the maximum value of the bending moment occurs at the end of a beam, or when there is a sudden change in the value of a bending moment due to the application of a couple.

Equation 1.10 can be integrated such that

$$M_2 - M_1 = \int_{x_1}^{x_2} -Q\,dx$$

Thus the change in moment between any two points is given by the area of the shear-force diagram between the two points. An exception to this would be if a couple were applied between the two points considered.

It is essential to acquire the skill of sketching bending-moment diagrams both rapidly and easily. The relations that have been derived can be of some assistance in achieving this, so also can the principle that is introduced in the next section.

1.10 Principle of superposition

This principle states that, for a linear elastic system in which changes in geometry are small, the effect m due to a cause M can be added to the effect n due to a cause N. Thus the result will be the same as the effect $m + n$ due to a cause $M + N$.

A simple example of this is a linear spring balance where the elongation of the spring due to a load $M + N$ could be found by summing the elongations resulting from the loads M and N applied separately.

The principle is not valid if the material of the system is non-elastic or has exceeded the yield point, or if the geometry of the structure changes appreciably as the load is applied.

Obviously the geometry of all structures must change slightly as loads are applied. Individual members will extend or contract resulting in points on the structure deflecting. The changes in the majority of cases are small and can be neglected in applying the principle.

An example of a structure in which the geometry is dependent upon the load is shown in figure 1.16. The two uniform members of the same length are made of linear elastic material, they are each pinned at one end to a rigid support and are joined together by a pin. A vertical load P is applied at the pin, and the dotted line shows the deflected form of the beams. It will be found that the deflection of the mid-point will not be proportional to the load.

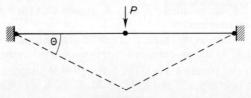

Figure 1.16

A small knowledge of elasticity is assumed in the following. Let the deflection be d and the original length of each member be l.

If $d \ll l, \tan \theta \approx \theta = d/l$, the tensile force F in each member is

$$\frac{P}{2 \sin \theta} \approx \frac{P}{2\theta} = \frac{Pl}{2d}$$

The strain ϵ is given by

$$\frac{(l^2 + d^2)^{1/2} - l}{l} \approx \frac{d^2}{2l^2}$$

Making use of Hooke's law

$$\epsilon = \frac{F}{AE}$$

where A is the cross-sectional area, therefore

$$\frac{d^2}{2l^2} = \frac{Pl}{2AEd}$$

or

$$P = \frac{AEd^3}{l^3}$$

So that when θ is small the relation between the load and displacement is of a cubic form.

The principle of superposition can sometimes be used to good effect when shear-force and bending-moment diagrams are required. In every case it would be perfectly possible to draw individual bending-moment diagrams for each load applied separately to a beam, and then finally sum the diagrams to get the total effect when all the loads are applied simultaneously. For the majority of cases this would be a rather tedious approach, and in general it is necessary to be selective. One particular application is in the case of statically indeterminate systems, where it is possible to get some idea of the shapes of the S.F. and B.M. diagrams. It is, of course, not possible to assign final numerical values to the diagrams.

Consider a beam built-in at each end and carrying a U.D.L. (figure 1.17a). Omitting any effects due to axial force the problem is statically indeterminate to the second degree, but may be reduced to a single degree by an equation of condition which enables us, from symmetry considerations, to state that the vertical reactions at each end are $pl/2$. There will however be an unknown moment M acting at each end. From symmetry it may be argued that the moments at each end will be of equal magnitude.

Consider the problem as two separate cases: a simply supported beam carrying a U.D.L. (figure 1.17b) and an unloaded simply supported beam with a moment M applied at each end (figure 1.17e). The S.F. and B.M. diagrams for the first case are shown in figures 1.17c and d. The B.M. diagram is parabolic with a maxi-

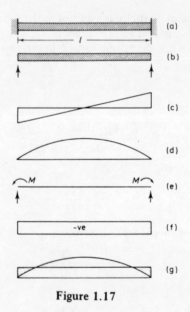

Figure 1.17

mum value of $pl^2/8$; a standard case that should be remembered. (For a central point load of P the maximum B.M. is at the centre and has a value of $Pl/4$.) Now consider the case of the beam simply supported with terminal moments $-M$. The sign is negative as these moments tend to hog the beam. The reactions at the ends will be zero, thus the shear force is everywhere zero. The bending moment will be constant at a value of $-M$ (figure 1.17f). The two bending-moment diagrams can now be combined to give the final result. The simplest way of doing this is to redraw figure 1.17d, invert figure 1.17f, and superimpose these giving figure 1.17g. The two diagrams now have to be subtracted, leaving the shaded area. From this it may be seen that the maximum bending moment can either occur at the centre of the beam, $pl^2/8 - M$, or at the ends of the beam, $-M$. The S.F. diagram does not have to be redrawn and is identical to figure 1.17c.

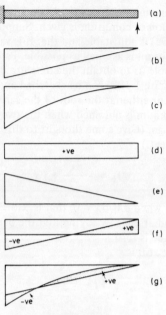

Figure 1.18

A further example is that of a propped cantilever (figure 1.18a). The value of the vertical reaction V cannot be determined by statics. Figures 1.18b and c show the S.F. and B.M. diagrams for the loaded cantilever without the prop. Figures 1.18d and e show the cantilever loaded with V at the right-hand end. The combined S.F. and B.M. diagrams are shown in figures 1.18f and g, the shaded area once again giving the final values. Note that the point of maximum positive bending-moment occurs when the shear force is zero. To find the value of V it is necessary to add some further information to the problem. In this case it might be stated that the prop is rigid, that is, the deflection at the end of the cantilever is zero. Deflection of beams will be found in a later chapter.

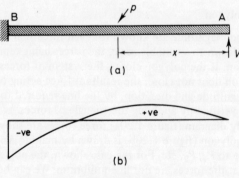

Figure 1.19

It is possible in certain circumstances to obtain a complete solution if further information or an equation of condition is given. Suppose that the height of the support at A (figure 1.19a) is adjustable and that it is varied until the maximum bending moment in the beam is as small as possible. Perhaps this is rather an artificial case but it will enable us to obtain the complete B.M. diagram.

The B.M. diagram (figure 1.19b) can be sketched and it is seen that the maximum moment could occur either at the root of the cantilever or somewhere in the positive region. The optimum is obtained when the two values are equal in magnitude but opposite in sign. (Give some thought to this statement and make sure you know why it is correct.)

$$M_x = -\frac{px^2}{2} + Vx$$

This will be a maximum when $dM/dx = 0$, that is, $x = V/p$. Actually this position could have been obtained at once by saying that the bending moment is a maximum when the shear force is zero.

The bending moment at B is

$$Vl - \frac{pl^2}{2}$$

therefore

$$\frac{px^2}{2} - Vx = Vl - \frac{pl^2}{2}$$

since

$$V = px \quad x^2 + 2lx - l^2 = 0 \quad x = (\sqrt{2} - 1)l$$

Hence

$$V = (\sqrt{2} - 1)pl$$

It is now possible to find the bending moment at any point in terms of p and l alone.

1.11 The force polygon

If the direction and magnitude are known for each force that is acting on a body, a polygon can be drawn to a suitable scale, the forces being represented by the sides of the polygon. If the polygon closes, the system of forces will be in equilibrium. If the polygon does not close, the resultant force acting on the body can be found, both in magnitude and direction, by the line required to close the polygon. This can easily be verified by applying the triangle of forces.

In the free-body diagram (figure 1.20a) three forces are shown acting on a body. The force polygon (figure 1.20b) is drawn by making AB, BC, etc., parallel and scaled in value to P_1, P_2, etc. For the case shown, the polygon does not close thus indicating that the forces are not in equilibrium. We can however close the polygon with a line DA. This will represent the magnitude and direction of the

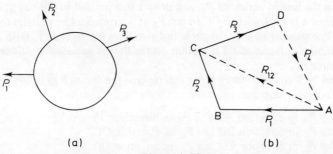

Figure 1.20

force P_4 that has to be applied to the body to keep it in equilibrium. This can be seen by applying the triangle of forces. The resultant of P_1 and P_2 is R_{12}, which when combined with P_3 gives P_4.

We must note at this stage that the construction gives the magnitude and direction of the resultant, but will not give the point of application, unless it so happens that all the forces pass through a single point. To find the point of application a further construction is required known as the funicular or string polygon.

1.12 The funicular polygon

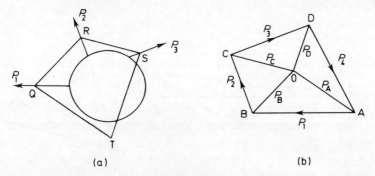

Figure 1.21

The same free-body diagram and force polygon are shown in figures 1.21a and b. Any point O is chosen on the force polygon diagram, this is known as the *pole*. It need not lie inside the force polygon but a point inside it has been chosen for convenience. Join OA, OB, etc.—these lines are termed *rays*. OAB now forms a force triangle, thus P_1 can be resolved into two other forces: P_A in direction AO and P_B in direction OB. Similarly P_2 is resolved into P_B in direction BO and P_C in direction OC. It is seen that by choosing a single point of intersection O, P_B, for example, will represent a common component of forces P_1 and P_2.

In the free-body diagram the rays must intersect the lines of action of the various forces, thus a line parallel to BO must intersect P_1 and P_2. Choose any point Q on the line of action of P_1, and draw a line parallel to OB to cut P_2 at R. From R draw a line parallel to OC to cut P_3 at S, from S a line parallel to OD and from Q a line parallel to OA. Let these last two lines intersect at T. QRST is known as the *funicular polygon* and T is a point on the line of application of the equilibriating force P_4.

This can be seen quite easily as we can replace the forces P as follows in the funicular polygon diagram.

P_1 by P_A in direction AO and P_B in direction OB
P_2 by P_B in direction BO and P_C in direction OC
P_3 by P_C in direction CO and P_D in direction OD.

The forces P_B and P_C will cancel out in pairs leaving only forces P_A and P_D acting at T. The resultant of P_A and P_D is P_4. So that T will be a point on the line of application of P_4.

Two conditions must be satisfied for a system of forces to be in equilibrium: (i) the force polygon must close; (ii) the funicular polygon must close.

A particular case can arise when the first condition is satisfied. However when the funicular polygon is drawn, the first and last lines are parallel, indicating that the resultant on the system is a couple.

1.13 Graphical determination of reactions

The force and funicular polygons can be used to determine the reactions that result when a beam is loaded.

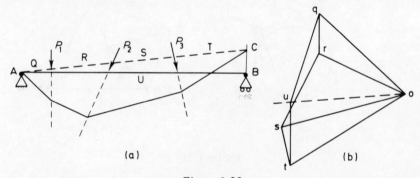

Figure 1.22

The beam in figure 1.22a will have a vertical reaction at B but the direction of the reaction at A is unknown. Draw lines on the force polygon (figure 1.22b) to represent P_1, P_2 and P_3. At this stage it will not be possible to close this diagram as there are two separate closing lines representing R_A and R_B. It is convenient to letter the spaces between the forces in figure 1.22a, this is known as *Bow's notation*. Thus Q is the space between R_A and P_1, R between P_1 and P_2, etc. Finally U is the space underneath the beam between the two reactions. This system of lettering is transferred to the force polygon, that is, P_1 is the force qr.

Next choose a suitable pole O and draw in the rays. The reason for the lettering system can now be seen as a line on the funicular polygon, across space Q for example, will be parallel to Oq. Using this system we are unlikely to make a mistake by drawing a line across the wrong space in a complicated system.

The funicular polygon can now be drawn. It is essential in this case to start the funicular polygon at A, as it is known that the reaction at A passes through this point. The various lines are shown in figure 1.22a. The reaction at B is known to be vertical; the line across space T cuts a vertical through B at C. AC is the closing line on the funicular polygon. A line parallel to AC is drawn through O on the force polygon, this is shown dotted. The force polygon is now closed by two lines representing R_A and R_B. R_B is known to be vertical and must pass through t. Draw tu to cut the dotted ray at u. The two reactions are then represented by qu and ut.

If we imagined a massless string fixed at A and C the funicular polygon would represent the shape that the string would take up if loads P were hung from it, hence the name *string polygon*. If the pole O had been chosen on the other side of the force polygon the funicular polygon would be inverted. We could now think of a series of massless rods, instead of strings, supporting the loads. Hence the term line arch.

1.14 Graphical construction of bending-moment diagrams

Figure 1.23 shows a light string fixed at the two end points A and B. The string carries two loads P_1 and P_2. Since the loads are vertical, $H_A = H_B$ and the horizontal component of tension in the string will be constant at H, say.

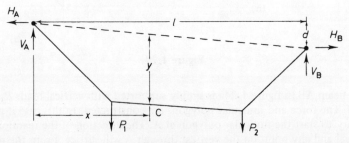

Figure 1.23

For a flexible string the bending moment at any point on the string must be zero. Then for ΣM_C

$$V_A x = H\left(y + \frac{xd}{l}\right) + \Sigma\, m_x \tag{1.12}$$

$\Sigma\, m_x$ represents the moment about C of all the applied loads P to the left of C. For ΣM_B

$$V_A l = Hd + \Sigma\, m_B \tag{1.13}$$

Σm_B represents the sum of the moments of the applied loads P about B.
Eliminating V_A from equations 1.12 and 1.13

$$Hy = \sum m_B \cdot \frac{x}{l} - \sum m_x \qquad (1.14)$$

Now $\Sigma m_B/l$ would be equal to V_A for the case when $H = 0$, thus the right-hand
side of the equation represents the bending moment at x for a simple beam of
length l with the same system of loads applied. Thus, considering the string, if we
multiply the horizontal force H by the vertical distance from a point on the string
to the line joining the two ends, we shall obtain a value for the bending moment
at the same point in a simple beam.

This analysis can be used for the graphical construction of bending-moment
diagrams for beams with applied vertical loads.

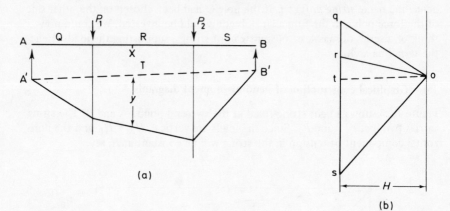

Figure 1.24

The beam AB in figure 1.24a is simply supported with vertical loads P_1 and P_2
applied. The force and funicular polygons are constructed. In this case it is not
necessary to start the funicular polygon at A. The direction of the reaction at A
is vertical and any point on the vertical through A will suffice. From the previous
discussion we can see that the funicular polygon will represent such a string fixed
at A' and B'. Thus the bending moment on the beam at X will be given by Hy.
The question that naturally arises is, what is the correct value of H? In the case of
the string, H was the value of the horizontal force at the supports. It can be seen
at once from figure 1.24b that this is represented by the horizontal distance from
O to the vertical line representing the loads.

It is of course necessary to take into account various scale factors when the
numerical value of the bending moment at a particular point is required in a
problem. If the force polygon has been drawn to a scale S_2 then the true value of
the horizontal thrust is S_2H, and if the line diagram has been drawn to a scale S_1,
the value of the bending moment will be given by S_1S_2Hy.

If some of the loads on the beam are not vertical the problem cannot be treated directly, the reason for this is that the value of H would not be constant. However, if only the vertical components of the loads are used in the construction, the resulting funicular polygon will represent the B.M. diagram, as the horizontal components do not affect the value of the bending moment.

When part of the loading system in a beam is uniformly distributed, it is best to consider it for the time being as a concentrated load placed at the centroid of the distributed load, and equal to the total value of the distributed load. The force and funicular polygon can be drawn. The portion of the funicular polygon between A and C in figure 1.25 will be incorrect as the variation of bending moment for a uniformly distributed load should be parabolic. The diagram can however be corrected by drawing in a parabola between A′ and C′ such that A′D′ and D′C′ are tangents to the parabola.

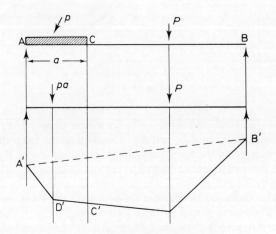

Figure 1.25

A non-uniform load can be plotted in graphical form and divided into a number of vertical elements or strips. The area and centroid of each strip can be found. The area will represent a small load which can be considered as acting through the centroid. The force and funicular polygon can be drawn, the latter will have to be slightly corrected by making the rays tangential to a curve, which can be drawn freehand.

1.15 Graphics applied to a three-pin arch

The problem of finding the reactions for a three-pin arch has already been dealt with theoretically in section 1.5. To solve the problem graphically it is necessary to consider the loads on each half of the arch separately.

Figure 1.26a shows the arch with the complete loading, and figure 1.26b shows it with the loads on the right-hand side removed. Considering this second case, the direction of the reaction R_B must pass through the points B and C as the

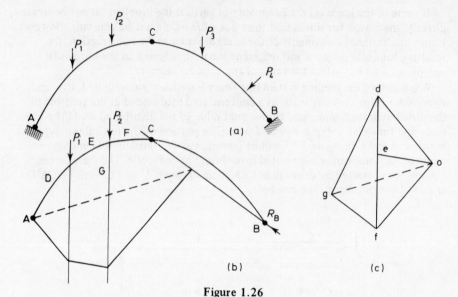

Figure 1.26

bending moment at both these points is zero and there is no other load on the right-hand side. The force and funicular polygons are drawn. The ray fo will pass between the lines of action of P_2 and R_B. The closing line from the funicular polygon is transferred to the force polygon. As the direction of R_B is known, a line parallel to this can be drawn from f to cut the final ray at g. R_A and R_B will be represented by dg and gf.

It will next be necessary to consider the arch with the loads on the right-hand half removed. The reactions can again be found. Finally the two sets of reactions can be combined to give the values when the arch is loaded on both sides.

The funicular polygons will not give a measure of the bending moment at any point as the horizontal thrust at the abutments has to be taken into account. However, once the reactions are known the moments can be found by statics.

Next consider a set of vertical loads applied to a three-pin arch (figure 1.27). It is essential that the funicular polygon, which also represents the line of thrust,

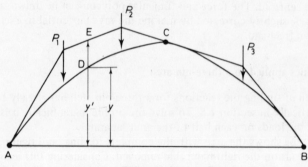

Figure 1.27

should pass through the three pins, that is, the bending moment at each pin is zero. This will mean that there is only one possible position for the pole on the force diagram. The location of this point is left as an exercise for the reader. It can be seen that the closing line on the funicular polygon must be horizontal.

We have already shown in section 1.14 that the horizontal distance H on the force polygon represents the horizontal thrust at the abutments and, for a vertical system of loads, will also represent the horizontal component of force at any point in the arch.

Any vertical distance y' on the funicular polygon multiplied by the polar distance H on the force polygon gives the bending moment at a particular point on a beam with the same vertical loading applied. So that in the case of the arch Hy' is the bending moment at a point D on the arch due to the vertical applied loads and the vertical component of the end reaction. To find the total moment the effect of the horizontal thrust must be taken into account.

$$M_D = Hy' - Hy = H(y' - y)$$

where y is the vertical coordinate of the arch at the point D. Note that $y' - y$ is the vertical distance between the arch and the funicular polygon.

If the arch had been designed such that its axis coincides with the funicular polygon, then the bending moment at every point on the arch would be zero, and the arch would only be subject to direct thrust. The funicular polygon represents the line of thrust in the arch, and is sometimes referred to as the pressure line.

We now have a very simple way of determining the position of the line of thrust. The moment at any point is divided by the value of the horizontal thrust, giving the value of $y' - y$.

1.16 The differential equation for a vertical load funicular polygon

Figure 1.28a shows the diagram of a non-uniform vertical load applied to a simply supported beam. To draw the force and funicular polygons (figures 1.28b and c) the load is split into a number of elements. It will be noted that the position of the pole O has been chosen such that the closing line on the funicular polygon is hori-

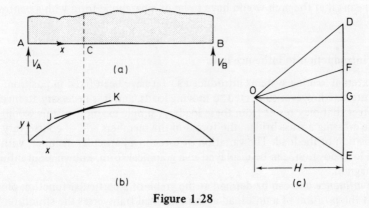

Figure 1.28

zontal. OF is a typical ray on the force polygon and is shown as the tangent JK to the funicular polygon. The closing ray on the force polygon is OG, thus DG and GE represent V_A and V_B.

The length DF is $\int_A^C p \, dx$ where p, the load, is some function of x.

Thus FG represents $V_A - \int_A^C p \, dx$. Now

$$\tan \angle FOG = \frac{V_A - \int_A^C p \, dx}{H}$$

and this is equal to the slope of JK on the funicular polygon. Thus

$$\frac{V_A - \int_A^C p \, dx}{H} = \left(\frac{dy}{dx}\right)_C$$

and since V_A is constant

$$\frac{d^2 y}{dx^2} = -\frac{p}{H} \tag{1.15}$$

This equation is known as the differential equation for a vertical load funicular polygon.

As a particular application let us find the shape of an arch that is carrying a load uniformly distributed over the span, such that the bending moment at any point on the arch is zero.

As the bending moment is zero it is essential that the funicular polygon and the centre line of the arch coincide.

In this case p is constant and equation 1.15 can be integrated twice to give

$$y = -\frac{px^2}{2H} + Ax + B$$

Applying suitable end conditions $y = 0$ when $x = 0$ or l

$$y = \frac{p}{2H}(lx - x^2)$$

This means that the arch would have to be of a parabolic form with a central rise of $pl^2/8H$.

1.17 Introduction to influence lines

The external loading systems introduced so far have been fixed in position. Many structures are however subjected to moving loads, and it is necessary to investigate the effect that may result from these loads. A simple example of this would be a vehicle crossing a truss bridge; the forces in the members will of course vary with the position of the load. The variation of force in a particular member, with respect to the load position, can be displayed in a graphical form, known as an influence-line diagram.

An *influence line* can be defined as the graph of a particular function plotted against the position of a unit load as the unit load transverses the structure. The function could be one of many different quantities, for example the shear force, bending moment or deflection at a particular point, or perhaps the force in a particular member.

It is absolutely essential from the outset to distinguish between, say, the S.F. diagram for a beam and the influence line for shear force at a particular point on the beam. The first diagram will give the value of the shear force at any point on the beam for a particular set of applied loads; whereas the influence line will give a diagram showing how the shear force varies at one particular point as a unit load moves over the beam.

1.18 Influence lines for reactions, shear force and bending moment

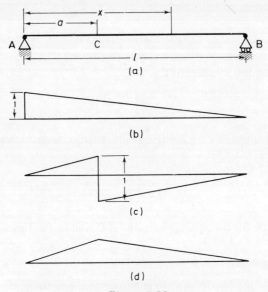

Figure 1.29

A simply supported beam is shown in figure 1.29a. The influence line for the reaction at A is first required. If the unit load is placed a distance x from A the value of V_A can be found by taking moments about B

$$V_A l = 1(l - x)$$

$$V_A = \frac{l - x}{l}$$

This value of V_A can now be plotted against x to form the influence line (figure 1.29b). V_A has a maximum value of unity when the applied load is at A and a value of zero when the applied load is at B, the variation being linear.

The influence lines are now required for shear force and bending moment at C, distant a from A.

Shear force

For $x < a$ the shear force at C is equal to the value of V_B which is x/l. The shear force $= x/l$, adopting the usual sign convention.

For $x > a$ the shear force at C is equal to the value of $-V_A$, that is, $-(l-x)/l$.

In both expressions there is a linear variation of shear force with position of the applied load. When $x = a$ there is a sudden change in value from a/l to $-(l-a)/l$, a total change of unity.

The influence line is shown in figure 1.29c. Note that the inclined lines must be parallel to one another and that there is a unit change at C.

Bending moment

For $x < a$ the bending moment at C is

$$V_B(l - a) = \frac{x}{l}(l - a)$$

For $x > a$ the bending moment at C is

$$V_A a = \left(\frac{l - x}{l}\right)a$$

Both of the expressions are positive as the bending moment is sagging and they also vary linearly with x.

At C when $x = a$

$$M = \frac{a}{l}(l - a)$$

The influence line for the bending moment at C is shown in figure 1.29d.

1.19 Loading systems

At this stage it might be pertinent to enquire into the possible use of an influence line. So long as the system is linearly elastic the principle of superposition can be invoked, and the influence line used to find the value of the function when any loading system is applied to the structure.

Let figure 1.30b represent the influence line at C for a particular function. This has of course been drawn for a unit load, so that if a load P crosses the beam, the value of the function will be given by Py, where y is the height of the influence-line diagram at the point under the load. The maximum value is of course given by Py_{max}. For a series of concentrated loads the value of the function will be given by ΣPy.

Next consider the case of a uniformly distributed load p. The load acting on a small element dx of the beam is $p\,dx$; as $dx \to 0$ the load can be considered as acting at a point and the value of the function would be $py\,dx$. The total value of the function due to the complete distributed load can be found by integrating this expression. That is $p\int y\,dx$ since p is a constant. $\int y\,dx$ is the area of the influence-line diagram underneath the uniformly distributed load.

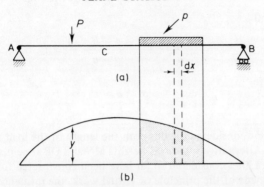

(a)

(b)

Figure 1.30

It is comparatively simple to find the maximum value of a function for a single point load, but the case of a uniformly distributed load requires further discussion.

Referring back to the S.F. diagram shown in figure 1.29c. If the U.D.L. is larger than the span of the beam and moving from left to right, the maximum positive shear will occur when the front of the load has just reached C, and is equal to p times the area above the base line. The maximum negative shear will occur when the rear of the load has reached C. If the load is shorter than the span, no particular difficulty arises; again there would be two cases to consider, the front and the rear of the load at C.

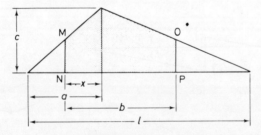

Figure 1.31

There is no need to discuss the case of bending moment when the load is longer than the span. If the length of the load is b and this is shorter than the span l an area such as that shown in figure 1.31 will have to be a maximum. The influence line for bending moment has been drawn for a point distant a from the left-hand support, the maximum height of the diagram being c, with the rear of the load at a distance x from the point considered.

The simplest approach is to find when the unshaded area is a minimum.

$$\text{Unshaded area } A = \frac{c(a-x)^2}{2a} + \frac{c(l-a-b+x)^2}{2(l-a)}$$

when $dA/dx = 0$

$$-\frac{(a-x)}{a} + \frac{(l-a-b+x)}{l-a} = 0.$$

This simplifies to

$$\frac{x}{b} = \frac{a}{l} \qquad\qquad (1.16)$$

Thus the point considered divides both the length of the load and the span in the same ratio. Alternatively the two heights MN and OP are equal.

The influence lines considered so far have been constructed for very simple cases. If use is made of the principle of virtual work, the influence line can often be sketched very rapidly for more complicated cases.

1.20 Application of virtual-work methods to influence lines

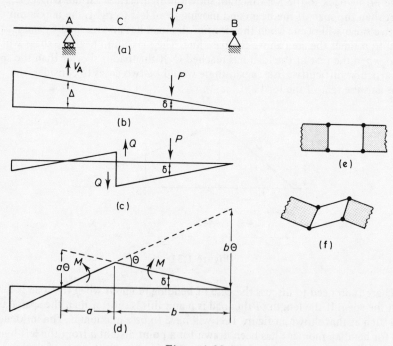

Figure 1.32

Reactions

We now meet our first real application of virtual work. The influence line for the vertical reaction at A is required. A virtual displacement Δ is applied to the beam at A. The corresponding displacement of P is δ (figure 1.32b). Since the system is in equilibrium

$$V_A \Delta - P\delta = 0$$

or

$$V_A = \frac{P\delta}{\Delta}$$

If we make both P and Δ equal to unity, then $V_A = \delta$. Thus an influence line has been constructed for the vertical reaction at A.

In general, to draw the influence line for a reaction, the point of application of the reaction is given a unit displacement in the direction of the reaction. The resulting beam displacement will give the influence line.

Shear force

The influence line for shear force at point C is required. It is necessary to remove resistance to shear force at this point, but the resistance to both bending moment and direct force must remain intact. We could imagine the beam cut through at C and a mechanism such as that shown in figure 1.32e introduced. When the shear forces Q are applied this would deform as shown in figure 1.32f, the final position being when both the links were vertical. The deformed beam is shown in figure 1.32c and the two portions are displaced through a relative vertical distance Δ at C, the corresponding displacement of P being δ.

By the principle of virtual work

$$Q\Delta - P\delta = 0$$

Again put Δ and P equal to unity. Thus $Q = \delta$ and the influence line for the shear force at C has been constructed.

It is essential that the two displaced portions of the beam should remain parallel. If not, there would be an extra term in the virtual-work equation involving the bending moment at C and the relative rotation of the two parts of the beam.

Bending moment

The influence line for bending moment at point C is required. The resistance to bending moment must be removed, but the beam must still be capable of carrying both shear and axial forces. This can be achieved by inserting a pin at C. Bending moments M in opposite directions are applied on each side of the pin (figure 1.32d), and a relative virtual rotation θ is imposed, θ of course being a very small angle. The corresponding displacement of P is δ.

By virtual work

$$M\theta - P\delta = 0$$

Thus

$$M = \frac{P\delta}{\theta}$$

Once again put $P = 1$, $\theta = 1$ then $M = \delta$ and the influence line for bending moment has been constructed.

As θ is assumed small, the distances shown at the supports can be taken as $a\theta$ and $b\theta$. Thus the height of the influence-line diagram at a distance d from the left-hand support is

$$\frac{db\theta}{a+b} = \frac{db\theta}{l} = \frac{db}{l}$$

when θ is unity. So that the maximum height of the diagram is ab/l and this occurs at C.

The application of these methods can probably best be illustrated by means of an example.

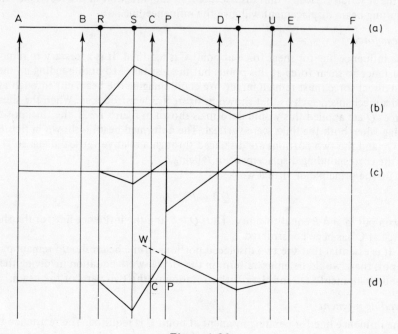

Figure 1.33

A bridge girder (figure 1.33a) has a total length of 100 m. It is supported at six points A, B, C, D, E and F, each span being 20 m. The girder is continuous over the supports, but has hinges at R and S in span BC and at T and U in span DE. BR = SC = DT = UE = 5 m.

Influence lines are required for the following: vertical reaction at C, and shear force and bending moment at P (situated 5 m from C in span CD).

The influence lines have all been sketched in figure 1.33. Figure 1.33b shows the reaction at C, c and d show the shear force and bending moment at P. In all cases the deformed beam, that is, the influence line, must remain in contact with the supports, except in b where unit displacement is applied at support C. The beam will pivot about the hinges S and T.

For the vertical reaction the height of the influence line at C is unity. For the shear force there is a unit change at P and in the deformed position SP is parallel to PT. For bending moment WC = CPθ, where θ is the relative rotation of the two parts of the beam at P. Now CP is 5 m and with $\theta = 1$, WC = 5 m.

The results can now be written down.

	A–R	S	C	P	D	T	U–F
V_C	0	5/4	1	3/4	0	−1/4	0
shear force	0	−1/4	0	1/4 to −3/4	0	1/4	0
bending moment	0	−15/4	0	15/4	0	−5/4	0

1.21 Multi-load systems

So far we have only considered the use of the influence-line diagram for single concentrated loads and uniformly distributed loads. The situation will often arise however, when a series of concentrated loads roll across the structure, and the maximum value of a particular function for a chosen point is required. Alternatively the absolute maximum value for a function for any point on the structure might be required.

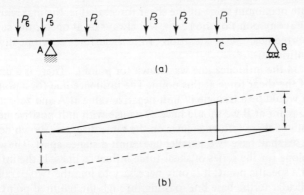

(a)

(b)

Figure 1.34

Shear force

The train load $P_1 - P_6$ (figure 1.34a) is travelling across the beam from left to right and the maximum value of the shear force at C is required. First the influence line for shear force at C is sketched as shown in figure 1.34b. As the loads move across the beam, the positive value of the shear force will steadily increase until P_1 is just to the left of C. When P_1 crosses C there will be a sudden decrease in shear force by an amount P_1. With the loads continuing to move to the right the value of the positive shear force will again increase while that of the negative shear force due to P_1 will decrease. This state of affairs will continue until

P_2 is just to the left of C, and the net shear force can be found. It can now be seen that the maximum value of the positive shear will occur when one of the point loads is just to the left of C. It also follows that the maximum value of negative shear will occur when one of the loads is just to the right of C.

The maximum value can be found by trial and error, but will always occur when one of the loads is just to the left or right of the point considered. Trial-and-error methods are often rather tedious, and it is possible to reduce the calculation involved.

Suppose that load P_n has just crossed C, the positive shear will have decreased by P_n. As the loads advance there will be a gradual rise in the positive shear equal to the change in value of the reaction V_B. This will increase until P_{n+1} reaches the point C. The total change in shear force dQ is given by

$$dQ = \Delta V_B - P_n \tag{1.17}$$

where ΔV_B represents the difference in the value of V_B when P_n and P_{n+1} are placed in turn at point C. If equation 1.17 is positive, the shear force will have been increased by moving the loads forward. The loads are thus advanced until there is a change of sign in equation 1.17. Once the sign becomes negative the maximum value of the shear force has been passed. The change in V_B can be found very simply if all the loads remain on the beam. Care must be exercised if a load comes onto or leaves the beam.

The absolute maximum value of the shear force, that is, the maximum value that can occur at any point of the beam, will always be at one of the supports for a simply supported beam. Consideration of the shear-force envelope discussed below will confirm this.

In figure 1.34 the influence line was drawn for point C. There is a unit change in the value of the shear force at this point. The influence line for a point next to the support A would be a triangle of unit negative value at A and zero at B. Similarly for a point at B we should have a triangle with unit positive height at B. If we draw both of these influence lines on the same diagram (shown dotted in figure 1.34b), we shall have two parallel lines, unit distance apart. These lines form the envelope for the series of shear-force influence lines. If the influence line is required for a specific point, it is only necessary to join the two parallel lines by a line perpendicular to the base line, passing through the required point.

Bending moment

The train load (figure 1.35a) is crossing the simply supported beam from left to right. The influence line for bending moment at point C has been drawn in figure 1.35b. The question arises, where should the loads be placed such that the bending moment at C is a maximum?

Let R_1 be the resultant of all the loads on the beam to the right of C, and R_2 the resultant of those on the left. The value of the bending moment at C will be given by

$$M = \frac{R_2 x_2 d}{a} + \frac{R_1 x_1 d}{b}$$

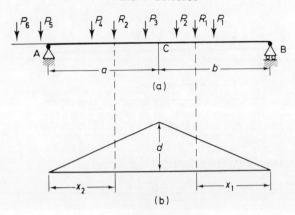

Figure 1.35

Let the loads now move a small distance dx to the right, such that $dx = dx_1 = dx_2$, but the values of R_1 and R_2 remain unchanged. There will be a change in the value of the bending moment

$$dM = \left(\frac{R_2 d}{a} - \frac{R_1 d}{b} \right) dx$$

Thus, as the loads move further onto the beam, the value of M will continue to increase as long as

$$\frac{R_2}{a} > \frac{R_1}{b} \tag{1.18}$$

Let us now assume that the inequality is true, and that P_5 comes onto the beam before P_3 has reached C. The value of R_2 will be increased while that of R_1 stays constant. Thus the inequality will not be affected by a load coming onto the beam. The same kind of argument will apply if P_1 leaves the beam.

If however P_3 reaches and crosses C the value of R_1 will increase suddenly and that of R_2 will decrease and the inequality may no longer be true.

We can now see that the maximum bending moment will occur when one of the loads is at the point C. The inequality (equation 1.18) may be used to determine which load will give the maximum value.

So far we have only considered a method for finding the maximum bending moment that can occur at a particular point. Let us next consider the problem of finding the absolute maximum bending moment as the train of loads crosses a simply supported beam. From our previous cases, it is clear that the maximum bending moment must occur underneath one of the loads.

Let us first find the position of P_4 such that the maximum bending moment occurs under P_4 (figure 1.36).

Let ΣP be the total load on the beam, and let a be the distance of the centroid of this load from P_4.

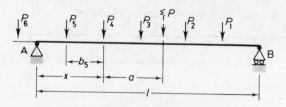

Figure 1.36

If x is the distance of P_4 from the left-hand support

$$V_A = \sum P \frac{(l - x - a)}{l}$$

The value of the bending moment under P_4 is

$$V_A x - P_5 b_5 = M$$

$$\frac{dM}{dx} = P \frac{(l - 2x - a)}{l} = 0 \text{ for a maximum}$$

then

$$x = \frac{l - a}{2} \qquad\qquad (1.19)$$

Thus the distance of P_4 from the support at A is $(l - a)/2$. The distance from the position of $\sum P$ to the support B is $l - x - a = (l - a)/2$.

Thus the maximum moment, under a particular concentrated load, will occur when the mid-point of the beam bisects the distance between the load being considered and the centroid of all the loads on the beam.

To find the absolute maximum it is necessary to proceed by trial and error. However in general the maximum bending moment will occur somewhere near the centre of a simply supported beam. Assuming that it occurs at the middle, the inequality 1.18 may first be used to find which load should be at the mid point, then the method just described can be used to make final adjustments. Hence the absolute maximum value can be found.

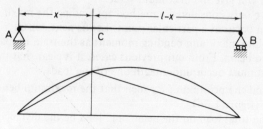

Figure 1.37

Before concluding this section let us determine the shape of the bending-moment envelope. The maximum height of the influence line for the bending moment at C (figure 1.37) has a value $x(l-x)/l$ (section 1.20). Thus the bending-moment envelope will be a parabola with maximum height $l/4$.

1.22 Influence lines for girders with floor beams

In the case of a fairly large span, the loading system is seldom applied directly to the beam. The usual form of construction is shown in figure 1.38a. Large beams, called girders, are first placed over the span, and then a number of floor beams a–e, are placed on top of, and at right angles to, the girders. These floor beams carry stringers, which are parallel to the girders and are simply supported on each floor beam. The decking system or roadway is placed on top of the stringers. The

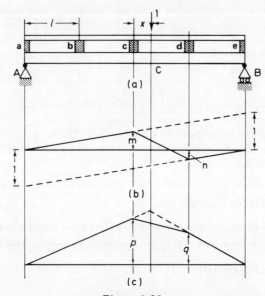

Figure 1.38

parts of the girder between a and b etc. are termed panels, and a, b, etc., are referred to as panel points. The effect of a load P placed on a panel will be transmitted to the girder via the two panel points. The floor beams are considered as being simply supported at the panel points.

The values of the reactions at A and B will not be affected by this type of construction, thus the influence line for reactions is identical to that of a simply supported beam.

For a load applied at any particular point on a stringer, the value of the shear force will be constant in any particular panel. Thus we can refer to the shear force in a particular panel rather than the shear force at a particular point on the girder.

We shall now construct the influence line for shear force in panel cd where the length cd is l. When a unit load is applied to the left of c the shear force will be equal to V_B, and with the load applied to the right of d, the shear force will be equal to V_A. These two portions of the influence line can be constructed and are identical to that of a simple beam. Consider the load applied between c and d at a distance x from c, with panel lengths all equal to l. The reaction at c will be $(l - x)/l$ and that at d will be x/l. The shear in the panel will be given by

$$\left(\frac{l-x}{l}\right)m - \left(\frac{x}{l}\right)n$$

where m and n are the heights of the influence line at c and d respectively. This is the equation of a straight line, joining the ends of the two lines that have already been constructed (figure 1.38b).

The influence line for bending moment has to be drawn for a particular point on the girder as the bending moment will not be constant in a panel. We shall draw the influence line for point C on the girder. First the influence line for a simple beam is drawn, this will be correct apart from the portion between the two panel points c and d (figure 1.38c).

The bending moment between c and d can be expressed as

$$M = \left(\frac{l-x}{l}\right)p + \frac{x}{l}q$$

where p and q are the heights of the influence line at c and d respectively. Again we have the equation of a straight line joining the two ends of the influence line already constructed.

1.23 Influence line for three-pin arch

The symmetrical three-pin arch (figure 1.39a) has a rise h and a span l. The influence lines at D (x, y) are required for bending moment, shear force and axial force.

For a unit load to the left of D

$$M_D = V_B(l - x) - Hy$$

For a unit load to the right of D

$$M_D = V_A x - Hy$$

It can be seen that the influence line for bending moment can be obtained from the influence lines for V_A, V_B and H. The influence lines for V_A and V_B are identical to those for a simply supported beam (figures 1.39b and c). A closer inspection of the first term in each of the expressions for M_D, reveals that this represents the influence line for the bending moment at D for a simply supported beam, that is, $H = 0$.

The influence line for H is shown in figure 1.39d. The shape of this can be determined by the method of virtual displacements; the maximum value is $l/4h$.

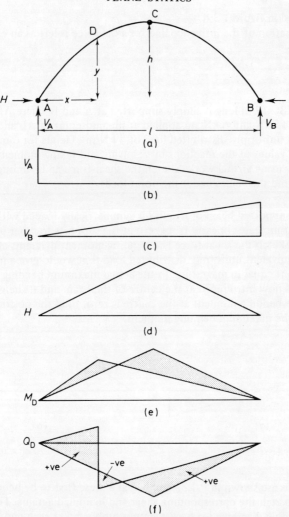

Figure 1.39

The influence line for the bending moment at D can now be drawn and consists of the difference between the influence line for M_D for a simply supported beam and a constant y times the influence line for H (figure 1.39e).

The influence line for the shear force at D can be found from the combination of two effects; the vertical load acting on a simply supported beam and the horizontal thrust. If θ is the slope of the arch at D, the influence line for shear force is given by

$$\begin{Bmatrix} \text{Influence line for shear at D,} \\ \text{for a simply supported beam} \end{Bmatrix} \cos\theta - \{\text{Influence line for } H\} \sin\theta$$

This is sketched in figure 1.39f.

The construction of the influence line for axial force is left as an exercise for the reader.

Problems

1.1 A beam ABC 12 m long is simply supported at A and B where AB = 8 m. Point loads of 15 kN and 6 kN are applied at the mid-point of AB and at C respectively. A uniformly distributed load of 4 kN/m extends for 6 m from A. Determine the values of the reactions by resolving and taking moments. Check your results by using virtual work. Sketch the shear-force and bending-moment diagrams and find the values under the point loads.

1.2 A simply supported beam of length l is symmetrically loaded with a distributed load that increases linearly from zero at the supports to p per unit length at the centre. Sketch the shear-force and bending-moment diagrams and find the value of the equivalent uniformly distributed load that would give a maximum bending moment equal in magnitude to the actual maximum bending moment.

A support is now introduced at the centre of the beam and its height is adjusted until the bending moment at the centre is zero. Find the position and magnitude of the maximum bending moment.

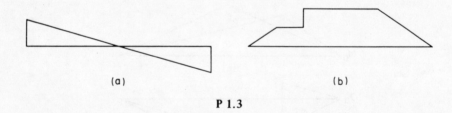

(a)   (b)

P 1.3

1.3 Two graphs are shown in P 1.3a and b. Take these first to be bending-moment diagrams and sketch the corresponding shear and loading diagrams. Then take them to be loading diagrams for beams simply supported at each end and sketch the corresponding shear-force and bending-moment diagrams.

1.4 A beam is simply supported at A, B, C and D. AB = CD = 8 m, BC = 12 m. Pins are inserted in span BC, 3 m from B and C respectively. AB has a uniformly distributed load of 4 kN/m, and CD a uniformly distributed load of 6 kN/m, with a point load of 10 kN applied at the centre of span BC. Show that the beam is statically determinate, and find the values of the reactions at the supports. Draw the shear-force and bending-moment diagrams.

1.5 Sketch the shear-force and bending-moment diagrams for the four beams shown in P 1.5. State whether each case is statically determinate or redundant.

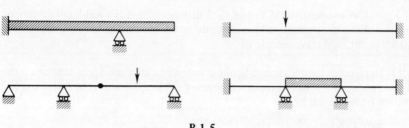

P 1.5

1.6 Show that the bending-moment diagram for a statically determinate beam with vertical loads applied is represented by a funicular polygon. Illustrate your answer with reference to a simply supported beam of length l with vertical loads of $P/2$ at the quarter points and a load P applied at the centre.

A symmetrical arch of varying cross-section is semicircular in form and has a span of l. It is pin-jointed at the abutments. A loading system identical to that on the beam is applied. The horizontal thrust is found to be $P/2$. Discuss how the line of thrust might be found. Sketch the bending-moment diagram for the arch, and determine the maximum value.

1.7 Show that the differential equation for a vertical load funicular polygon is given by the expression $d^2y/dx^2 = -p/H$. Discuss what is meant by H.

An arch is pinned at each abutment and has a span of l. A third pin is placed at mid-span where the rise is $l/2$. Define the shape of the arch for the following load system if no bending action is permitted.

Horizontal distance from left-hand abutment	$\dfrac{l}{8}$	$\dfrac{l}{4}$	$\dfrac{3l}{4}$	$\dfrac{7l}{8}$	$\dfrac{l}{4} - \dfrac{3l}{4}$	
Load	$\dfrac{P}{2}$	P	$2P$	P	distributed load of $\dfrac{4P}{l}$ per unit horizontal length	

1.8 For the beam described in problem 1.1 when no loading is present, sketch the influence line for the reactions at A and B, and the influence lines for shear force and bending moment at the mid-point of AB.

1.9 For the beam described in problem 1.4 when no loading is present, sketch the influence lines for the reactions at A and B, and the influence line for shear force and bending moment at a point 5 m from A. Determine the maximum bending moment and shear force at this point when two loads of 6 kN and 3 kN spaced 3 m apart traverse the beam.

1.10 A symmetrical parabolic arch rib is hinged at both abutments and also at the crown. The span is 40 m and the rise 10 m. Draw the influence line for the horizontal thrust at the abutments, and the influence line for the bending moment at a point whose horizontal and vertical distances from the left-hand abutment are x and y. Find the position and magnitude of the maximum bending moment as a 50 kN point load traverses the rib.

1.11 Determine the maximum bending moment and shear force that can occur at a third point of a simply supported beam of 24 m span when the following load system traverses the beam.

load (kN)	5	10	15	10
spacing (m)		2	2	2

Also determine the maximum possible bending moment and shear force that can occur in the beam.

2 STATICALLY DETERMINATE STRUCTURES

2.1 Simple plane trusses

A structure of this type consists of a number of members that are connected by pins at the ends, all the members lying in a single plane. Of course most structures that exist are three-dimensional, but in a large number of cases it is possible to treat the problem from a two-dimensional point of view. An example of this would be a bridge consisting of two parallel trusses connected by transverse members carrying the roadway. If the loads that are transmitted by the transverse members to the trusses are known, then each truss can be analysed as a two-dimensional system.

As the discussion is being limited to pin-connected members at this stage, no bending moments will be developed in the members if the loads are applied through the joints of the structure—they will only be subject to tensile and compressive forces. If however a load is applied between the ends of a particular member, that member will have bending moment and shear force developed in it, in addition to an axial force.

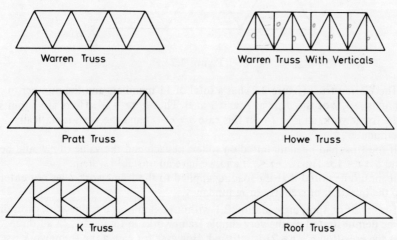

Warren Truss

Warren Truss With Verticals

Pratt Truss

Howe Truss

K Truss

Roof Truss

Figure 2.1

Figure 2.1 shows five different types of bridge truss and one typical roof truss. The top and bottom members, referred to as chords, need not necessarily be parallel. The two inclined members at the ends are called end-posts and the rest of the members are referred to as diagonals and verticals.

2.2 Stability and determinacy

In the previous chapter we discussed the question of determination of reactions without any reference to the structure itself. It will now be necessary to decide whether the complete structure is unstable, statically determinate or redundant.

For pin-jointed structures the unknown quantities will be the axial force in the members and the reactions at the supports. Thus if there are b members and r reactions the total number of unknowns will be $b + r$.

For a plane pin-jointed framework it is possible to resolve for the forces in the framework in two directions at every joint in the system. Thus for j joints a total of $2j$ equations can be formed relating the unknowns. If

$b + r = 2j$ the structure will be statically determinate   (2.1)

$b + r < 2j$ the structure will be unstable or a mechanism   (2.2)

$b + r > 2j$ the structure will be statically indeterminate or redundant (2.3)

It is necessary to qualify these statements to some extent, as it is quite possible to conceive a structure that might be redundant in one part and unstable in another, and yet a count of members, reactions and joints would satisfy equation 2.1. We can state here and now that it is no good blindly applying the above relationships; it is always necessary to inspect a structure carefully at the same time.

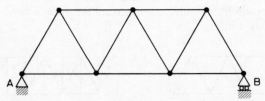

Figure 2.2

The Warren truss (figure 2.2) has a total of 11 members and the number of reactions is 3, 2 of these are at A and 1 at B. Thus $b + r = 14$. There are 7 joints in the framework, so $2j = 14$. In this case $b + r = 2j$ and the truss is statically determinate.

If the truss had been mounted on rollers at each end the value of r would be 2 and $b + r = 13$. Thus $b + r < 2j$ and we have an unstable system.

It should be noted that the loading applied to the framework does not enter into the question of stability or redundancy.

The result of applying equation 2.1 without examining a framework carefully can be demonstrated by the very simple frameworks in figure 2.3. In all three cases the equation $b + r = 2j$ is satisfied. However for case a the framework itself

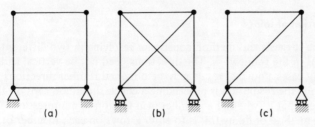

Figure 2.3

is unstable (consider a small lateral displacement) and the reactions are redundant. For b the framework is redundant and the reactions are unstable. Case c will be a suitable statically determinate structure.

Let us consider case b in a little more detail. It has been stated that the framework itself is redundant. The simplest framework that can be formed from a minimum number of bars is that of a triangle (figure 2.4). This will be statically determinate as long as the support conditions are correct. Suppose that it is required to add one more joint D to the basic triangle ABC. If the problem is to remain statically determinate we can only connect D to the triangle by two bars.

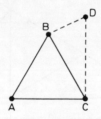

Figure 2.4

This can be seen very simply if we consider a load applied in any direction at D. It would be possible to resolve the load in two directions at D and hence find the forces in the members meeting at D. If we add a third bar DA the frame becomes statically indeterminate.

Thus if a framework is built up from a simple triangle and each joint added is only connected by two members to the framework, the frame will be statically determinate. This will lead to the following relationship for the members in a statically determinate frame

$$b = 2j - 3 \tag{2.4}$$

A number of texts on structures make use of this relationship but in certain cases it leads to difficulties and the relationship given by equation 2.1 is preferred.

There are a number of different methods for finding the forces in the members of a framework and it is often a matter of experience and practice to decide which one should be used. The sign convention used throughout will be that tensile forces are considered positive.

2.3 Resolution at joints

As the name suggests this method consists of resolving in two different directions
at each joint in the framework. The directions need not be vertical and horizontal
and in many cases a quicker result may be obtained if other directions are considered. While the resolution is proceeding a check should be kept on the direction
of the forces, that is, whether a member is in tension or compression. The most
convenient method of doing this is to enter arrows on each member of a line
drawing of the framework. Thus ←——→ would indicate compression and →——←
would indicate tension. With practice, if the angles between members happen to
be simple ones, the forces can be written down straight away on the line drawing.

  The frame should first be checked to see if it is statically determinate. It is
always worthwhile spending a few moments on an inspection of the frame. Sometimes certain members may not be loaded and if this can be seen straight away,
time may be saved later on. Often it will be necessary to determine the values of
the reactions before resolution can start.

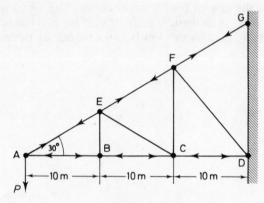

Figure 2.5

  For the framework in figure 2.5 we need not attempt to determine the
reactions at G and D as resolution can start straight away at joint A. However
after checking that the frame is statically determinate, we can see that the force
in EB is zero (vertical resolution at B). This will mean that the force in BC is
equal to that in AB. Again as the force in EB is zero, by resolving perpendicular
to EF at E, the force in EC is zero. In fact the forces in all members will be zero
except for those lying along AG or AD. (Consider joint C and then joint F.) If
this is so the forces in AE, EF and FG must all be equal as are those in AB, BC
and CD.

  Resolution of forces at joint A
Vertical

$$F_{AE} \sin 30° = P \quad \text{therefore } F_{AE} = 2P \text{ tensile}$$

Horizontal

$$F_{AB} = F_{AE} \cos 30° \quad \text{therefore } F_{AB} = \sqrt{3}P \text{ compressive}$$

Results

$$F_{AB} = F_{BC} = F_{CD} = -\sqrt{3}P$$
$$F_{AE} = F_{EF} = F_{FG} = 2P$$

We might question the necessity of having the vertical and diagonal members in this frame as the forces in these members are zero. Basically the frame could be constructed out of two members AG and AD. If these were continuous the frame would be stable and the two member forces could be found by resolution at A. If AG and AD are made up of several shorter members connected together, the extra members will be required even though the forces in the members are zero. The framework has to be stable and this stability is completely independent of the actual loading system applied.

In our analysis we have ignored the dead weight of all the members themselves. If this were taken into account there would be a force, for example, in member FD.

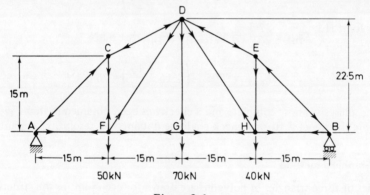

Figure 2.6

A somewhat more complicated example is shown in figure 2.6. The frame is statically determinate and by taking moments about A and vertical resolution for the whole frame, the two reactions at A and B are

$$V_A = 82 \cdot 5 \text{ kN} \quad V_B = 77 \cdot 5 \text{ kN}$$

By inspection $F_{GD} = 70$ kN and $F_{FG} = F_{GH}$.
The lengths of members CD and FD will be required

$$CD = 16 \cdot 8 \text{ m}, \quad FD = 27 \text{ m}$$

In a case of this kind where the framework is symmetrical but the loading system is not, it is probably best to start resolution at each end of the frame.

Joint A $\uparrow$ $F_{AC} = -82 \cdot 5\sqrt{2} = -116 \cdot 5$ kN
  $\rightarrow F_{AF} = 82 \cdot 5$ kN

Joint B $\uparrow$ $F_{BE} = -77 \cdot 5\sqrt{2} = -109 \cdot 5$ kN
  $\rightarrow F_{BH} = 77 \cdot 5$ kN

Joint C →
$$F_{CD} \times \frac{15}{16\cdot8} = \frac{-116\cdot5}{\sqrt{2}} \qquad F_{CD} = -92\cdot3 \text{ kN}$$

↑
$$F_{CF} = \frac{116\cdot5}{\sqrt{2}} - \left(92\cdot3 \times \frac{7\cdot5}{16\cdot8}\right) = 41\cdot2 \text{ kN}$$

Joint E →
$$F_{ED} \times \frac{15}{16\cdot8} = \frac{-109\cdot5}{\sqrt{2}} \qquad F_{ED} = -86\cdot5 \text{ kN}$$

↑
$$F_{EH} = \frac{109\cdot5}{\sqrt{2}} - \left(86\cdot5 \times \frac{7\cdot5}{16\cdot8}\right) = 38\cdot8 \text{ kN}$$

Joint F ↑
$$F_{FD} \times \frac{22\cdot5}{27} = 50 - 41\cdot2 \qquad F_{FD} = 10\cdot6 \text{ kN}$$

→
$$F_{FG} = 82\cdot5 - \left(10\cdot6 \times \frac{15}{27}\right) = 76\cdot6 \text{ kN}$$

Joint H ↑
$$F_{DH} \times \frac{22\cdot5}{27} = 40 - 38\cdot7 \qquad F_{DH} = 1\cdot6 \text{ kN}$$

→
$$F_{GH} = 77\cdot5 - \left(1\cdot16 \times \frac{15}{27}\right) = 76\cdot6 \text{ kN}$$

The analysis is now complete and a check has been obtained in that F_{FG} and F_{GH} are the same. A further check could be obtained by resolution at joint D.

2.4 Graphical method

A series of force triangles or polygons are drawn for each joint in the structure. They are connected together in one diagram to avoid the repetition that would result if a separate polygon were drawn for each joint. In connection with this method it is useful to employ Bow's notation such that the forces in the members of the structure are easily identified.

It is first necessary to make an accurate scale drawing of the framework. The spaces between all external and internal forces are lettered. The force diagram, drawn to a suitable scale can then be started. It is essential to commence the drawing from a joint at which there are only two unknowns.

The truss in figure 2.7a has all members the same length. By inspection $V_A = V_B = P$ and $H_A = H_B = 0$. The spaces between the forces have been lettered; note that space U extends right round the top outside edge of the truss and separates V_A and V_B. The force diagram could be started from either A or B, as at both these points there are one known and two unknown forces.

Starting from B and proceeding round this joint in a clockwise manner, we see that from U to S there is a vertical upwards force of $V_B = P$. This is drawn on the force diagram and lettered such that s is above u, indicating that R_B acts upwards. The rest of the force triangle for joint B can now be drawn and is represented by rsu.

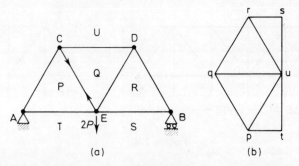

Figure 2.7

We can now proceed to joint D, again working round this joint in a clockwise manner urq will be the force triangle. Proceeding systematically round the truss the force diagram can be completed (figure 2.7b).

The values of the forces can be scaled off the force diagram and their directions found making use of Bow's notation. As an example the force in member CE is represented by the distance qp. We proceed round joint C in a clockwise direction from Q to P—this is represented by the direction qp. An arrow is entered on the line diagram, close to C, in this direction. If we repeat the process for joint E we shall obtain an arrow in the opposite direction. Thus it can be seen that the member CE is in tension.

The accuracy of the final result is entirely dependent on the accuracy of the two scale drawings. In the majority of problems some idea of the accuracy can be obtained when the last few lines are drawn on the force diagram. These should meet at a point, but a small triangle or error may very well result. If the errors are too large to be acceptable, it will be necessary to check both diagrams most carefully to find any major source of error.

2.5 Method of sections

In this method, ascribed to Ritter, an imaginary cut is made through the truss and forces are applied to each part of the structure to keep it in equilibrium. These applied forces will have the same values as the forces in the members that have been sectioned. As there are only three equations of equilibrium it will not be possible to find the values of the forces if more than three members are cut, in separating the two parts of the truss, unless the forces in some of the members are already known.

A great advantage with this method arises when forces are only required in certain members of a truss; very often a section can be made and the forces determined at once. In both the methods that have been discussed previously, it would be necessary to start at one end of the truss and work steadily through the joints until the required forces were found.

The forces in members AB, CD, AD and AC are required for the Pratt truss (figure 2.8a).

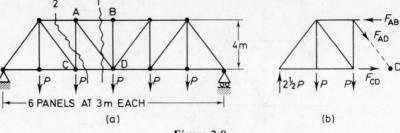

Figure 2.8

A section 1 is made cutting AB, AD and CD. The isolated left-hand part of the frame is shown in figure 2.8b with the necessary forces for equilibrium.

ΣM_A

$$4F_{CD} + 3P = 6 \times 2\tfrac{1}{2}P$$

therefore

$$F_{CD} = 3P \quad \text{tensile}$$

ΣM_D

$$4F_{AB} + 3P + 6P = 9 \times 2\tfrac{1}{2}P$$

therefore

$$F_{AB} = 3\tfrac{3}{8}P \quad \text{compressive}$$

The force in AD can be found very simply by considering the vertical equilibrium of the sectioned portion. The vertical component of the force in member AD must balance the applied vertical loading and reactions on the left-hand portion of the truss.

Thus the vertical component of F_{AD} is

$$2\tfrac{1}{2}P - 2P = \frac{P}{2}$$

therefore

$$F_{AD} = \frac{P}{2} \times \frac{5}{4} = \frac{5P}{8} \quad \text{tensile}$$

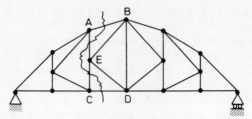

Figure 2.9

We can proceed in the same manner to find the force in AC, but a new section, 2, is required.

Vertical equilibrium for the left-hand portion of the frame gives

$$F_{AC} = \frac{P}{2} \quad \text{compressive}$$

A particular case arises with a K truss, where it is possible to section four members. If the section is made in the manner shown in figure 2.9, the force in CD can be found by taking moments about A, and that in AB by moments about C. Also note that by horizontal resolution $F_{BE} = -F_{ED}$.

2.6 Compound trusses

In the case of simple trusses it is possible to make a complete analysis of the structure by any of the methods previously discussed. For compound trusses however it is very often necessary to make use of the method of sections in addition to resolution at joints.

Several examples of compound frameworks are shown in figure 2.10. These are often formed by connecting two or more simple trusses with sufficient members to make the structures rigid. In figure 2.10b, for example, there are two simple triangles connected by three members. Figure 2.10c shows two smaller roof trusses, each consisting of half the span, again interconnected by three members.

A brief discussion follows on a possible method of solution for each problem illustrated in figure 2.10.

Figure 2.10a: This problem can be started by resolving at A then D. Difficulties will arise however if we now attempt to resolve at C—five members meet at this point, and the forces in only two of these have so far been found. If a section is put through as shown, the forces in EG and FH and CH can be found. It is now possible to return to C and resolve as only two unknowns remains. The rest of the analysis can then proceed.

Figure 2.10b: A section is put through the three interconnecting members. Moments can then be taken about the point of intersection X of two of the connecting members—this will give the force in the third member, and resolution can then proceed normally.

Figure 2.10c: This can proceed on similar lines to a.

Figure 2.10d: In this case there are three unknowns at every joint and it will not be possible to make a start by resolving. Again it is not possible to put a section through without cutting more than three members. If we examine the problem it consists of two basic triangles AEF and BCD connected by three members AC, DE, and FB. The triangle AEF is shown isolated in figure 2.10e. The values of F_{AC}, F_{DE} and F_{FB} can be found by resolving and taking moments. No further difficulties will arise.

2.7 Complex trusses

In this type of truss it will not be possible to solve directly using resolution at joints and method of sections. Indeed it is often difficult to determine whether

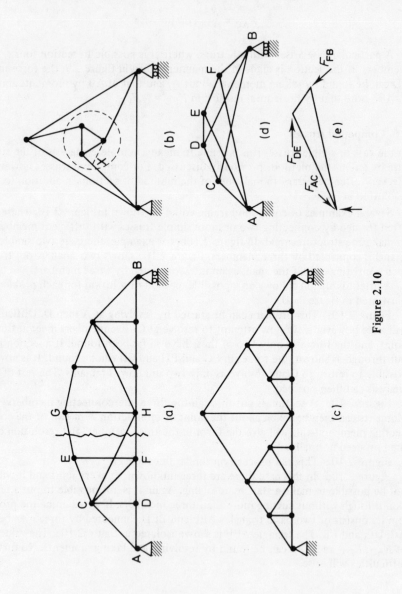

Figure 2.10

or not a truss of this type is statically determinate. The usual test—by satisfying equation 2.1—can be applied, and this will check that the number of members is correct. It is much more difficult to decide whether these are in the correct positions and it might well be that one part of the truss is statically indeterminate and another part unstable.

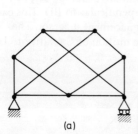
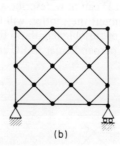
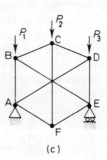

(a)   (b)   (c)

Figure 2.11

Three examples of complex trusses are given in figure 2.11. In all these cases there are at least three members meeting at every joint. It would of course be possible to resolve at every joint and form a total of $2j$ equations in terms of all the unknown bar forces—these equations could then be solved. This approach would be somewhat tedious, but if used it would be best to employ the method of tension coefficients which will be dealt with in the section on three-dimensional statically determinate structures.

An alternative approach would be to proceed from one of the supports and resolve as far as possible. Having reached a point where there are three unknowns, let one of these be T. The resolution can now proceed, and the value of T found when resolving at the last joint in the structure. Once T is known the rest of the forces in the members will be known.

2.8 Virtual work

This is a useful method to assist the analysis of a complex truss. If the principle is applied once to the structure we can determine the force in one of the members. Having determined this force, the rest of the analysis can generally proceed by resolution at joints. It may in certain cases be necessary to apply the principle of virtual work a second time to find the force in a second member.

The method of application is very similar to that demonstrated in section 1.7. Basically a member of the truss is removed and replaced by an unknown force applied at each end of the member. The frame will now have turned into a mechanism and a virtual displacement is applied in the direction of the removed member. The resulting displacements of all the applied forces have to be calculated. The total work done can then be summed to zero.

In certain simple cases it may be easy to calculate the displacement of the applied forces in terms of the applied virtual displacement. If the problem is complicated it is probably best to use a graphical approach—termed a *displacement diagram.*

As an introduction let us consider the very simple example shown in figure 2.12. The force in member AC is required, so this member is removed and a force equal to F_{AC} is applied—the frame will become a mechanism pivoting about B. We start the displacement diagram from a pole o which also represents any fixed points, in this case A and B. The point D can now describe a circular arc of radius BD, but for a very small initial movement the motion will be perpendicular to BD. This can be represented by a line od of suitable length on the displacement diagram. The motion of C relative to B is perpendicular to BC and the motion of C relative to D is perpendicular to CD.

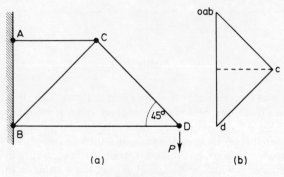

(a)   (b)

Figure 2.12

These two directions can be represented by lines on the displacement diagram drawn from b and d respectively which must intersect at c. Note that in this case we have a triangle bcd which will be similar to BCD. Let od equal x. The point c has moved horizontally and vertically through $x/2$. We require the displacement in the direction of AC, that is, $x/2$.

The virtual-work equation can now be applied. Assuming that AC is in tension, the horizontal displacement of C is in the opposite sense to the force. Thus

$$-F_{AC}\frac{x}{2} + Px = 0$$

or

$$F_{AC} = 2P$$

As a more complicated case we shall consider the same truss that was discussed in section 2.7. The force is required in member BE, this has been removed in figure 2.13a. It is probably best to give E a known horizontal displacement oe; note E is on rollers and can only move horizontally. The point f can next be fixed; d must coincide with o as the point D will not move for a small initial movement of E; c can now be fixed and this can be followed by b and the displacement diagram (figure 2.13b) is complete.

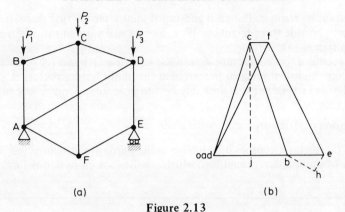

Figure 2.13

The movement of E relative to B is given by be—we require the component in the direction of BE, that is, bh. Neither B nor D move vertically and the vertical movement of C is given by cj. This implies that there is no virtual work term for P_1 or P_3, and that for P_2 is given by $-P_2$cj.

The complete virtual-work equation assuming BE in tension is

$$F_{BE} \, \text{bh} - P_2 \text{cj} = 0$$

therefore

$$F_{BE} = P_2 \, \frac{\text{cj}}{\text{bh}}$$

2.9 Statically determinate space structures

The type of structure to be considered in this section will consist of members that are jointed at their ends such that complete rotational freedom exists and only an axial force will act in the members. It will be necessary to determine whether a three-dimensional structure is stable and statically determinate. We can proceed on similar lines to the test for a plane structure.

The total number of unknown forces in the members will be equal to the number of members b in the structure; in addition to this there are the unknown reactions r, making a total of $b + r$ unknowns. If

$b + r = 3j$ the structure will be statically determinate   (2.5)
$b + r < 3j$ the structure will be unstable
$b + r > 3j$ the structural will be statically indeterminate

It is quite possible for equation 2.5 to be satisfied, but part of the structure could be unstable and another part statically indeterminate. This is more difficult to determine by inspection than in a plane structure, one reason for this being that 'thinking' in three dimensions does not come automatically to the majority of people. The simplest basic element for a space structure is a triangle—if a further joint has to be fixed in space, it will be necessary to provide three more members connected to the triangle. For each additional joint after the first three, it is necessary to provide three members. We can often make use of this when checking through a framework.

When solving a space structure we can use equations 1.1 and 1.2 to determine the reactions, and then proceed to resolve at the joints in the structure. Unless the problem is very simple it is probably best to proceed in a systematic manner.

2.10 Tension coefficients

Basically this method consists of resolving at the joints in a structure and forming a number of simultaneous equations, which are then solved to find the unknown forces.

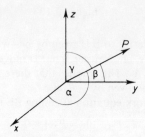

Figure 2.14

Consider the vector force P (figure 2.14). This can be expressed as

$$P = P_x i + P_y j + P_z k$$

where P_x, etc. are the resolved parts of P in the directions of the three axes. These can be found by multiplying P by its direction cosines.

$$P_x = P \cos \alpha \quad P_y = P \cos \beta \quad P_z = P \cos \gamma$$

In a structure the values of the direction cosines for a particular member can be expressed as the resolved length of the member in the x, y or z direction divided by the length of the member. So that if a plan and elevation for the structure are given it will be a very simple matter to determine the resolved length.

The term *tension coefficient, T,* for a member is defined as the force in the member divided by the length of the member. So that if the tension coefficient is multiplied by the resolved length of the member in a particular direction, the resolved part of the force in that direction is obtained. (The term 'tension' is used since at the start of an analysis all members are assumed to be in tension.)

The elevations and plan of a space structure are shown in figure 2.15. The forces are required in all the members.

$$b = 12 \quad r = 9 \quad j = 7$$

therefore

$$b + r = 3j$$

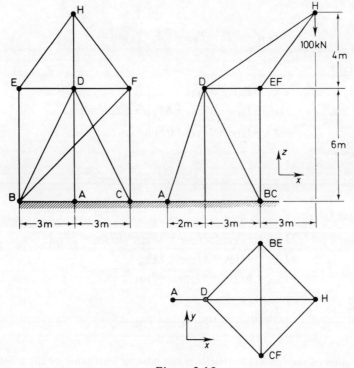

Figure 2.15

and the frame is statically determinate. A further inspection of the structure shows that members are correctly placed such that no part is a mechanism.

It is probably best to start the resolution at joint H. There are only three unknowns at this joint and the tension coefficients for three members can be found straight away.

Joint H

$$x \quad 3T_{HE} + 3T_{HF} + 6T_{HD} = 0$$
$$y \quad 3T_{HE} = 3T_{HF}$$
$$z \quad 4T_{HE} + 4T_{HF} + 4T_{HD} + 100 = 0$$

therefore

$$T_{HE} = T_{HF} = -25 \quad T_{HD} = 25$$

Joint E

$$x \quad 3T_{HE} = 3T_{ED}$$
$$y \quad 3T_{HE} + 3T_{ED} + 6T_{EF} = 0$$
$$z \quad 4T_{HE} = 6T_{EB}$$

therefore

$$T_{ED} = -25 \quad T_{EF} = 25 \quad T_{EB} = -\frac{50}{3}$$

Joint F

$$x \quad 3T_{HF} = 3T_{FD}$$
$$y \quad 3T_{HF} + 3T_{FD} + 6T_{FB} + 6T_{EF} = 0$$
$$z \quad 4T_{HF} = 6T_{FC} + 6T_{FB}$$

therefore

$$T_{FD} = -25 \quad T_{FB} = 0 \quad T_{FC} = -\frac{50}{3}$$

Joint D

$$x \quad 3T_{DB} + 3T_{ED} + 3T_{DC} + 3T_{FD} + 6T_{HD} = 2T_{DA}$$
$$y \quad 3T_{DB} + 3T_{ED} = 3T_{DC} + 3T_{FD}$$
$$z \quad 6T_{DA} + 6T_{DC} + 6T_{DB} = 4T_{DH}$$

therefore

$$T_{DA} = 10 \quad T_{DB} = T_{DC} = \frac{20}{6}$$

The values of the tension coefficients can now be multiplied by the lengths of the members to obtain the forces in kN.

Member	HD	HF	HE	ED	DF	FE	FC	EB	AD	BD	CD	BF
length	7·2	5·84	5·84	4·23	4·23	6	6	6	6·31	7·35	7·35	8·9
T.C.	25	−25	−25	−25	−25	25	−16·67	−16·67	10	3·33	3·33	0
force	180	−145	−145	−100	−100	150	−100	−100	63·2	24·5	24·5	0

2.11 Rigid jointed structures

The majority of structures of this type are highly redundant and are dealt with in later chapters. In certain cases it is possible to make assumptions that will enable an approximate solution to be obtained by statics. It is useful at this stage to have some idea of the problems involved in solving rigid jointed structures. Three examples are shown in figure 2.16.

The majority of engineering structures in everyday use have their members rigidly connected at the ends. In steelwork joints are made by welding, rivetting or bolting. There will probably be a small amount of slip at the joint in the latter two cases, but usually the joint can be assumed to be rigid. In reinforced concrete, where there is a joint between a column and a beam, part of the reinforcement in the column is bent round into the beam so that a continuous joint is made.

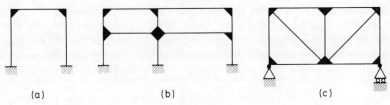

Figure 2.16

Since the joints are rigid the members will develop both bending moment and shear force in addition to axial load. Thus for each member in the structure there will be three unknowns, in addition to the support reactions being unknown. It is possible to form three equations at each joint relating the unknown quantities by resolving in two directions and taking moments. If

$3b + r = 3j$ the structure is statically determinate   (2.6)
$3b + r < 3j$ the structure is unstable
$3b + r > 3j$ the structure is statically indeterminate

Consider the structures in figure 2.16.

(a) $b = 3$ $j = 4$ $r = 6$ $3b + r = 15$ $3j = 12$

therefore the structure is three times redundant.

(b) $b = 10$ $j = 9$ $r = 9$ $3b + r = 39$ $3j = 27$

therefore the structure is 12 times redundant.

(c) $b = 9$ $j = 6$ $r = 3$ $3b + r = 30$ $3j = 18$

therefore the structure is 12 times redundant.

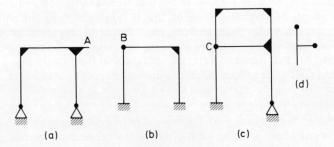

Figure 2.17

If we have a structure with a cantilever portion (figure 2.17a) the end A of the cantilever should be considered as a joint and the cantilever as a member

$$b = 4 \quad j = 5 \quad r = 4 \quad 3b + r = 16 \quad 3j = 15$$

therefore the structure is singly redundant.

Alternatively the cantilever portion could be completely ignored.

A structure that has a mixture of pinned and rigid joints can probably best be treated by first stiffening up all the pinned joints in the structure. The degree of redundancy can then be found. We can then subtract the number of releases or equations of condition that result from turning the rigid joints back into pinned ones.

The portal-type frame (figure 2.17b) is the same as that in figure 2.16a except for the pin joint at B. The original frame was three times redundant. By inserting a pin at B we provide one equation of condition, that is, the moment at B is zero. Thus the degree of redundancy is reduced to two.

For the frame in figure 2.17c the joint at C is first stiffened—there is no need to alter the pinned support since the number of reactions can be calculated. The stiffened frame is five times redundant. Re-introducing the pin at C will not provide three releases, that is, one for each member meeting at C. The effect of the pin is shown in the subsidiary sketch (figure 2.17d). From this it is quite clear that there are only two releases, thus the structure is three times redundant.

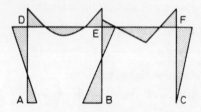

Figure 2.18

To produce a complete analysis of a statically determinate structure with some rigid joints, we shall finally require the values of axial force, shear force and bending moment for every member in the structure. The sign convention that was originally adopted for bending moment, that is, sagging bending-moment positive, was quite satisfactory for beam analysis, but this will no longer suffice when dealing with frameworks consisting of beams and columns. The usual convention to be adopted is that if part of a framework is viewed from the inside, a positive bending moment will produce tension in the near edges of the member. This requires some more explanation and will be discussed in conjunction with the two-bay portal-type structure in figure 2.18.

An assumed bending-moment diagram has been drawn using the line drawing of the frame as a base line. Consider ABED—if the bending-moment diagram lies on the inside of this frame it will be positive and will produce tension on the inside of the members.

For member DE the bending moment is positive over the centre portion and negative towards the ends. Thus considering the lower edge it will be in tension over the centre and in compression towards the ends. For member BE the bending moment is negative at end E of the member but changes sign along the member and is positive at B. So that considering ABED the inside edge of BE is in compression at E and in tension at B.

Now let us consider the member BE from the point of view of BEFG. The bending moment is positive at E and negative at B and is in tension at E and compression at B. This means that the sign of the bending-moment diagram is dependent upon the side of the member from which it is viewed, but the convention is consistent as it produces tension, say, on one edge of the member and compression on the other edge. We can give numerical values to points on a bending-moment diagram but must be careful about the sign.

The convention we shall adopt is such that if we view part of a framework from the inside, when the near edge of a member is in tension then the bending-moment diagram will be drawn on the inside edge of the member, and can be considered as positive.

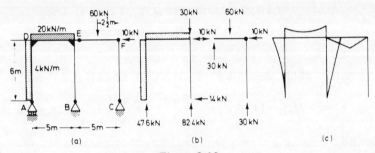

Figure 2.19

An analysis of the framework and a sketch of the bending-moment diagram are required for the two-bay structure (figure 2.19a).

A check is made first to see if the framework is statically determinate. It is necessary to make joints E and F rigid.

$$b = 5 \quad r = 5 \quad j = 6 \quad 3b + r = 20 \quad 3j = 18$$

The framework is twice redundant and two releases will be required. These are provided by the pins at E and F.

ΣM_F gives $H_C = 0$. ΣM_E gives

$$5V_C = 60 \times 2\tfrac{1}{2}$$
$$V_C = 30 \text{ kN}$$

Next it is worthwhile splitting the frame into two parts at joint E (figure 2.19b). The forces required for equilibrium are entered on each part. Thus the left-hand portion requires a vertical load of 30 kN and a horizontal load of 10 kN at E.

Resolve horizontally for the left-hand portion

$$H_B = (4 \times 6) - 10 = 14 \text{ kN}$$

ΣM_B

$$5V_A = \frac{20 \times 5^2}{2} + (10 \times 6) - \frac{4 \times 6^2}{2}$$

therefore

$$V_A = 47 \cdot 6 \text{ kN}$$

Resolve vertically for the left-hand portion

$$V_B = (20 \times 5) + 30 - 47 \cdot 6 = 82 \cdot 4 \text{ kN}$$

A check can be obtained by taking moments about A.

The axial and shear forces and the bending moments can now be found for each member.

Member AD

Axial $-47 \cdot 6$ kN; shear force zero at A, 24 kN at D—linear

$$M_{DA} = -72 \text{ kNm}, \quad M_{AD} = 0\text{—parabolic}$$

Member DE

Axial -24 kN; shear force $-47 \cdot 6$ kN at D, $52 \cdot 4$ kN at E—linear

$$M_{DE} = 72 \text{ kNm}, \quad M_{ED} = (47 \cdot 6 \times 5) - \frac{20 \times 5^2}{2} - \frac{4 \times 6^2}{2} = -84 \text{ kNm—parabolic}$$

Member BE

Axial $-82 \cdot 4$ kN; shear force 14 kN—constant

$$M_{BE} = 0, \quad M_{EB} = -(14 \times 6) = -84 \text{ kNm—linear (viewed from ADEB)}$$

Member EF

Axial -10 kN; shear force 30 kN from E to G, -30 kN from G to F

$$M_{EF} = M_{FE} = 0; \quad M_{max} \text{ at centre} = 75 \text{ kNm}$$

Member CF

Axial -30 kN; shear force = bending moment = 0

It should be noted that a check can be made at all the joints by summing the various components of forces and moments; the resultant should be zero in all cases. This is shown for joint E in figure 2.20.

The bending-moment diagram (figure 2.19c) can now be drawn and salient values attached. The maximum or minimum moment should be found for DE; this occurs when the shear force is zero, that is at $2 \cdot 38$ m from A. The bending moment at this point is

$$-72 + (47 \cdot 6 \times 2 \cdot 38) - \left(20 \times \frac{2 \cdot 38^2}{2}\right) = -16 \text{ kNm}$$

An example of a structure that is statically indeterminate but that can be solved by making an assumption is shown in figure 2.21a. The beams are very much more

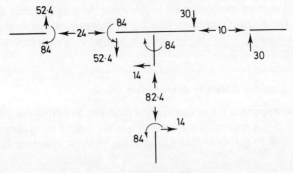

Figure 2.20

rigid than the columns. Bending moments are required at the ends of all the members.

If the beams are assumed to be infinitely stiff, they will not deform under the action of bending moments, hence there will not be any joint rotation when the load is applied. If in addition the effect of axial forces is neglected the frame will deform in the manner shown in figure 2.21b. The columns in a particular storey must deform in an identical fashion, hence they must be identically loaded. If we look at b we can see that the point of contraflexure or zero bending moment must occur at the mid-point of each column; we have in fact established an equation of condition based on the approximations stated above. The column ACE, shown in c, is broken into two parts at C with all the horizontal forces for equilibrium entered on the diagram.

For column CE, taking moments about the mid-point H

$$M_{EC} = 30 \times 1\tfrac{1}{2} = 45 \text{ kNm} = M_{CE}$$

For column CA, taking moments about the mid-point G

$$M_{CA} = 70 \times 2 = 140 \text{ kNm} = M_{AC}$$

It is now possible to draw the complete bending-moment diagram. The moments in the right-hand column are the same as the corresponding ones in the left-hand column. As the sum of the bending moments at joint C is zero, M_{CD} = 185 kNm.

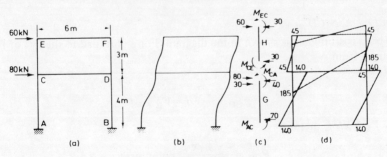

Figure 2.21

The same convention has been used to draw the bending-moment sketch (figure 2.21d). Bending moments are positive if they produce tension in the inside edge of a member and positive bending moments have been drawn on the inside of the frame.

2.12 Influence lines for statically determinate trusses

The virtual-work approach is the simplest method for sketching the influence line for the axial force in a particular member of a truss. Before discussing this in detail it is advisable to get some idea of the shape of these influence lines.

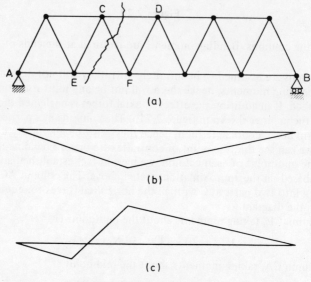

Figure 2.22

The influence lines for the forces in various members are required when a unit load crosses from A to B (figure 2.22a). All the members of the pinjointed truss have the same length l.

Chord member CD

The truss is sectioned as shown in the diagram. For a unit load to the left of F and distant x from A

ΣM_F

$$F_{CD} \times \frac{\sqrt{3}}{2} l = V_B \times 3l$$

Now $V_B = x/5l$, therefore

$$F_{CD} = \frac{2\sqrt{3}}{5} \frac{x}{l} \quad \text{compressive}$$

For the unit load to the right of F

ΣM_F

$$F_{CD} \times \frac{\sqrt{3}}{2} l = V_A \times 2l$$

Now $V_A = (5l - x)/5l$, therefore

$$F_{CD} = \frac{4}{\sqrt{3}} \frac{(5l - x)}{5l} \quad \text{compressive}$$

Both of the expressions for F_{CD} are linear and give a maximum value of $4\sqrt{3}/5$ when the unit load is at F (figure 2.22b). The influence line for other chord members can be found in a similar manner.

Inclined member CF

Making use of the same section, the vertical component of the force in CF will be equal to the shear force in panel CDEF. The influence line for shear in a panel has already been discussed in section 1.22. To obtain the influence line for the axial force in CF, the influence line for shear force will have to be multiplied by $2/\sqrt{3}$ (figure 2.22c). It will be noted that the member is in compression while the unit load moves from A to E and reaches a maximum value of $2/5\sqrt{3}$ at E. From E to F the force changes from compression to tension and has a maximum tensile value of $6/5\sqrt{3}$. At F. Thereafter the axial force decreases linearly, remaining in tension.

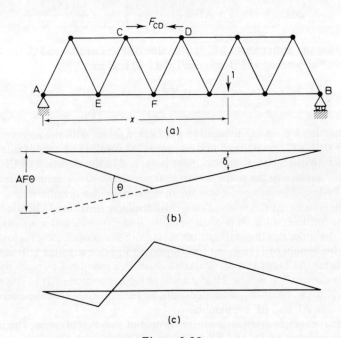

Figure 2.23

If the influence line for the axial force in CE were required, it would have exactly the same shape and numerical values as that for CF except that all signs would be reversed.

If a vertical member exists in a truss, a suitable section can usually be chosen. The influence line would have the same shape and numerical value as the influence line for shear force in an adjacent panel.

The virtual-work method will now be applied to the same problem.

The member CD is removed and equal and opposite forces F_{CD} are applied at C and D. These two points are given a virtual displacement Δ and approach one another. Now both ACF and BDF are rigid bodies connected together at the pin F. Thus the base line AEFB would deform as shown in figure 2.23b. The unit load being displaced by δ.

Using virtual work

$$F_{CD} \times \Delta + 1 \times \delta = 0 \quad F_{CD} = -\frac{\delta}{\Delta}$$

The negative sign indicates that CD is in compression. The force is proportional to the vertical displacement of the unit load, so that the displaced shape of AEFB represents the shape of the influence line. It will only be necessary to find the ratio δ/Δ to fix the scale of the influence line.

Let θ be the relative rotation of AF and BF. When the unit load is to the right of F

$$\Delta = \frac{\sqrt{3}}{2} l\theta \quad \delta = AF \times \theta \left(\frac{5l - x}{5l}\right) \quad \frac{\delta}{\Delta} = \frac{4}{\sqrt{3}} \frac{(5l - x)}{5l}$$

Note that the term $(5l - x)/5l$ is the value of V_A. For the load to the left of F, the value of δ becomes FB $\times \theta (x/5l)$ and δ/Δ becomes

$$\frac{2\sqrt{3}}{5} \frac{x}{l}$$

For member CF we can proceed in a similar manner. With the member removed ACE and DBF will remain rigid bodies connected together by two parallel links CD and EF. When a virtual displacement is applied in the direction of CF the line AEFB will displace to the shape shown in figure 2.23c. Once again the ratio of δ/Δ can be found in terms of x and the dimensions of the members.

It is suggested that the method of virtual displacements should be used to sketch the influence lines. It is often easier to find a spot value for a particular point on the influence line, rather than trying to fix a general scale factor. For example for member CD it would be convenient to place the load at F and hence find the force in CD for this particular position. For member CF, place the load at E, the value of F_{CF} will be $2V_B/\sqrt{3}$ where V_B has a value of $1/5$. The maximum positive value can be obtained immediately by remembering that the parts of the influence line AE and BF are parallel.

As a final example consider a bridge formed of two Pratt trusses. The influence lines for members CD, FG and EF are required (figure 2.24a).

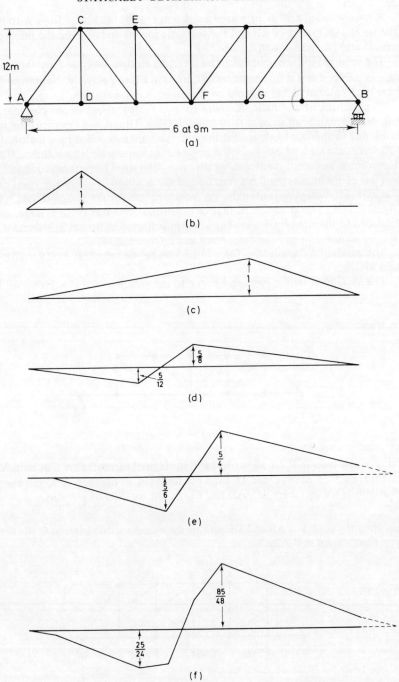

Figure 2.24

A four-wheeled vehicle of wheelbase 4·5 m crosses the bridge from A to B. The front-axle load is 20 kN and the rear-axle load is 40 kN. Find the maximum tension and compression in member EF.

The required influence lines can be sketched straight away and are shown in figures 2.24b, c and d for members CD, FG and EF respectively. Numerical values have been calculated and entered on the diagram at salient points.

Next consider the vehicle. As there are two Pratt trusses, we can consider one-half of each axle load to be carried by each truss. Thus d will represent the force in EF due to the front-axle load if the ordinates are multiplied by a factor of 10 kN. The rear-axle load can be represented to the same scale if an influence line is drawn for a 'two-unit' load crossing the truss. The total force is required in EF and this could be sketched as a function of the position of either the front or the rear axle. Choosing the front axle, the influence line for the two-unit load is drawn (figure 2.24e). This is similar to that of d, but has ordinates twice as large, and is displaced to the right by 4·5 m. The graphs illustrated in figures 2.24d and e can next be summed to give the total force in EF (figure 2.24f).

The maximum compressive force in EF due to the vehicle is 25/24 × 10 = 10·4 kN.

The maximum tensile force in EF due to the vehicle is 85/48 × 10 = 17·7 kN.

Problems

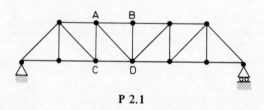

P 2.1

2.1 For the truss in P 2.1 all vertical and horizontal members are 5 m long. Vertical loads of 5 kN act at C and D. Find the forces in all the members, and check your results for members AC, AD and CD by the method of sections.

2.2 Find the forces in AB and DE and all the members that meet at C for the K-type truss shown in P 2.2.

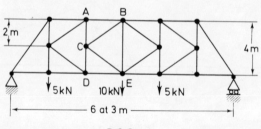

P 2.2

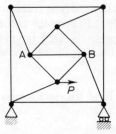

P 2.3

2.3 For a statically determinate framework, derive relations between the number of bars, reactions and joints for the following cases.

(a) two-dimensional pin-jointed;
(b) two-dimensional rigid-jointed;
(c) three-dimensional pin-jointed.

For a and b sketch examples that are not completely statically determinate but for which the relation holds.

The plane framework P 2.3 is pin-jointed and basically consists of an inner square of side $a/\sqrt{2}$ connected to an outer square of side $2a$. The diagonal bar of the inner square is horizontal. Find the force in AB when the horizontal load P is applied.

2.4 A plane pin-jointed structure consists of 9 nodes, 4 of which are pinned to a rigid foundation. Sketch a suitable framework such that the complete structure is statically determinate. Repeat for the case of a three-dimensional pin-jointed structure.

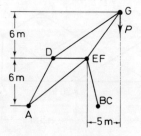

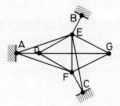

P 2.4

The plan and elevation of a derrick-type crane are shown in P 2.4. All members are pin-jointed. A, B, C, and D, E, F, lie at the corners of equilateral triangles of 10 m and 5 m side lengths respectively. DEF lies in a plane parallel to the ground and AD = BE = FC. Find the forces in all the members when the vertical load P is applied at G.

2.5 The framework shown in P 2.5 has rigid joints.

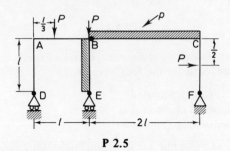

P 2.5

(a) Show that the framework is statically determinate.
(b) Find the reactions at D, E and F when $pl = P$.
(c) Sketch the bending-moment diagram stating the sign convention adopted.

2.6 In the structure shown in P 2.6 the members are vertical, horizontal or inclined at 30° to the horizontal, and they cross without being connected. Show that the frame is statically determinate and find the force in the member AD. Hence find the forces in the other members.

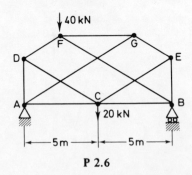

P 2.6

2.7 For the truss shown in P 2.1 draw the influence lines for the forces in members AB, AC, AD, and CD. A 7 m long, uniformly distributed load of 5 kN/m crosses the truss. Find the maximum force in member CD.

3 ELEMENTARY ELASTICITY, PLASTICITY AND BENDING OF BEAMS

3.1 Sign conventions

It is necessary to adopt a sign convention when using a rectangular or cartesian set of axes. A right-handed system will be adopted here. This means that if the axes are taken in the sequence Ox, Oy, Oz, Ox etc. and the directions of any two of the axes are chosen, then the direction of the third axis will be given by the right-handed-screw rule.

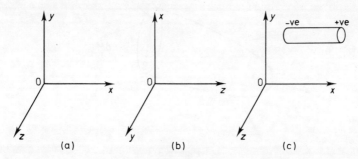

Figure 3.1

Two examples of right-handed sets are shown in figures 3.1a and b. Suppose that the directions of x and z have been chosen in a. The sequence would be z, x, y, so that Oy must point vertically upwards.

Next consider the section of a body in figure 3.1c. The exposed face will be considered positive if the outward normal to the face acts in a positive direction when referred to the coordinate system. This means that the right-hand face in figure 3.1c is positive while the left-hand face is negative.

A force or component of a force or moment will be considered as positive if it acts on a positive face in a positive direction, or on a negative face in a negative direction.

This will lead to a tensile force as positive (figure 3.2a) and a compressive force as negative (3.2b). Positive shear force is shown at c and a positive bending moment at d. These conventions are identical with those adopted in chapter 1, where for example a sagging bending moment was considered as positive.

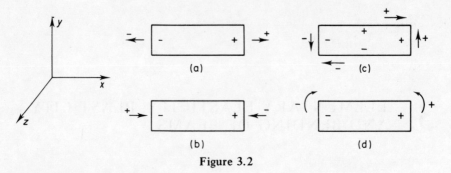

Figure 3.2

3.2 Stress

If a body that is under the action of external forces is imagined to be sectioned and equilibrium maintained, then there must be internal forces acting on the cross-section (figure 3.3).

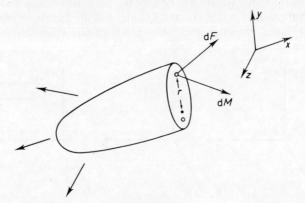

Figure 3.3

On the small element of cross-section dA there is a force dF and a moment dM. There is of course no reason why the force should act normal to the surface.

Total force acting $F = \Sigma \, dF$

The moment of the force about O could be written as

$$\Sigma \, dM + \Sigma \, r \times dF$$

If a set of axes is now introduced such that the x direction is normal to the sectioned face, as dA tends to zero the value of dF/dA is termed the stress. This stress is in fact a tensor. A vector has three properties, direction, magnitude and sense. This would not be sufficient in the case of stress as it is also necessary to know the direction of the normal relative to the element face.

It would be perfectly possible to resolve the small force into three components dF_x, dF_y and dF_z. Since dF_x is in the direction of the normal to the surface it is called the normal or direct force. The other two forces act at right angles to the normal and lie in the plane of the section, these are both referred to as shear forces.

We now write

$$\sigma_{xx} = \frac{\mathrm{d}F_x}{\mathrm{d}A} \qquad \tau_{xy} = \frac{\mathrm{d}F_y}{\mathrm{d}A} \qquad \tau_{xz} = \frac{\mathrm{d}F_z}{\mathrm{d}A} \tag{3.1}$$

where σ is a normal stress and τ is a shear stress. The first subscript denotes the direction of the normal to the face and the second subscript the direction in which the stress is acting. It is usual to express a stress in terms of N/mm^2.

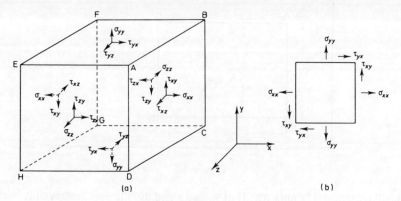

Figure 3.4

A small parallelepiped of sides $\mathrm{d}x$, $\mathrm{d}y$ and $\mathrm{d}z$ is shown in figure 3.4a, its centroid is at x, y, z, and it has been cut from the body such that the face ABCD lies in the plane of the original section. The block is in equilibrium. All possible stresses that can act have been shown on the diagram acting in a positive direction. It will be seen that there are nine different stress components—three normal and six shear. If we look directly at the front face EADH, we see the stresses indicated in figure 3.4b.

shear forces on face ABFE and DCGH are $\tau_{yx}\,\mathrm{d}x\,\mathrm{d}z$
the moment about the centre of EADH is $\tau_{yx}\,\mathrm{d}x\,\mathrm{d}z\,\mathrm{d}y$
shear forces on faces ABCD and EFGH are $\tau_{xy}\,\mathrm{d}y\,\mathrm{d}z$
the moment about the centre of EADH is $\tau_{xy}\,\mathrm{d}y\,\mathrm{d}z\,\mathrm{d}x$

Now as the parallelepiped is in equilibrium there cannot be any rotation, hence the sum of the moments must be zero. Thus

$$\tau_{yx} = \tau_{xy} \tag{3.2}$$

similarly $\tau_{xz} = \tau_{zx}$ and $\tau_{yz} = \tau_{zy}$

It can now be seen that the original nine different stresses have been reduced to six. It should also be noted that it is impossible for τ_{xy}, say, to exist without τ_{yx}—a shear stress is always accompanied by a complementary shear stress.

3.3 Strain

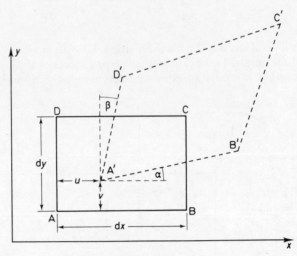

Figure 3.5

A small rectangular lamina ABCD of sides dx and dy, the coordinates of A being x, y, is shown in figure 3.5. The lamina is loaded in some manner such that it displaces to A'B'C'D', the displacement of A being u in the x direction and v in the y direction. We now want to try to write down expressions for the new co-ordinates of A', B' and D'. There is certainly no difficulty about A' and it has coordinates $x + u, y + v$.

A convenient way of representing the horizontal displacement of B would be to say that it was equal to the horizontal displacement of A plus the rate of change of displacement in the x direction times the length, that is

$$u + \frac{\partial u}{\partial x} dx$$

The partial derivative is used as we only require the rate at which u has changed with x.

Similarly the vertical displacement of B could be written

$$v + \frac{\partial v}{\partial x} dx$$

Thus the coordinates of B' are

$$x + dx + u + \frac{\partial u}{\partial x} dx, \quad y + v + \frac{\partial v}{\partial x} dx$$

In a similar manner it is easily shown that the coordinates of D′ are

$$x + u + \frac{\partial u}{\partial y}\, dy, \quad y + dy + v + \frac{\partial v}{\partial y}\, dy$$

The horizontal component of length of A′B′ is

$$dx + \frac{\partial u}{\partial x}\, dx$$

and the vertical component

$$\frac{\partial v}{\partial x}\, dx$$

So that the length that was originally dx has become

$$\left\{ \left(dx + \frac{\partial u}{\partial x}\, dx \right)^2 + \left(\frac{\partial v}{\partial x}\, dx \right)^2 \right\}^{1/2} = dx \left\{ 1 + 2\frac{\partial u}{\partial x} + \left(\frac{\partial u}{\partial x} \right)^2 + \left(\frac{\partial v}{\partial x} \right)^2 \right\}^{1/2}$$

Now $\partial u/\partial x$ is in any case a very small term so that it will be in order to neglect terms such as $(\partial u/\partial x)^2$ and the length becomes $dx(1 + \partial u/\partial x)$. The increase in length of AB is

$$dx \left(1 + \frac{\partial u}{\partial x} \right) - dx = \frac{\partial u}{\partial x}\, dx$$

If we define the normal or direct strain as

$$\epsilon = \frac{\text{increase in length}}{\text{original length}}$$

$$\epsilon_{xx} = \left(\frac{\partial u}{\partial x}\, dx \right) \Big/ dx = \frac{\partial u}{\partial x}$$

In a similar manner we can show that $\epsilon_{yy} = \partial v/\partial y$. The analysis could be extended to a three-dimensional case and

$$\epsilon_{xx} = \frac{\partial u}{\partial x} \quad \epsilon_{yy} = \frac{\partial v}{\partial y} \quad \epsilon_{zz} = \frac{\partial w}{\partial z} \tag{3.3}$$

It should be noted that the displacements u, v and w are in the x, y and z directions respectively.

The angle DAB, after the deformation has taken place, is no longer a right angle. The shear strain γ_{xy} is defined as the change in value of the right angle.

$$\tan \alpha = \frac{\dfrac{\partial v}{\partial x}\, dx}{dx + \dfrac{\partial u}{\partial x}\, dx} = \frac{\dfrac{\partial v}{\partial x}}{1 + \epsilon_{xx}} \approx \frac{\partial v}{\partial x}$$

similarly

$$\tan \beta = \frac{\partial u}{\partial y}$$

The angles are very small hence $\tan \alpha = \alpha$ and $\tan \beta = \beta$.

$$\alpha + \beta = \gamma_{xy} = \frac{\partial v}{\partial x} + \frac{\partial u}{\partial y}$$

In a three-dimensional case it can be shown that

$$\gamma_{xy} = \frac{\partial v}{\partial x} + \frac{\partial u}{\partial y} \quad \gamma_{xz} = \frac{\partial w}{\partial x} + \frac{\partial u}{\partial z} \quad \gamma_{yz} = \frac{\partial w}{\partial y} + \frac{\partial v}{\partial z} \qquad (3.4)$$

The only assumptions that have been made in deriving equations 3.3 and 3.4 is that all the displacements are small.

3.4 Relations between stress and strain

The fundamental experimental law on which linear elastic theory is based was discovered by Robert Hooke in 1678. When subjected to uniaxial loading it was found that for certain materials the extension was proportional to the applied load. If a rod of linear elastic material with a uniform cross-section is loaded axially, the stress is uniform over the cross-section and would be equal to the load divided by the area of the cross-section. The axial strain would be the extension divided by the original length, so that Hooke's law may be written

$$\sigma_{xx} \propto \epsilon_{xx}$$

A constant of proportionality, E, which will vary from one elastic material to another can be introduced—this is known as Young's modulus, or the modulus of elasticity.

$$\sigma_{xx} = E \epsilon_{xx} \qquad (3.5)$$

If careful measurements are made on the material under test it will be found that the dimensions of the cross-section also change as the load is applied. The strain in both the y and z directions is proportional to the strain in the x direction but opposite in sign. The constant of proportionality ν is referred to as Poisson's ratio.

$$\epsilon_{yy} = \epsilon_{zz} = -\nu \epsilon_{xx} \qquad (3.6)$$

Typical values of E and ν for three different materials are given below.

	Mild steel	Brass	Duralumin
$E\,(\text{kN/mm}^2)$	210	100	70
ν	0·3	0·35	0·33

If P is the load applied to the rod and A the cross-sectional area $\sigma_{xx} = P/A$. Suppose that the rod is of length l and extends by an amount e, then $\epsilon_{xx} = e/l$.

Substituting into equation 3.5

$$e = \frac{Pl}{AE}$$

It has been assumed that the direct stress is uniformly distributed over the cross-section. If the load is applied to the rod by means of a testing machine the stress distribution at the ends will certainly not be uniform. A section that was plane in the unloaded state will no longer be plane when the load is applied. However at a small distance from the ends it will be found that the stress distribution is sensibly uniform and it may be assumed that plane sections remain plane. This is a particular application of what is known as St Vernant's principle.

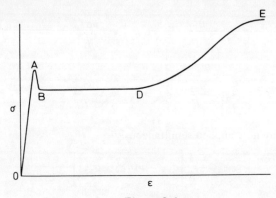

Figure 3.6

Not all materials exhibit this linear stress–strain relationship and those that do will only behave in this manner for a limited value of strain. Mild steel for example has a curve similar to that shown in figure 3.6. The portion OA is linear; at A there is a yield point with a sudden drop in stress to a lower yield point B. This is followed by what is termed a *plastic* range, BD, where there is a large increase in strain for a small increase in stress. Thereafter the stress increases more rapidly with work hardening until the ultimate stress is reached at E. It is of interest to note that the strain at D is about twelve times that at A.

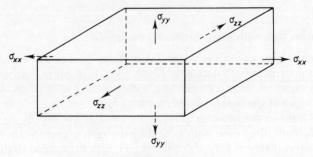

Figure 3.7

A rectangular parallelepiped is shown in figure 3.7 with stresses σ_{xx}, σ_{yy}, and σ_{zz} applied. The material is linearly elastic, homogeneous and isotropic, that is, it has the same properties in all directions. Hence we can use the method of super-position: apply each stress separately, determine its effect, and then add all the effects together as follows

σ_{xx} applied alone

$$\epsilon_{xx} = \frac{\sigma_{xx}}{E} \qquad \epsilon_{yy} = \epsilon_{zz} = -\frac{\nu\sigma_{xx}}{E}$$

σ_{yy} applied alone

$$\epsilon_{yy} = \frac{\sigma_{yy}}{E} \qquad \epsilon_{xx} = \epsilon_{zz} = -\frac{\nu\sigma_{yy}}{E}$$

σ_{zz} applied alone

$$\epsilon_{zz} = \frac{\sigma_{zz}}{E} \qquad \epsilon_{xx} = \epsilon_{yy} = -\frac{\nu\sigma_{zz}}{E}$$

If all stresses are now applied simultaneously

$$\epsilon_{xx} = \frac{1}{E}\left[\sigma_{xx} - \nu(\sigma_{yy} + \sigma_{zz})\right] \tag{3.7}$$

two similar expressions can be derived for ϵ_{yy} and ϵ_{zz}.

These expressions are the generalized form of Hooke's law. A very common error is to assume that the uniaxial case of equation 3.5 holds on all occasions—in other words the Poisson-ratio effect is ignored. A further relationship exists between shear stress τ and shear strain γ.

$$\tau_{xy} = G\gamma_{xy} \qquad \tau_{yz} = G\gamma_{yz} \qquad \tau_{xz} = G\gamma_{xz} \tag{3.8}$$

where G is another elastic constant called the *modulus of rigidity* or sometimes the *shear modulus.* These linear relationships again only apply to certain materials and will hold only for a limited value of shear strain.

3.5 Composite rods with axial tension or compression

The type of problem we shall consider is that of two rods of the same length, but of different materials which have been rigidly fixed together at both ends. A load P is applied such that there is no bending in either rod. The question arises as to what proportion of the load is carried by each rod.

Figure 3.8 shows the two rods of materials 1 and 2. As the rods are rigidly con-nected at the ends, they must both extend by the same amount when the load is applied or alternatively as they are of the same length the normal strains must be the same.

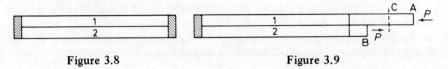

Figure 3.8   Figure 3.9

Let P_1 and P_2 be the loads carried by each rod of areas A_1 and A_2 and Young's moduli E_1 and E_2.

$$\epsilon_1 = \epsilon_2$$

thus

$$\frac{P_1}{A_1 E_1} = \frac{P_2}{A_2 E_2}$$

also

$$P_1 + P_2 = P$$

From these

$$P_1 = \frac{P A_1 E_1}{A_1 E_1 + A_2 E_2}$$

$$P_2 = \frac{P A_2 E_2}{A_1 E_1 + A_2 E_2}$$

Note that

$$\sigma_1 = \frac{P}{A_1 + A_2(E_2/E_1)} \tag{3.9}$$

It is worthwhile examining the expression for σ_1 in a little more detail. The stress is given by the total load applied divided by the area of material 1 together with a modified area of material 2. This is the same as converting material 2 to an equivalent amount of material 1. This process is referred to as transforming the section: that is, converting it all to one material.

Now consider the effect of a temperature variation on the composite rod. Again the restriction is made that axial effects only are considered. This could affect the configuration of the material forming the rods as a symmetrical arrangement would have to be used to avoid secondary bending effects. One possible arrangement would be to have a circular rod of material 1 inside a tube of material 2. However for the analysis we shall consider the two rods side by side—figure 3.9 with the fixing removed at one end.

If a temperature rise T is applied and the rods have coefficients of linear expansion α_1 and α_2 where $\alpha_1 > \alpha_2$, rod 1 expands to A while rod 2 only expands to B. In fact the rods must finish with the same length as they are really fixed at both ends. The only way of achieving this is to introduce two forces each having a magnitude P, to bring the internal force system to equilibrium. A compressive force P acts on material 1 and a tensile force P on material 2 such that the final position at both ends is at C. Both rods have an original length l.

Due to the temperature change, T, the extension of rod 1 is $l\alpha_1 T$ and due to the force P the extension is $-Pl/A_1E_1$. Note that the original length l is used for this second extension and not the length $l(1 + \alpha_1 T)$—the reason for this being that changes in length are very small compared with the original length.

$$\text{Total extension of } 1 = l\alpha_1 T - \frac{Pl}{A_1 E_1}$$

$$\text{Total extension of } 2 = l\alpha_2 T + \frac{Pl}{A_2 E_2}$$

These extensions are equal—hence the strains are equal, and

$$\alpha_1 T - \frac{P}{A_1 E_1} = \alpha_2 T + \frac{P}{A_2 E_2}$$

therefore

$$P = \frac{A_1 A_2 E_1 E_2 (\alpha_1 - \alpha_2) T}{A_1 E_1 + A_2 E_2} \tag{3.10}$$

As an example a steel tie rod 20 mm in diameter is placed concentrically in a brass tube of 3 mm wall thickness and 50 mm mean diameter. Nuts and washers are fitted to the tie rod such that the ends of the tube are closed by the washers. The nuts are first tightened to give a compressive stress of 30 N/mm² in the tube. A tensile load of 20 kN is then applied to the tie rod. Find the resulting stresses in both the tie rod and the tube (a) with no change in temperature, (b) when the temperature rises by 60 °C. $E_S = 210$ kN/mm², $E_B = 100$ kN/mm², $\alpha_S = 12 \times 10^{-6}/°C$, $\alpha_B = 20 \times 10^{-6}/°C$.

Area of steel = 314 mm², area of brass = 471 mm². Initial load in brass = initial load in steel therefore

$$\text{stress in steel} = \frac{30 \times 471}{314} = 45 \text{ N/mm}^2$$

Now consider the effect of a 20 kN load applied to the unloaded system assuming that the tube and rod are fixed together at their ends. They would therefore extend by the same amount, the 20 kN being shared between them.

$$\sigma_S = \frac{20 \times 10^3}{314 + 471 \times \frac{100}{210}} = 37 \cdot 6 \text{ N/mm}^2$$

$$\sigma_B = 37 \cdot 6 \times \frac{100}{210} = 18 \cdot 8 \text{ N/mm}^2$$

therefore

$$\text{resultant stress in steel} = 45 + 37\cdot6 = 82\cdot6 \text{ N/mm}^2 - \text{tensile}$$

$$\text{resultant stress in brass} = 30 - 18\cdot8 = 11\cdot2 \text{ N/mm}^2 - \text{compressive}$$

Considering the temperature effect only, from equation 3.10

$$P = \frac{314 \times 471 \times 210 \times 100(20 - 12) \times 10^{-6} \times 60}{(210 \times 314) + (100 \times 471)} = 13\cdot1 \text{ kN}$$

The stress in the steel is $41\cdot7$ N/mm^2 tensile, and in the brass $27\cdot8$ N/mm^2 compressive. Therefore final stresses are

$$\text{steel } 82\cdot6 + 41\cdot7 = 124\cdot3 \text{ N/mm}^2 - \text{tensile}$$

$$\text{brass } 11\cdot2 + 27\cdot8 = 39\cdot0 \text{ N/mm}^2 - \text{compressive}$$

3.6 Composite rods loaded beyond the yield point

When a tensile test is carried out on a mild steel specimen, a stress–strain diagram similar to that shown in figure 3.6 will be obtained. If the load is removed after the yield point has been reached, that is, in the range BDE, the unloading path will be a straight line parallel to the elastic part of the original loading curve. It may then be assumed that the unloading process is linear elastic with a modulus of E. This indicates, for a tensile loading process exceeding the elastic limit, that there will be a residual strain in the specimen when the applied stresses reach zero. In other words the specimen will not regain its original length but will remain extended; the extension is referred to as a permanent set.

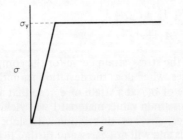

Figure 3.10

If the strains are not too large the stress–strain diagram can be idealised without too much error to the form indicated in figure 3.10. It will be noted that the upper yield point has been suppressed.

Let us now examine the case of a composite system consisting of a rod of material 1 inside a tube of material 2. The two members are of the same length and are rigidly fastened together at each end. The idealised stress–strain curve for each material is shown in figure 3.11. It will be assumed that the yield stresses

σ_{y1} and σ_{y2}, and the moduli E_1 and E_2 are all known, as are the cross-sectional areas A_1 and A_2.

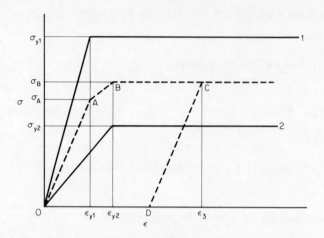

Figure 3.11

When both materials are in the elastic range, from equation 3.9

$$\epsilon_1 = \epsilon_2 = \frac{P}{A_1 E_1 + A_2 E_2}$$

The equivalent value of Young's modulus E_c for the composite can be found from the stress $\sigma = P/(A_1 + A_2)$ and the strain $\epsilon = \epsilon_1$, hence

$$E_c = \frac{E_1 A_1 + E_2 A_2}{A_1 + A_2}$$

The initial portion of the stress–strain curve for the composite may now be drawn. The question arises, when does the deviation from the straight line occur? Material 1 reaches a value of σ_{y1} at a strain of ϵ_{y1}, so that as soon as the strain in the composite reaches this strain value, material 1 will yield. This means that the composite will have a linear stress–strain relation with slope E_c in the region OA. At A the material 1 yields and will not carry any further load in excess of $\sigma_{y1} A_1$. Material 2 is, however, still behaving in a linear elastic manner and will continue to do so until the strain reaches a value of ϵ_{y2}. It would appear that there is a further linear relation between σ and ϵ; at first sight the slope of this line might well be thought to be E_2. This is, however, incorrect. For an increment of strain ϵ beyond ϵ_{y1} the increase in stress in 2 is ϵE_2 and the increase in the load P is $\epsilon E_2 A_2$. The average stress in the composite is $\epsilon E_2 A_2/(A_1 + A_2)$. This gives an equivalent value of E of $E_2 A_2/(A_1 + A_2)$. The stress–strain relation will be linear from A to B with this slope. At B the material yields and there is no further increase in stress with increasing strain. The values of the stresses at which there is a change of slope on the composite stress–strain curve will be given by

$$\sigma_A = \left(\frac{E_1 A_1 + E_2 A_2}{A_1 + A_2} \right) \frac{\sigma_{y1}}{E_1} \text{ and } \sigma_B = \left(\frac{\sigma_{y1} A_1 + \sigma_{y2} A_2}{A_1 + A_2} \right)$$

We shall next examine the situation when the composite is strained to a value ϵ_3 which is greater than ϵ_{y2} and then unloaded. It has already been stated that unloading takes place in a linear elastic manner. Since the initial value for E was E_c, this must be the elastic constant for unloading and the line CD will be followed, giving a permanent set with strain value OD.

The decrease in strain while unloading is

$$\frac{\sigma_B}{E_C} = \frac{\sigma_{y1} A_1 + \sigma_{y2} A_2}{E_1 A_1 + E_2 A_2}$$

The change in stress in the individual materials can next be found from this value of strain and the Young's modulus for each material, that is

$$\sigma_1 = \left(\frac{\sigma_{y1} A_1 + \sigma_{y2} A_2}{E_1 A_1 + E_2 A_2} \right) E_1$$

This is obviously not equal to σ_{y1} and there must be a residual stress in material 1 given by

$$\sigma_{y1} - \sigma_1 = \frac{(\sigma_{y1} E_2 - \sigma_{y2} E_1) A_2}{E_1 A_1 + E_2 A_2}$$

and that in material 2 is

$$\frac{(\sigma_{y2} E_1 - \sigma_{y1} E_2) A_1}{E_1 A_1 + E_2 A_2}$$

Examining the two results it can be seen that they are of opposite sign: one material will have a residual tensile stress and the other a residual compressive stress. This result is to be expected because the residual load in the composite is zero. This can easily be checked by multiplying the residual stresses by the corresponding cross-sectional areas.

3.7 Principal stresses in two dimensions

The rectangular element in figure 3.12a is of unit thickness and is acted on by the stress system shown. This is referred to as a plane stress system, since $\sigma_{zz} = \tau_{xz} = \tau_{yz} = 0$. The triangular element in figure 3.12b has been generated by sectioning the rectangular element perpendicular to the xy plane. We now want to find the stresses that must act on plane AC to maintain the triangular element in equilibrium. The obvious way of proceeding is to resolve forces along and normal to AC. A fundamental point arises here in that it is essential to resolve forces and not stresses.

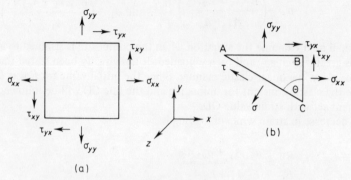

(a)

(b)

Figure 3.12

Resolving forces normal to AC

$$\sigma \times AC = \sigma_{xx} \times BC \cos \theta + \sigma_{yy} \times AB \sin \theta + \tau_{xy} \times BC \sin \theta + \tau_{yx} \times AB \cos \theta$$

In section 3.2 we showed that $\tau_{xy} = \tau_{yx}$ therefore

$$\sigma = \sigma_{xx} \cos^2 \theta + \sigma_{yy} \sin^2 \theta + \tau_{xy} \sin \theta \cos \theta + \tau_{xy} \sin \theta \cos \theta$$

or

$$\sigma = \tfrac{1}{2}(\sigma_{xx} + \sigma_{yy}) + \tfrac{1}{2}(\sigma_{xx} - \sigma_{yy}) \cos 2\theta + \tau_{xy} \sin 2\theta \qquad (3.11)$$

If we resolve in the direction AC we shall find that

$$\tau = \tfrac{1}{2}(\sigma_{xx} - \sigma_{yy}) \sin 2\theta - \tau_{xy} \cos 2\theta \qquad (3.12)$$

If τ is zero then

$$\tan 2\theta = \frac{2\tau_{xy}}{\sigma_{xx} - \sigma_{yy}} \qquad (3.13)$$

therefore

$$\theta = \tfrac{1}{2} \tan^{-1} \frac{2\tau_{xy}}{\sigma_{xx} - \sigma_{yy}}$$

or

$$\theta = \tfrac{1}{2} \tan^{-1} \frac{2\tau_{xy}}{\sigma_{xx} - \sigma_{yy}} + \frac{\pi}{2}$$

There are therefore two planes inclined at $90°$ to each other on which the value of the shear stress is zero. These are known as principal planes—the stress acting is normal to these planes and is called a principal stress.

We can find the planes on which the direct stress is a maximum by differentiating equation 3.11 and setting the result equal to zero.

$$-(\sigma_{xx} - \sigma_{yy}) \sin 2\theta + 2\tau_{xy} \cos 2\theta = 0$$

or

$$\tan 2\theta = \frac{2\tau_{xy}}{\sigma_{xx} - \sigma_{yy}}$$

This result was also obtained for the case when $\tau = 0$. So the principal stresses are also the maximum and minimum direct stresses that can occur. To find the values of these stresses it is only necessary to substitute the value of tan 2θ into equation 3.11. The arithmetic is somewhat tedious and an alternative method exists.

In figure 3.10b assume that σ is a principal stress, that is, τ will be zero. Resolve normal and along the direction of BC.

$$\sigma \times AC \cos \theta = \sigma_{xx} BC + \tau_{xy} AB$$

or

$$\tan \theta = \frac{\sigma - \sigma_{xx}}{\tau_{xy}}$$

$$\sigma \times AC \sin \theta = \sigma_{yy} AB + \tau_{xy} BC$$

or

$$\cot \theta = \frac{\sigma - \sigma_{yy}}{\tau_{xy}}$$

Eliminating θ

$$(\sigma - \sigma_{xx})(\sigma - \sigma_{yy}) - \tau_{xy}^2 = 0$$

This is a quadratic in σ which solves to give

$$\sigma_1, \sigma_2 = \frac{\sigma_{xx} + \sigma_{yy}}{2} \pm \tfrac{1}{2}[(\sigma_{xx} - \sigma_{yy})^2 + 4\tau_{xy}^2]^{1/2} \qquad (3.14)$$

these are the two principal stresses. Note that

$$\sigma_1 + \sigma_2 = \sigma_{xx} + \sigma_{yy} \qquad (3.15)$$

We should also determine the maximum value of the shear stress and the plane on which it acts. The maximum value will occur when $d\tau/d\theta = 0$ in equation 3.12 or

$$\tan 2\theta = \frac{\sigma_{yy} - \sigma_{xx}}{2\tau_{xy}} \qquad (3.16)$$

The plane on which the maximum shear stress acts is inclined at $45°$ to the planes of principal stress (compare equation 3.16 with equation 3.13).

If the value of $\tan 2\theta$ from equation 3.16 is substituted into equation 3.12 the maximum value of the shear stress can be found.

$$\tau_{max} = \tfrac{1}{2}\left[(\sigma_{xx} - \sigma_{yy})^2 + 4\tau_{xy}^2\right]^{1/2} \tag{3.17}$$

From the values of σ_1 and σ_2 given by equation 3.14 it can be seen at once that

$$\tau_{max} = \tfrac{1}{2}(\sigma_1 - \sigma_2) \tag{3.18}$$

This is a most important relationship—it is very simple to remember that the maximum shear stress is equal to one half of the difference between the principal stresses.

If a knowledge of matrix algebra is assumed, an elegant method for determining the values of the principal stresses is as follows.

In figure 3.12, define the plane AC as a principal plane, that is, a plane on which the shear stress is zero. Resolving forces in the x and y directions will give

$$\sigma_{xx}\,\text{BC} + \tau_{yx}\,\text{AB} = \sigma\text{AC}\cos\theta$$

$$\sigma_{yy}\,\text{AB} + \tau_{xy}\,\text{BC} = \sigma\text{AC}\sin\theta$$

or

$$\sigma_{xx}\cos\theta + \tau_{yx}\sin\theta = \sigma\cos\theta$$

$$\tau_{xy}\cos\theta + \sigma_{yy}\sin\theta = \sigma\sin\theta$$

In matrix form

$$\begin{bmatrix} \sigma_{xx} & \tau_{yx} \\ \tau_{xy} & \sigma_{yy} \end{bmatrix} \begin{Bmatrix} \cos\theta \\ \sin\theta \end{Bmatrix} = [\sigma] \begin{Bmatrix} \cos\theta \\ \sin\theta \end{Bmatrix} = [\sigma]\,[\mathbf{I}] \begin{Bmatrix} \cos\theta \\ \sin\theta \end{Bmatrix}$$

the reason for the last step being that $[\sigma]$ is a diagonal matrix, since there are no shear stresses.

$$\begin{bmatrix} \sigma_{xx} - \sigma & \tau_{yx} \\ \tau_{xy} & \sigma_{yy} - \sigma \end{bmatrix} \begin{Bmatrix} \cos\theta \\ \sin\theta \end{Bmatrix} = 0$$

This is an eigenvalue problem and for a non-trivial solution the determinant of the matrix must be zero

$$(\sigma_{xx} - \sigma)(\sigma_{yy} - \sigma) - \tau_{xy}^2 = 0$$

Using the numerical values for stresses from the problem in figure 3.14a

$$(50 - \sigma)(-30 - \sigma) - 900 = 0$$

$$\sigma^2 - 20\sigma - 2400 = 0$$

$$\sigma = 60 \text{ or } -40$$

It is now possible to determine the modal matrix and hence the orientation of the principal planes. However if we go back to our original matrix, it can be seen

at once that

$$\tan \theta = \frac{\sigma - \sigma_{xx}}{\tau_{yx}} \text{ or } \frac{\tau_{xy}}{\sigma - \sigma_{yy}}$$

In the given case $\theta = 18\frac{1}{2}°$ or $-71\frac{1}{2}°$.

3.8 The Mohr circle diagram for stress

The theory that has been derived so far leads to fairly lengthy expressions for principal stresses. An alternative method of presentation is by what is termed the Mohr circle diagram; once the significance of this is appreciated it will be seen to be a much easier method to use.

First consider the problem illustrated in figure 3.13a. Given σ_1 and σ_2 what are the values of σ and τ? It can very quickly be shown by resolution of forces that

$$\sigma = \tfrac{1}{2}(\sigma_1 + \sigma_2) + \tfrac{1}{2}(\sigma_1 - \sigma_2) \cos 2\theta \tag{3.19}$$

$$\tau = \tfrac{1}{2}(\sigma_1 - \sigma_2) \sin 2\theta \tag{3.20}$$

If a set of orthogonal axes are taken with σ horizontal and τ vertical. Two points $(\sigma_1, 0)$, $(\sigma_2, 0)$ are entered on the diagram represented by OE and OF. A circle with centre D is then drawn with FE as diameter. The length OD is given by $(\sigma_1 + \sigma_2)/2$ and the radius r of the circle by $(\sigma_1 - \sigma_2)/2$. It is now possible to find the coordinates σ, τ of point P where DP is inclined at 2θ to the horizontal. The horizontal coordinate is OD $+ r \cos 2\theta$ and the vertical $r \sin 2\theta$. So that

$$\sigma = \tfrac{1}{2}(\sigma_1 + \sigma_2) + \tfrac{1}{2}(\sigma_1 - \sigma_2) \cos 2\theta$$

and $\tau = \tfrac{1}{2}(\sigma_1 - \sigma_2) \sin 2\theta$. These expressions are identical to equations 3.19 and 3.20, thus indicating that the construction is correct.

The circle can also be constructed given values of σ_{xx}, σ_{yy} and τ_{xy}. Hence values of σ_1, σ_2 and θ can be found.

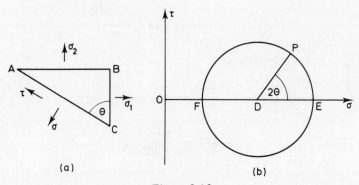

(a)   (b)

Figure 3.13

The sign convention adopted for direct stresses will apply when constructing a Mohr circle diagram, that is, tensile stress is positive. Our previous shear-stress convention will not be satisfactory. However, let us now consider shear stresses acting on a pair of opposite faces of the rectangular element in figure 3.12a. If the effect of the shear stresses is to form a clockwise couple when viewed from the front then the shear stress will be considered as positive. This would mean that σ_{xx} in the diagram is accompanied by a negative shear and σ_{yy} by a positive shear.

As an example, let us determine the principal stresses and their directions when $\sigma_{xx} = 50$ N/mm^2, $\sigma_{yy} = -30$ N/mm^2, $\tau_{xy} = -30$ N/mm^2. The stresses are shown on the rectangular element figure 3.14a. Using the theoretical result

$$\sigma_1, \sigma_2 = \tfrac{1}{2}(50 - 30) \pm \tfrac{1}{2}[(50 + 30)^2 + (4 \times 900)]^{1/2} = 10 \pm 50$$

therefore

$$\sigma_1 = 60 \text{ N/mm}^2 \quad \sigma_2 = -40 \text{ N/mm}^2$$

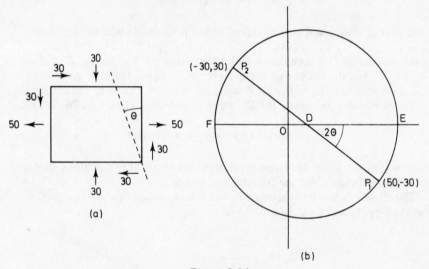

(a)

(b)

Figure 3.14

The maximum shear stress is

$$\frac{60 + 40}{2} = 50 \text{ N/mm}^2$$

$$\tan 2\theta = \frac{2 \times 30}{(50 + 30)} = \tfrac{3}{4}$$

therefore

$$\theta = 18\tfrac{1}{2}°$$

To construct the Mohr circle diagram, the two axes are first drawn (figure 3.14b. Points P_1 and P_2 with coordinates $(50, -30)$ and $(-30, 30)$ can now be found. These are joined by a line which cuts the abscissa at D. The circle with radius DP_1 or DP_2 can now be constructed. By measurement OE = 60, OF = -40 and $2\theta = 37°$. The maximum shear stress is of course equal to the radius of the circle. Note that starting from P_1 on the circle and proceeding through an angle of $37°$ anti-clockwise we come to a principal stress of 60 N/mm^2. So that starting from the plane on which the 50 N/mm^2 stress is acting figure 3.14a and proceeding anti-clockwise through an angle of $18\tfrac{1}{2}°$ we shall come to a plane on which a principal stress of 60 N/mm^2 is acting.

3.9 Principal strain in two dimensions

Suppose that a line MN was scribed on a piece of material that is strained and displaced such that its new position is $M'N'$ (figure 3.15a). The coordinates of M being x, y and those of N $x + dx$, $y + dy$. The displacements of M' are u and v,

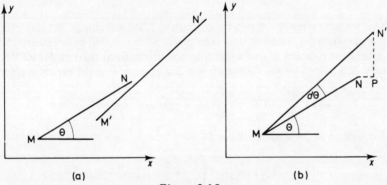

Figure 3.15

shown greatly exaggerated in the diagram. It is possible by making use of Taylor's theorem to express the displacements of N' as

$$u + \frac{\partial u}{\partial x}\,dx + \frac{\partial u}{\partial y}\,dy \quad v + \frac{\partial v}{\partial y}\,dy + \frac{\partial v}{\partial x}\,dx$$

Now move the line MN so that it remains parallel to its original position, such that M and M' are coincident (figure 3.15b). The horizontal and vertical displacement of N' relative to N are NP and PN' where

$$NP = \frac{\partial u}{\partial x}\,dx + \frac{\partial u}{\partial y}\,dy$$

$$PN' = \frac{\partial v}{\partial y}\,dy + \frac{\partial v}{\partial x}\,dx$$

The component of displacement along MN' is $PN \cos \theta + PN' \sin \theta$ and if the original length of MN was dl the strain is

$$\epsilon = \left[\left(\frac{\partial u}{\partial x} \, dx + \frac{\partial u}{\partial y} \, dy \right) \cos \theta + \left(\frac{\partial v}{\partial y} \, dy + \frac{\partial v}{\partial x} \, dx \right) \sin \theta \right] \bigg/ dl$$

$$= \epsilon_{xx} \cos^2 \theta + \epsilon_{yy} \sin^2 \theta + \gamma_{xy} \sin \theta \cos \theta$$

or

$$\epsilon = \tfrac{1}{2}(\epsilon_{xx} + \epsilon_{yy}) + \tfrac{1}{2}(\epsilon_{xx} - \epsilon_{yy}) \cos 2\theta + \frac{\gamma_{xy}}{2} \sin 2\theta \qquad (3.21)$$

The angle $d\theta$ through which MN' has rotated is given by the normal component of the displacement of N divided by the length.

$$d\theta = \frac{PN' \cos \theta - PN \sin \theta}{dl}$$

$$= \left[\left(\frac{\partial v}{\partial y} \, dy + \frac{\partial v}{\partial x} \, dx \right) \cos \theta - \left(\frac{\partial u}{\partial x} \, dx + \frac{\partial u}{\partial y} \, dy \right) \sin \theta \right] \bigg/ dl$$

$$= \frac{\partial v}{\partial x} \cos^2 \theta - \frac{\partial u}{\partial y} \sin^2 \theta + (\epsilon_{yy} - \epsilon_{xx}) \sin \theta \cos \theta$$

Now we have already defined shear strain as being the change in value of a right angle when deformation has taken place. So that it will be necessary to find the equivalent rotation $d\theta_1$ of a line that was originally at right angles to MN. This is easily determined by substituting $\theta + \pi/2$ in place of θ in the expression for $d\theta$

$$d\theta_1 = \frac{\partial v}{\partial x} \sin^2\theta - \frac{\partial u}{\partial y} \cos^2 \theta - (\epsilon_{yy} - \epsilon_{xx}) \sin \theta \cos \theta$$

The change in the angle is $d\theta_1 - d\theta$

$$\gamma = (\epsilon_{xx} - \epsilon_{yy}) \sin 2\theta - \left(\frac{\partial v}{\partial x} + \frac{\partial u}{\partial y} \right)(\cos^2 \theta - \sin^2 \theta)$$

therefore

$$\frac{\gamma}{2} = \tfrac{1}{2}(\epsilon_{xx} - \epsilon_{yy}) \sin 2\theta - \frac{\gamma_{xy}}{2} \cos 2\theta \qquad (3.22)$$

If equations 3.21 and 3.22 are compared with equations 3.11 and 3.12 they will be found to be identical in form if normal stress is replaced by normal strain and shear stress by half the shear strain. This means that any method used for the analysis of principal stress can also be used to analyse principal strains. So that from equation 3.14 the principal strains can be written down as

$$\epsilon_1, \epsilon_2 = \tfrac{1}{2}(\epsilon_{xx} + \epsilon_{yy}) \pm \tfrac{1}{2}[(\epsilon_{xx} - \epsilon_{yy})^2 + \gamma_{xy}^2]^{1/2} \qquad (3.23)$$

3.10 The Mohr circle diagram for strain

The two equations 3.21 and 3.22 for ϵ and $\gamma/2$ may. be represented on a circle diagram in a similar manner to the stress circle. ϵ is plotted horizontally and $\gamma/2$ vertically.

If a Mohr circle diagram has been obtained for stresses it is a straightforward matter to draw the equivalent strain circle. Once the principal stresses have been found, the principal strains can be determined by using Hooke's law in a two-dimensional form. The strain circle can then be drawn.

The strain circle is however of more use when constructed from experimental readings of strain. A number of different experimental devices exist for measuring normal strain but there is no method for measuring the shear strain directly. This means that the circle could not be constructed in the usual manner from two sets of coordinates.

A very convenient method of measuring strain is by means of an electrical resistance strain gauge. This relies on the fact that if a wire is extended its electrical resistance increases. For particular materials the ratio of electrical strain to mechanical strain is a constant, where the electrical strain is the change in resistance divided by the original resistance. It is not the purpose of this book to describe experimental methods and it will suffice to say that by using resistance strain gauge rosettes it is possible to determine the strain in three different known directions.

It is assumed that the three strain values ϵ_a, ϵ_b, ϵ_c have been found and that the strains are measured at relative angles α and β (figure 3.16a). Equation 3.21 can be modified for principal strains by putting $\epsilon_{xx} = \epsilon_1$, $\epsilon_{yy} = \epsilon_2$ and $\gamma_{xy} = 0$, so that

$$\epsilon_a = \tfrac{1}{2}(\epsilon_1 + \epsilon_2) + \tfrac{1}{2}(\epsilon_1 - \epsilon_2) \cos 2\theta$$

$$\epsilon_b = \tfrac{1}{2}(\epsilon_1 + \epsilon_2) + \tfrac{1}{2}(\epsilon_1 - \epsilon_2) \cos 2(\theta + \alpha)$$

$$\epsilon_c = \tfrac{1}{2}(\epsilon_1 + \epsilon_2) + \tfrac{1}{2}(\epsilon_1 - \epsilon_2) \cos 2(\theta + \alpha + \beta)$$

These three equations can be solved for ϵ_1, ϵ_2 and θ. In rosette strain gauges the angles α and β between the gauges are equal and generally have values of $45°$, $60°$ or $120°$.

To construct the Mohr strain circle directly from the three strain readings it is first necessary to arrange the gauge readings such that the one with the middle value lies between the gauges with the highest and lowest values; the angles between the gauges must be less than $90°$. This rearrangement will be demonstrated in a following example. A vertical line YY is drawn as the strain origin; additional verticals are then drawn at scaled distances equal to ϵ_a, ϵ_b and ϵ_c (figure 3.16b). A point P is chosen on the vertical line representing the median value of strain. Lines inclined at α and β are drawn to cut the verticals representing ϵ_a and ϵ_c respectively at Q and R. A circle can now be drawn to pass through P, Q and R. This is in fact the Mohr circle. The circle cuts the vertical representing ϵ_b at P and P′. We shall show that P′, Q and R represent the correct points on the Mohr circle diagram. These points most certainly have the true values of strain but it will still be necessary to show that they are correctly orientated. The angle between ϵ_a and ϵ_b is α, so that on the Mohr circle diagram the two gauge points P′ and Q

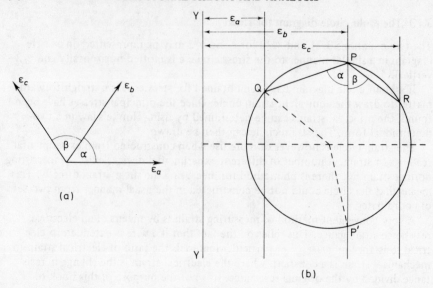

Figure 3.16

should subtend an angle of 2α at the centre of the circle. This is so, as the chord P'Q subtends an angle at the centre of the circle equal to twice the angle at the circumference. Similarly it can be shown that the angle subtended by P' and R at the centre of the circle is 2β. This means that the gauge points have the correct orientation and that the constructed circle is indeed the required Mohr circle.

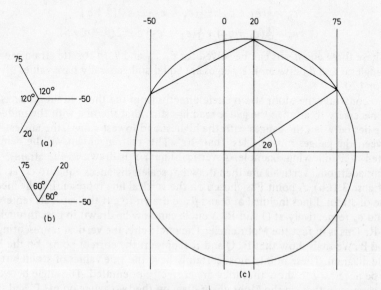

Figure 3.17

As an example of the construction of a Mohr circle diagram for strain, the readings taken from a $120°$ rosette strain gauge are 75×10^{-5}, 20×10^{-5} and -50×10^{-5}. The gauge was mounted on a piece of sheet steel and the principal stresses and their directions are required. The gauge readings (figure 3.17a) are rearranged as in b. Note that the angles between the gauges are now less than $90°$ and the median reading lies between the other two. A suitable scale is chosen and the Mohr diagram constructed as previously described. From the diagram, figure 3.17c, the principal strains are found to be 87.5×10^{-5} and -57.5×10^{-5}; the angle 2θ is $34°$; so that the direction of the principal strain of 87.5×10^{-5} is at an angle of $17°$ measured clockwise to the direction of the strain gauge reading of 75×10^{-5}.

The general form of Hooke's law for the case of plane stress when principal stresses are used is

$$\epsilon_1 = \frac{1}{E}(\sigma_1 - \nu\sigma_2)$$

$$\epsilon_2 = \frac{1}{E}(\sigma_2 - \nu\sigma_1)$$

If the two equations are solved for σ_1 and σ_2 we have expressions for principal stresses in terms of principal strains

$$\sigma_1 = \frac{E}{1-\nu^2}(\epsilon_1 + \nu\epsilon_2) \quad \sigma_2 = \frac{E}{1-\nu^2}(\epsilon_2 + \nu\epsilon_1) \tag{3.24}$$

Assuming values of $E = 210 \text{ kN/mm}^2$ and $\nu = 0.3$ we find in this case that

$$\sigma_1 = 162 \text{ N/mm}^2 \text{—tensile} \quad \sigma_2 = 72 \text{ N/mm}^2 \text{—compressive}$$

3.11 Relationships between elastic constants

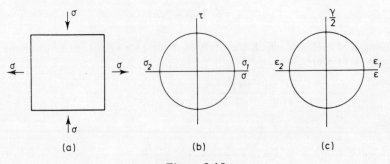

Figure 3.18

A particular case of plane stress is shown in figure 3.18a where $\sigma_{xx} = \sigma$, $\sigma_{yy} = -\sigma$ and $\tau_{xy} = 0$. The Mohr circle for stress is drawn in figure 3.18b. Note that $\sigma_1 = \sigma$, $\sigma_2 = -\sigma$ and $\tau_{max} = \sigma$, so that the maximum shear strain is σ/G. The strain diagram is shown in figure 3.18c and from this it can be seen that

$$\epsilon_1 = \frac{\gamma_{max}}{2} = \frac{\sigma}{2G}$$

From Hooke's law

$$\epsilon_1 = \frac{1}{E}(\sigma + \nu\sigma)$$

Thus

$$\frac{\sigma}{2G} = \frac{\sigma}{E}(1 + \nu)$$

or

$$G = \frac{E}{2(1 + \nu)} \tag{3.25}$$

So that if two of the three constants E, G or ν are known, the third may be determined theoretically.

A body that is subjected to hydrostatic pressure has the same compressive stress or pressure acting on it from all directions. Due to this the volume of the body, V will decrease by an amount dV. The volumetric strain is defined as dV/V. If the compressive stress is σ and the material behaves in an elastic manner the bulk modulus K of the material is defined as

$$K = \frac{-\sigma}{(dV/V)}$$

Let us consider a small cube of the material having unit edge length. Using Hooke's law for a three-dimensional case, the strain along the directions of three adjacent edges is identical and has a value

$$\epsilon = \frac{1}{E}(-\sigma + 2\nu\sigma)$$

The length of each side which was unity becomes $(1 + \epsilon)$ and the new volume of the cube is $(1 + \epsilon)^3$. So that

$$\frac{dV}{V} = \frac{(1 + \epsilon)^3 - 1}{1} \approx 3\epsilon = -\frac{3\sigma}{E}(1 - 2\nu)$$

Hence

$$K = \frac{\sigma}{\dfrac{3\sigma}{E}(1 - 2\nu)} = \frac{E}{3(1 - 2\nu)} \tag{3.26}$$

Both G and K are positive quantities so that from equations 3.25 and 3.26 the following limits for ν can be found

$$-1 < \nu < \tfrac{1}{2}$$

We shall conclude this section with a discussion on the solution of a particular problem. This may help to fix some of the ideas that have been developed so far in this chapter.

Consider the case of two thin-walled circular tubes. One is of brass and the other of steel. Both are the same length. The brass tube is inserted inside the steel tube and is an exact fit at room temperature. The tubes are closed by rigid end plates. The wall thicknesses of the tubes are t_B and t_S. These dimensions are small compared with the mean radius r of the tubes. The vessel is filled with water. How much more water has to be pumped in to raise the pressure by an amount p?

Two effects will have to be considered: the compression of the water due to the increase in pressure; the changes in dimensions of the tubes due to the pressure increase.

If the bulk modulus for water is K

$$V = \pi r^2 l$$

$$\frac{-p}{(dV/V)} = K$$

Thus

$$dV = -\frac{\pi r^2 l p}{K}$$

This is a decrease in volume.

It is assumed that both materials remain elastic. When the pressure is applied the internal tube will tend to expand and a radial stress σ_R will be developed between the two tubes (figure 3.19a).

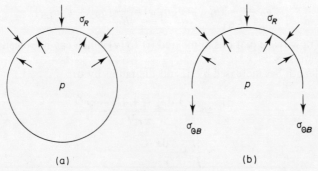

(a)   (b)

Figure 3.19

Consider the inner brass tube, in particular the equilibrium about a diametral cut figure 3.19b. For a unit length of tube where $\sigma_{\theta B}$ is the hoop stress.

$$\sigma_{\theta B} \times 2t_B = 2r(p - \sigma_R)$$

thus

$$\sigma_{\theta B} = \frac{r(p - \sigma_R)}{t_B} \qquad (i)$$

for the steel tube

$$\sigma_{\theta S} = \frac{r\sigma_R}{t_S} \qquad (ii)$$

For equilibrium in the longitudinal direction with longitudinal stresses σ_{LB} and σ_{LS}

$$\sigma_{LB} \times 2\pi r t_B + \sigma_{LS} \cdot 2\pi r t_S = \pi r^2 p$$

or

$$\sigma_{LB} t_B + \sigma_{LS} t_S = \frac{rp}{2} \qquad (iii)$$

So far there are a total of five unknown stresses σ_R, $\sigma_{\theta B}$, $\sigma_{\theta S}$, σ_{LB} and σ_{LS} and there are only three equations relating them. This means that two further equations have to be found.

The length of each tube and the radius of each tube must extend by the same amount. Otherwise they would not fit together after the pressure is applied. This is referred to as compatibility of strain. Working in terms of strains, the longitudinal strain is

$$\epsilon_L = \frac{1}{E_B}(\sigma_{LB} - \nu_B \sigma_{\theta B}) = \frac{1}{E_S}(\sigma_{LS} - \nu_S \sigma_{\theta S}) \qquad (iv)$$

Hoop strain

$$\epsilon_\theta = \frac{1}{E_B}(\sigma_{\theta B} - \nu_B \sigma_{LB}) = \frac{1}{E_S}(\sigma_{\theta S} - \nu_S \sigma_{LS}) \qquad (v)$$

There are now five equations numbered (i) to (v) which can be solved for the unknown stresses.

If the length has increased by dl and the radius by dr

$$\frac{dV}{V} = \frac{\pi(r + dr)^2(l + dl) - \pi r^2 l}{\pi r^2 l}$$

$$\approx \frac{dl}{l} + \frac{2\,dr}{r}$$

$$= \epsilon_L + 2\epsilon_\theta$$

Thus $dV = (\epsilon_L + 2\epsilon_\theta)\pi r^2 l$. Both ϵ_L and ϵ_θ are known.

The total change in volume can now be found and hence the volume of water pumped in is this value plus the bulk modulus effect.

3.12 Strain energy

If a spring is extended it is obvious that a certain amount of external work has been done. Assume that the process is adiabatic, that is, there is no heat exchange with the surroundings, and that the force is applied sufficiently slowly so that there is no change in temperature of the spring and hence no change in internal energy. If the change in kinetic energy is zero then the external work done must be equal to the increase in potential energy of the spring. This particular form of potential energy is usually referred to as strain energy and is denoted by the symbol U.

In the case of a spring that remains linearly elastic the relationship between the applied force P and the extension e can be written $P = ke$ where k is the spring constant. As the load increases from 0 to P_0 the extension increases from 0 to e_0 (figure 3.20).

$$dU = P\,de = ke\,de$$

therefore

$$U = k \int_0^{e_0} e\,de = \tfrac{1}{2}ke_0^2 = \tfrac{1}{2}P_0e_0 = \frac{1}{2}\frac{P_0^2}{k} \tag{3.27}$$

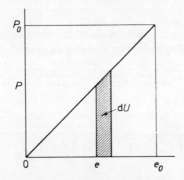

Figure 3.20

Exactly the same analysis will apply for the case of a rod of linearly elastic material under the action of an axial tensile or compressive force.

The spring constant k will be replaced by AE/l where A is the cross-sectional area and l the length. Thus

$$U = \frac{1}{2}\frac{AEe_0^2}{l} = \tfrac{1}{2}P_0e_0 = \frac{1}{2}\frac{P_0^2l}{AE} \tag{3.28}$$

Now consider the case of a rectangular parallelepiped of dimensions dx, dy, dz. A uniform stress σ_{xx} is applied, so that the dimension dx will change to $dx(1 + \epsilon_{xx})$. The work done by the force $\sigma_{xx}\,dy\,dz$ during a small change in the length dx of

$d(\epsilon_{xx})\, dx$ is

$$\sigma_{xx}\, dy\, dz\, dx\, d\epsilon_{xx}$$

or the strain energy stored per unit volume.

$$\frac{dU}{dV} = \int \sigma_{xx}\, d\epsilon_{xx}$$

Now

$$\epsilon_{xx} = \frac{\sigma_{xx}}{E}$$

Thus

$$\frac{dU}{dV} = \frac{\sigma_{xx}^2}{2E} = \frac{\sigma_{xx}\epsilon_{xx}}{2}$$

By similar reasoning the strain energy stored per unit volume for a case of pure shear will be given by

$$\frac{dU}{dV} = \frac{\tau_{xy}^2}{2G} = \frac{\tau_{xy}\gamma_{xy}}{2}$$

This can be extended to a case of plane stress and indeed to the general case of a three-dimensional stress system. For this general case

$$\frac{dU}{dV} = \tfrac{1}{2}(\sigma_{xx}\epsilon_{xx} + \sigma_{yy}\epsilon_{yy} + \sigma_{zz}\epsilon_{zz} + \tau_{xy}\gamma_{xy} + \tau_{xz}\gamma_{xz} + \tau_{yz}\gamma_{yz})$$

Now

$$\epsilon_{xx} = \frac{1}{E}[\sigma_{xx} - \nu(\sigma_{yy} + \sigma_{zz})] \quad \text{etc.}$$

Thus

$$\frac{dU}{dV} = \frac{1}{2E}[\sigma_{xx}^2 + \sigma_{yy}^2 + \sigma_{zz}^2 - 2\nu(\sigma_{xx}\sigma_{yy} + \sigma_{xx}\sigma_{zz} + \sigma_{yy}\sigma_{zz})]$$

$$+ \frac{1}{2G}[\tau_{xy}^2 + \tau_{xz}^2 + \tau_{yz}^2] \tag{3.29}$$

3.13 Pure bending of beams

A straight prismatic beam of homogeneous isotropic material is loaded in such a way that the bending moment at any cross-section has a constant value of M. This immediately implies that the shear force is zero. Also there is no axial force applied. As the bending moment is applied the beam will bend into a curved shape.

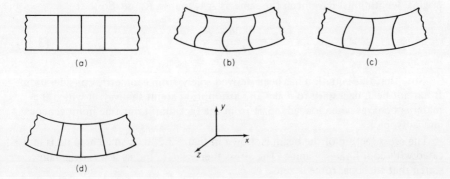

Figure 3.21

A small element of the beam is shown in figure 3.21a with three vertical lines indicating plane sections. After the application of the bending moment it is assumed that the straight lines could be deformed as in b or c. For b if a rotation of 180° about the y axis is considered it will be seen from symmetry arguments that the assumed deformation is not possible. c is also impossible because the two adjacent elements on the beam although subjected to the same bending moment have deformed in an entirely different manner. The only deformation that is suitable is that shown in figure 3.21d. From this it may be concluded that plane sections remain plane when pure bending is applied.

In figure 3.22a a small portion of a beam is shown bent, such that the radius of AA in the xy plane is R_y. The two inclined lines were originally vertical and a distance dx apart. It will be seen that the bending effect will compress some fibres such as BB, and extend others. There must however be some position AA where the fibres have not changed length. We shall refer to this particular position as the neutral axis.

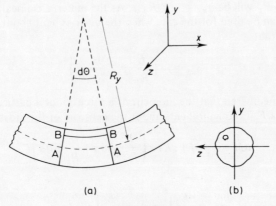

Figure 3.22

The length of the line BB distant y from AA can be written as $(R_y - y)\,d\theta$. Its original length must have been the same as AA, that is, $R_y\,d\theta$. So that

$$\epsilon_{xx} = \frac{(R_y - y)\,d\theta - R_y\,d\theta}{R_y\,d\theta} = -\frac{y}{R_y} \tag{3.30}$$

Note that this equation has been derived solely from geometric considerations. It has not been necessary to make any assumption about the way in which the material behaves when loaded, apart from the fact that it must be homogeneous and isotropic.

The cross-section of the beam is shown in figure 3.22b. A small area da is indicated with coordinates y and z. The stress there is σ_{xx}, and as it was originally stated that the axial force is zero

$$\int \sigma_{xx}\,da = 0$$

Now make the assumption that the material of the beam is linearly elastic, so that $\sigma_{xx} = E\epsilon_{xx}$. Thus

$$\int E\epsilon_{xx}\,da = 0$$

or

$$-\frac{E}{R_y}\int y\,da = 0$$

and

$$\int y\,da = 0$$

If the beam had been bent such that the radius of curvature was R_z, by similar reasoning it would be found that $\int z\,da = 0$. So the position of the neutral axis is defined—it must pass through the centroid of the cross-section.

The stress due to the radius of curvature R_y can be written $\sigma_{xx} = -(E/R_y)y$ and that due to R_z will be $\sigma_{xx} = -(E/R_z)z$. As the material is linearly elastic, superposition can be used for the case when the beam is bent about two axes at the same time, that is

$$\sigma_{xx} = -\left(\frac{E}{R_y}y + \frac{E}{R_z}z\right) \tag{3.31}$$

It will be remembered that the moment of a force about a particular point can be written as $r \times F$. Taking moments about the centroid of the cross-section

$$dM = -(yj + zk) \times \left(\frac{E}{R_y}y\,da + \frac{E}{R_z}z\,da\right)i$$

or

$$M_z = \frac{E}{R_y}\int y^2\,da + \frac{E}{R_z}\int yz\,da \tag{3.32}$$

and

$$M_y = - \frac{E}{R_y} \int yz \, \mathrm{d}a - \frac{E}{R_z} \int z^2 \, \mathrm{d}a$$

It should be noted from these equations that even if $1/R_z$ is zero it is necessary in this general case to apply moments about both the z and y axes to produce a curvature in the xy plane. In fact M_z and M_y can be combined and a moment applied about an inclined axis.

The reader may possibly recognise some of the integrals in equations 3.32. Integrals of this type are used for finding moments of inertia. However we shall refer to $\int y^2 \, \mathrm{d}a$ and $\int z^2 \, \mathrm{d}a$ as being second moments of area and denote them by I_{zz} and I_{yy} respectively. $\int yz \, \mathrm{d}a$ is called a product second moment of area and is denoted by I_{yz} so that

$$M_z = \frac{EI_{zz}}{R_y} + \frac{EI_{yz}}{R_z} \qquad (3.33)$$

$$M_y = - \frac{EI_{yz}}{R_y} - \frac{EI_{yy}}{R_z}$$

If I_{yz} is zero it will only be necessary to apply a moment M_z to produce a radius of curvature R_y. For this case equations 3.33 are much simplified and I_{zz} and I_{yy} are then referred to as principal second moments of area and y and z are called principal axes. The requirement is that $\int yz \, \mathrm{d}a$ should be zero. This is certainly true if a cross-section has an axis of symmetry.

All the sections shown in figure 3.23 have at least one axis of symmetry, hence the directions of the principal axes are known at once. For bending about the z axis we may write

$$- \frac{\sigma_{xx}}{y} = \frac{M_z}{I_{zz}} = \frac{E}{R_y} \qquad (3.34a)$$

Similarly for bending about the y axis

$$\frac{\sigma_{xx}}{z} = \frac{M_y}{I_{yy}} = - \frac{E}{R_z} \qquad (3.34b)$$

We shall confine our attention for the time being to bending about a principal axis so that the expressions given by equations 3.34a and b may be used.

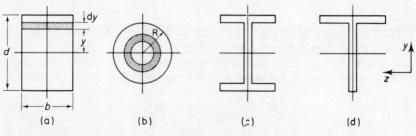

Figure 3.23

It is necessary to be able to determine values of I_{zz} and I_{yy}. For the rectangle in figure 3.23a

$$I_{zz} = \int y^2 \, \mathrm{d}a = \int_{-d/2}^{d/2} by^2 \, \mathrm{d}y = \frac{b \, d^3}{12}$$

For the circular cross-section in figure 3.23b it is easiest to determine I_{xx}.

$$I_{xx} = \int_0^R 2\pi r^3 \, \mathrm{d}r = \frac{\pi R^4}{2}$$

Now $I_{yy} + I_{zz} = I_{xx}$ and in this case $I_{yy} = I_{zz}$, so that

$$I_{yy} = I_{zz} = \frac{\pi R^4}{4}$$

For the Universal beam (figure 3.23c) one suggested method of finding I_{zz} is to take the value for a rectangle equal to the maximum dimensions of the section and then to subtract the values for two smaller rectangles each of the same size.

The derivation of the bending formula was based on a basic assumption that the bending moment was constant. This would mean that the theory would have a very limited application in practice. However the formula is still used when the bending moment changes and the shear force is thus no longer zero. If the length of the beam is long when compared with the dimensions of the cross-section the errors introduced by the simple theory will not be great. The negative sign is often omitted in equations 3.34, mathematically it should be present but in most simple cases the sign of the stresses can be determined by inspection. For the particular case when M_z, say, is constant, R_y will be constant and the beam will bend into the arc of a circle.

As a fairly straightforward example of the application of the bending theory, the maximum value of p, the uniformly distributed load, is required for the cantilever in figure 3.24. The cross-section is rectangular and the maximum stress is limited to σ_1.

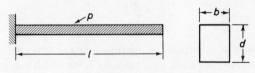

Figure 3.24

From equation 3.34a it can be seen that for a given cross-section the stress will be a maximum when both y and M have their greatest values. The maximum value of y in this case is $d/2$ and the worst bending moment occurs at the support and has a value of $pl^2/2$. The value of I is $bd^3/12$, so that

$$\frac{pl^2}{2} = \tfrac{1}{12} bd^3 \sigma_1 \frac{2}{d} = \frac{bd^2 \sigma_1}{6}$$

and

$$p = \frac{bd^2\sigma_1}{3l^2}$$

In a large number of cases we are interested in the maximum value of y and often the value of I/y_{max} is required for a particular section. This is referred to as the section modulus and it is denoted by Z. It would of course be perfectly possible to have two values of Z for a particular section, for example, consider the Tee section in figure 3.23d, for bending about the z axis.

For a given maximum value of stress the bending moment will be a maximum when Z is a maximum. For a symmetrical cross-section of given area the material should be placed as far away from the neutral axis as possible. There are limitations to this as other effects such as shear and stability have to be considered. The Universal beam (figure 3.23c) makes good use of material and for a given cross-sectional area would carry much greater bending moments than a rectangular beam. For a beam of this type, the web (the vertical part) does not contribute much to the bending resistance. A fairly close approximation to the moment is obtained by only considering the flanges (the horizontal portion) and assuming that the longitudinal stress is constant within them.

3.14 Bending of a composite beam

A cross-section of one type of composite beam is shown in figure 3.25a. It is formed from a piece of material 2 sandwiched between two similar pieces of material 1. Assume that there is perfect bond between the two materials. To make the problem general, a bending moment M is applied about an axis inclined at θ to the y axis. The maximum resulting tensile stress is required in material 1. $E_1 > E_2$.

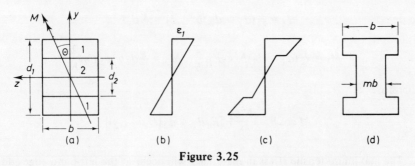

Figure 3.25

It will be necessary to work about the principal axes. These will be the y and z axes as shown. The bending moment can be resolved into two components

$$M_z = M \sin \theta \quad M_y = M \cos \theta$$

For bending about the z axis the strain and stress distributions will be of the forms shown in figures 3.25b and c. The strain must be linear over the section and has a maximum value of ϵ_1 say. At the junction of the two materials the strain must have the same value in each material and as $E_1 > E_2$ there will be a sudden change in stress at the junction.

The maximum stress in material 1 is $\epsilon_1 E_1$

$$I_1 = \tfrac{1}{12} b(d_1^3 - d_2^3)$$

So that the moment carried by material 1 is

$$M_1 = \tfrac{1}{12} b(d_1^3 - d_2^3)\epsilon_1 E_1 \frac{2}{d_1}$$

The maximum stress in material 2 is $\epsilon_1(d_2/d_1)E_2$ so that

$$M_2 = \tfrac{1}{12} bd_2^3 \epsilon_1 \frac{d_2}{d_1} E_2 \frac{2}{d_2}$$

The two moments must sum to M_z and the value of ϵ_1 can be found

$$M \sin \theta = \frac{b\epsilon_1}{6 d_1} \{E_1 d_1^3 - d_2^3(E_1 - E_2)\}$$

The maximum tensile stress in material 1 will occur at the bottom edge and has a value $E_1\epsilon_1$, therefore

$$\sigma_{max} = \frac{6E_1 d_1 M \sin \theta}{b [E_1 d_1^3 - d_2^3(E_1 - E_2)]} \qquad (3.35)$$

For bending about the y axis the strain distribution and the stress distribution for each material will be linear. If the maximum strain is ϵ_2

$$I_1 = \tfrac{1}{12}(d_1 - d_2)b^3 \qquad I_2 = \tfrac{1}{12}d_2 b^3$$

$$M_1 = \tfrac{1}{12}(d_1 - d_2)b^3 \epsilon_2 E_1 \frac{2}{b} \qquad M_2 = \tfrac{1}{12}d_2 b^3 \epsilon_2 E_2 \frac{2}{b}$$

therefore

$$M \cos \theta = \frac{b^2 \epsilon_2}{6} [E_1 d_1 - d_2(E_1 - E_2)]$$

The maximum tensile stress in material 1 will occur at the left-hand edge and has a value $E_1\epsilon_2$

$$\sigma_{max} = \frac{6E_1 M \cos \theta}{b^2 [E_1 d_1 - d_2(E_1 - E_2)]} \qquad (3.36)$$

The maximum value of the total tensile stress will occur at the bottom left-hand corner of the section and has a value given by the sum of the stresses given by equations 3.35 and 3.36.

An alternative approach to this type of problem is to try and find an equivalent section that consists entirely of material 1. This is referred to as transforming the section (see section 3.5).

If the strain in material 2 at a particular distance y from the neutral axis is ϵ the stress is $E_2\epsilon$ then the force carried by an element of width b and height dy is $E_2\epsilon b\, dy$. If this material is replaced by an equivalent amount of material 1 the strain value would have to remain at ϵ but the stress would be $E_1\epsilon$. It is essential that the force transmitted by an element of height dy should remain the same, otherwise the bending moment would be altered. Hence the only possible modification is to the breadth, changing it to b_1 say

$$E_1\epsilon b_1\, dy = E_2\epsilon b\, dy$$

$$b_1 = b\frac{E_2}{E_1} = bm$$

where m is referred to as the modular ratio.

For bending about the z axis the transformed section for the composite beam we have been discussing is shown in figure 3.25d, where material 2 has been turned into an equivalent amount of material 1.

The value of the equivalent I is

$$\tfrac{1}{12}bd_1^3 - \tfrac{1}{12}(b - mb)\, d_2^3$$

This will lead to the same expression for ϵ_1 as the one already obtained.

The transformed section for bending about the y axis is left as an exercise for the reader.

3.15 Bending of unsymmetrical sections

The bending problems considered so far have been confined to sections which have had an axis of symmetry and hence the directions of the principal axes have been known. We shall now proceed to make use of equations 3.33 which were derived from a general case of bending. These two equations can be solved simultaneously to give values of E/R_y and E/R_z

$$\frac{E}{R_y} = \frac{M_z I_{yy} + M_y I_{yz}}{I_{zz}I_{yy} - I_{yz}^2}$$

$$\frac{E}{R_z} = \frac{-M_y I_{zz} - M_z I_{yz}}{I_{zz}I_{yy} - I_{yz}^2}$$

If these values are now substituted into equation 3.31 we shall obtain a general expression for the longitudinal stress σ_{xx}

$$\sigma_{xx} = \frac{-M_z(yI_{yy} - zI_{yz}) + M_y(zI_{zz} - yI_{yz})}{I_{yy}I_{zz} - I_{yz}^2} \tag{3.37}$$

This rather cumbersome expression will give the value of the longitudinal stress for any chosen pair of axes. It will be seen to simplify to the expressions 3.34a and b for the case when $I_{yz} = 0$.

The alternative approach as we have seen is to resolve the applied bending moment into components acting in the directions of the principal axes. This is a perfectly simple operation so long as the directions of the principal axes are known; however in the case of an asymmetrical section we have no idea of the position of the principal axes. It will be assumed that a particular set of axes y, z have been chosen for an asymmetrical section (figure 3.26), also that it has been possible to determine I_{yy}, I_{zz} and I_{yz}.

The principal axes Y and Z are assumed to be at an angle θ to the y, z axes. Consider a small area da with coordinates y, z or Y, Z.

$$Y = y \cos \theta - z \sin \theta$$

$$Z = z \cos \theta + y \sin \theta$$

$$I_{YY} = \int Z^2 \, da = \int z^2 \cos^2 \theta \, da + \int 2yz \sin \theta \cos \theta \, da + \int y^2 \sin^2 \theta \, da$$

Now

$$\int z^2 \, da = I_{yy} \quad \int y^2 \, da = I_{zz} \quad \int yz \, da = I_{yz}$$

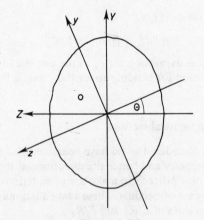

Figure 3.26

Thus

$$I_{YY} = \tfrac{1}{2}(I_{yy} + I_{zz}) + \tfrac{1}{2}(I_{yy} - I_{zz}) \cos 2\theta + I_{yz} \sin 2\theta \qquad (3.38)$$

A similar expression to this can be obtained for I_{ZZ}

$$I_{YZ} = \int YZ \, da$$

$$= \int y^2 \sin \theta \cos \theta \, da - \int z^2 \sin \theta \cos \theta \, da + \int yz (\cos^2 \theta - \sin^2 \theta) \, da = 0$$

since Y and Z have been defined as principal axes. Hence

$$\frac{I_{zz} - I_{yy}}{2} \sin 2\theta + I_{yz} \cos 2\theta = 0$$

or

$$\tan 2\theta = \frac{2I_{yz}}{I_{yy} - I_{zz}} \qquad (3.39)$$

We should by now recognise the form of the two equations 3.38 and 3.39—they are in fact similar to the expressions obtained for principal stress and strain. So that once again the Mohr circle diagram can be used to good effect. Suitable axes y and z are chosen for the section, I_{yy}, I_{zz} and I_{yz} are determined. Axes are now set up so that second moments of area are plotted horizontally and product second moments of area vertically. Points (I_{yy}, I_{yz}) and $(I_{zz}, -I_{yz})$ are located. These lie at the ends of a diameter of the circle, which can then be drawn. The principal second moments of area are given by the values where the circle cuts the horizontal axis.

To indicate the various calculations the principal second moments of area for the unequal angle section in figure 3.27a will be found. The dimensions are in mm and the thickness of the material is 6 mm.

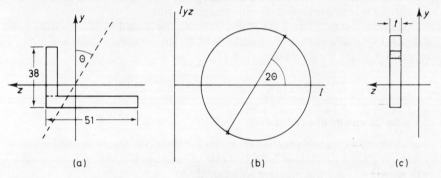

Figure 3.27

It will first be necessary to locate the centroid and we shall consider the angle made up of the two rectangles indicated.

The area of the section is $192 + 306 = 498$ mm^2. Taking moments about the bottom edge

$$y_c = \frac{(192 \times 22) + (306 \times 3)}{498} = 10\cdot3 \text{ mm}$$

and moments about the left-hand edge

$$z_c = \frac{(192 \times 3) + (306 \times 25\cdot5)}{498} = 16\cdot8 \text{ mm}$$

Having located the centroid suitable axes as shown are chosen.

$$I_{zz} = (\tfrac{1}{12} \times 6 \times 32^3) + (192 \times 11\cdot7^2) + (\tfrac{1}{12} \times 51 \times 6^3) + (306 \times 7\cdot3^2) = 59\,900 \text{ mm}^4$$

$$I_{yy} = (\tfrac{1}{12} \times 32 \times 6^3) + (192 \times 13\cdot8^2) + (\tfrac{1}{12} \times 6 \times 51^3) + (306 \times 8\cdot7^2) = 126\,600 \text{ mm}^4$$

It is also necessary to determine the value of I_{yz} which is of the form $\int yz \, da$.

Suppose that the product second moment of area is required for the rectangle in figure 3.27c. An elemental area is shown in the figure as $t \, dy$

$$I_{yz} = \int yzt \, dy$$

Now z is constant for all similar small areas and is the horizontal distance from the axis to the centroid of the rectangle, that is, $\bar{z}$

$$I_{yz} = \bar{z} \int yt \, dy \qquad \int yt \, dy = A\bar{y}$$

where $\bar{y}$ is the vertical distance from the axis to the centroid of the rectangle. So that $I_{yz} = A\bar{y}\bar{z}$. For the given case taking account of signs. *Note* it is perfectly possible for I_{yz} to be negative for a particular case.

$$I_{yz} = (192 \times 11\cdot7 \times 13\cdot8) + 306(-7\cdot3)(-8\cdot7) = 50\,400 \text{ mm}^4$$

The coordinates $(126\,600, 50\,400)$ and $(59\,900, -50\,400)$ are plotted and the Mohr circle diagram drawn. From this the principal second moments of area are found to be $154\,000$ and $33\,000 \text{ mm}^4$, and the angle 2θ is $56°$. This means that if we measure clockwise from the y axis through an angle of $28°$ we shall come to an axis of the major principal second moment of area.

An alternative to drawing the Mohr circle diagram would be to substitute into a similar expression to that of equation 3.14 for the principal values, and into equation 3.39 for the directions.

3.16 Strain energy due to bending

For an elastic material a general expression for the strain energy stored per unit volume is given by equation 3.29. In a case of pure bending the only stress that will arise is σ_{xx} and the equation will simplify to

$$\frac{dU}{dV} = \frac{\sigma_{xx}^2}{2E}$$

Assuming that bending takes place about a principal axis and that a bending moment M has been applied

$$\sigma_{xx} = -\frac{yM_z}{I_{zz}}$$

For a small area of cross-section dA and unit length of beam

$$U = \int \frac{M_z^2 y^2 \, dA}{2EI_{zz}^2}$$

Now

$$\int y^2 \, \mathrm{d}A = I_{zz}$$

so that

$$U = \int \frac{M_z^2}{2EI_{zz}}$$

The total strain energy in the beam of length l is then

$$U = \int_0^l \frac{M_z^2}{2EI_{zz}} \, \mathrm{d}x = \frac{M_z^2 l}{2EI_{zz}}$$

since M_z has been assumed constant.

If M_z is a variable, that is, we are no longer considering a case of pure bending, a similar expression can be used for the bending strain energy

$$U = \int_0^l \frac{M_z^2 \, \mathrm{d}x}{2EI_{zz}} \tag{3.40}$$

3.17 Combined bending and axial force

The cross-section of a short column is shown in figure 3.28a. Oy and Oz are principal axes. A vertical 'compressive' load P is applied at the top end of the column so that its position is defined by e, the eccentricity, and θ.

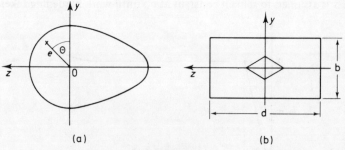

(a)   (b)

Figure 3.28

The load P can be replaced by an axial load P and a moment of value Pe. The moment can be resolved into two components

$$M_y = -Pe \sin \theta \qquad M_z = Pe \cos \theta$$

At any point y, z the stress due to the axial load is $\sigma_{xx} = -P/A$. From equations 3.34a and b the stress due to M_y is $-Pez \sin \theta / I_{yy}$ and that due to M_z is $-Pey \cos \theta / I_{zz}$.

The total stress can be found by superposition.

$$\sigma_{xx} = -\frac{P}{A} - \frac{Pez\sin\theta}{I_{yy}} - \frac{Pey\cos\theta}{I_{zz}}$$

Certain constructional materials are weak in tension; concrete and brickwork are notable examples. If it is assumed in a particular design of a column made of one of these materials that no tensile stress is allowed to develop, the position of the applied load P will be limited and must lie within a certain area called the core. We shall find the core for a rectangular section (figure 3.28b).

$A = bd$; $I_{yy} = bd^3/12$; $I_{zz} = db^3/12$. The worst stresses arise when $z = -d/2$, $y = -b/2$, so that

$$\frac{1}{bd} - \frac{6e\sin\theta}{bd^2} - \frac{6e\cos\theta}{b^2 d} = 0$$

$$be\sin\theta + de\cos\theta = \frac{bd}{6}$$

This represents the equations of straight lines that cut the z axis at $\pm d/6$ and the y axis at $\pm b/6$, and the core is as sketched in figure 3.28b.

3.18 Plastic bending of beams

The stress–strain curve for mild steel in both tension and compression will be idealized as shown in figure 3.29. The upper yield point has been neglected and the stress is assumed to remain constant at σ_y until work hardening takes place.

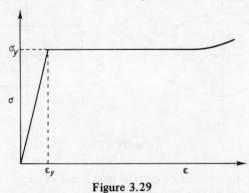

Figure 3.29

We are not interested beyond this 'plastic' region as in this case large deflections of the beam would take place.

The following assumptions will be made when a beam is bent

(1) the lower yield stress and the modulus of elasticity have the same value in tension and compression

(2) the stress remains constant in the plastic range
(3) plane sections remain plane
(4) the material is isotropic and homogeneous

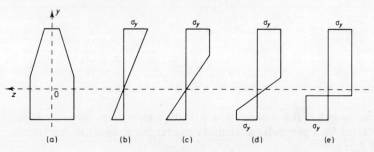

Figure 3.30

The section of a beam in figure 3.30a is symmetrical about one axis and a steadily increasing bending moment is applied. If the stresses are below σ_y, the yield stress, the neutral axis will pass through O, the centroid of the section. In figure 3.30b the top edge has just reached σ_y. The elastic section modulus $Z_e = I/y_{max}$, and the moment at yield is given by

$$M_y = \frac{\sigma_y I}{y_{max}} = \sigma_y Z_e$$

The effect of increasing bending moment is shown in figures 3.30c, d and e. In c, part of the upper portion of the section has become plastic, but the whole of the section below the neutral axis is still elastic. Yield has taken place over part of the bottom portion of the section in d. Finally in e the beam has reached what is termed the fully-plastic bending moment. This last diagram does not quite correspond to the stress–strain curve (figure 3.29) since a small area of the section either side of the neutral axis remains elastic. This however will have a negligible effect on the value of the fully-plastic bending moment. It should be noted that there has been a steady shift of the neutral axis towards the bottom edge of the beam. This would not occur if the beam was of a completely symmetrical section. The neutral axis will finally divide the section into two equal areas.

The value of the fully-plastic bending moment M_p will be required. If y_1 and y_2 are the distances from the neutral axis to the centroids of the areas above and below the neutral axis respectively, and A is the area of the cross-section, then

$$M_p = \frac{\sigma_y A}{2}(y_1 + y_2) \tag{3.41}$$

The plastic section modulus is denoted by Z_p where $M_p = \sigma_y Z_p$, then

$$Z_p = \frac{A}{2}(y_1 + y_2) \tag{3.42}$$

Since $M_y = \sigma_y Z_e$ we can obtain the ratio

$$\frac{M_p}{M_y} = \frac{Z_p}{Z_e} = \alpha$$

The ratio is only dependent on the dimensions of the section and is called the 'shape factor'. It is also sometimes referred to as the 'form factor'.

For the case of a rectangular beam of breadth b and depth d

$$Z_e = \frac{bd^2}{6} \qquad Z_p = \frac{bd^2}{4}$$

therefore

$$\alpha = 1 \cdot 5$$

For universal beams, that is, I-sections, an average value of α about the major axis is $1 \cdot 15$.

It can be seen that it will not be a difficult matter to calculate the value of the shape factor for a particular section. As an example consider the tee section (figure 3.31).

It is first necessary to find the position of the centroid. Taking moments of areas about the lower edge of the section

$$\bar{y} = \frac{(10t^2 \times 5t) + (10t^2 \times 10 \cdot 5t)}{20t^2} = 7 \cdot 75t$$

$$I = (\tfrac{1}{12} \times 10t \times t^3) + (10t^2 \times 2 \cdot 75^2 t^2) + (\tfrac{1}{12}t \times 1000t^3) + (10t^2 \times 2 \cdot 75^2 t^2)$$

$$= 235 \cdot 4t^4$$

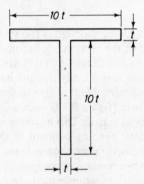

Figure 3.31

so that

$$Z_e = \frac{235 \cdot 4t^4}{7 \cdot 75t} = 30 \cdot 3t^3$$

For the fully plastic case the neutral axis will be at the junction of the flange and web. $y_1 = t/2$, $y_2 = 5t$, $A = 20t^2$, then

$$Z_p = \frac{20t^2}{2} \left(5t + \frac{t}{2}\right) = 55t^3$$

therefore

$$\alpha = \frac{Z_p}{Z_e} = \frac{55}{30 \cdot 3} = 1 \cdot 82$$

If a particular stress–strain relationship is assumed it is perfectly possible to calculate a theoretical moment–curvature relationship for a given cross-section. We shall do this for the rectangular cross-section shown in figure 3.32a. The

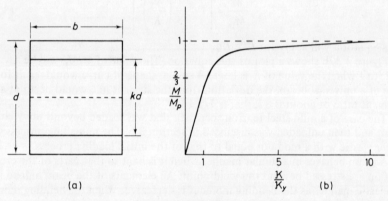

Figure 3.32

material is ideal elasto-plastic, with a yield stress of σ_y. This means that the stress–strain relation is similar to figure 3.29. Assuming that the yield stress has been exceeded and that a total depth of beam kd remains elastic. The moment carried by the elastic portion is

$$M = \frac{bk^2 d^2 \sigma_y}{6}$$

The moment carried by the plastic portion is

$$M = \frac{bd^2}{4}(1 - k^2)\sigma_y$$

Thus the total moment is

$$M = \frac{\sigma_y bd^2}{12}(3 - k^2)$$

This can be expressed in a slightly different form if we make use of the fully plastic bending moment $M_p = \sigma_y bd^2/4$

$$\frac{M}{M_p} = \left(1 - \frac{k^2}{3}\right) \tag{3.43}$$

From the consideration that plane sections remain plane $\epsilon = y/R$, where y is the distance from the neutral axis to any fibre, and R is the radius of curvature, $1/R = \epsilon/y = K$, the curvature. Therefore $K = 2\sigma_y/Ekd$. If we put $K_y = 2\sigma_y/Ed$—the maximum fully elastic curvature

$$\frac{K}{K_y} = \frac{1}{k}$$

therefore

$$\frac{M}{M_p} = \left[1 - \frac{1}{3} \left(\frac{K_y}{K} \right)^2 \right] \qquad (3.44)$$

This equation will apply for $K > K_y$.

Figure 3.32b shows a plot of this equation. The value of M/M_p will be asymptotic to 1 when the value of K is large. A similar shape of curve would result for the case of a universal beam, the deviation from the straight line would occur at a moment ratio of about 0·87, that is $1/1·15$.

The case of a mild steel tension specimen that was loaded beyond the yield point and then unloaded was discussed in section 3.6. The unloading process was linear elastic with a modulus equal to that of the initial loading process.

A beam behaves in a similar manner when it is bent so that parts of the cross-section are stressed beyond the yield point. All elements of the beam unload in an elastic manner as the bending moment is decreased. When the bending moment is finally removed it will be seen that there are residual stresses locked in the cross-section rather like the composite member that was discussed in section 3.6. The distribution of stresses must be such that the resultant axial force and bending moment are zero. The beam which was initially straight will be permanently curved. It is a straightforward matter to calculate the residual stress distribution and the final value of the curvature. This will be demonstrated with the help of the following example.

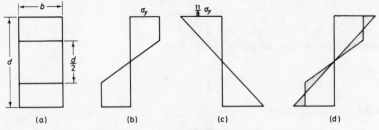

Figure 3.33

A beam of rectangular section has a moment applied such that half the cross-section becomes plastic with stress σ_y (figure 3.33a). The moment is then removed. It is first necessary to calculate the bending moment applied. This can either be found from first principles or from equation 3.43.

$$M = \frac{\sigma_y bd^2}{4} (1 - \tfrac{1}{12}) = \tfrac{11}{48} \sigma_y bd^2$$

If the bending moment is now reduced to zero the unloading process will be elastic. A simple way of considering this type of problem is first to determine the stress distribution when the bending moment M is applied (known in this case). Next find the stresses that would result if a bending moment $-M$ were applied to the beam and it is assumed that the material in the beam behaves in an elastic manner. If the two bending moments are summed, the result is of course zero. The resulting stress distribution can be found by adding the two stress distributions together. In our given case the bending moment $-M$ will produce a linear stress distribution giving a maximum stress

$$\sigma_{max} = \frac{M(d/2)}{bd^2/12}$$

where

$$M = \tfrac{11}{48}\, \sigma_y\, bd^2$$

so that

$$\sigma_{max} = \tfrac{11}{8}\, \sigma_y$$

The distribution is shown in figure 3.33c. At first sight it might appear somewhat worrying that the stresses are greater than the yield stress of the material; they do not in fact exist by themselves and are of opposite sign to the initial loading stresses. Let us consider the stress at the top of the beam. Under the action of M it had a value of σ_y. As the bending moment is reduced to zero this stress will steadily reduce, it will go through zero and end up of opposite sign, the final value is $\tfrac{3}{8}\, \sigma_y$. The final stress distribution is given in figure 3.33d where the two stress distributions, one with a sign change, have been superimposed. The shaded portion gives the resultant stress distribution.

To find the final curvature it is necessary to calculate the initial curvature for the loading condition and then subtract the change in curvature that takes place when unloading.

For loading at $y = \pm d/4$ the stress has just reached a value of $\pm\sigma_y$, hence the strain at this position is $\pm\sigma_y/E$ therefore

$$\frac{\sigma_y}{E} = \frac{d}{4R_1}$$

or

$$\frac{1}{R_1} = \frac{4\sigma_y}{dE}$$

The change of curvature when unloading is found from the fact that the value of the stress is $(11/8)\sigma_y$ at $y = d/2$

$$\frac{d}{2R_2} = \frac{11\sigma_y}{8E}$$

or

$$\frac{1}{R_2} = \frac{11}{4}\frac{\sigma_y}{dE}$$

The final curvature is

$$\frac{\sigma_y}{dE}\left(4 - \tfrac{11}{4}\right) = \frac{5}{4}\frac{\sigma_y}{dE}$$

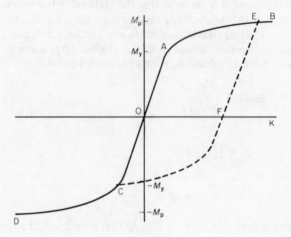

Figure 3.34

In figure 3.34 OAB is the curve representing the moment–curvature relation when a positive bending moment is applied to a beam. OCD is the curve for a moment of opposite sign. EF represents an unloading process and the residual curvature will be given by OF. Once the point F has been reached let us consider what will happen if a bending moment of opposite sign is applied. Initially the relation between moment and curvature will be a straight line, a continuation of EF. Can a bending moment of $-M_y$ be applied before the material yields again? If the stress distribution corresponding to F is considered, similar to figure 3.33d, there is a stress at the top of the beam that is of opposite sign to that resulting from the original loading. The effect of now applying a negative bending moment is to increase the value of this stress, and the material will yield well before a moment of $-M_y$ is applied. It can be seen at once that, starting from point E with a stress of $+\sigma_y$, an elastic bending moment of value M_y would reduce the stress to zero and a further bending moment of value M_y would give a stress of $-\sigma_y$. This implies that there is an elastic range of bending moment of value $2M_y$. So that starting from any point beyond M_y on the loading curve the maximum change of bending moment before yield occurs is $2M_y$.

3.19 Combined plastic bending and axial force

We shall consider the case of a rectangular section (figure 3.35a) which has an axial load P applied. A bending moment M is then applied until the section becomes fully plastic (figure 3.35b). It is obvious that when this stress distribution is integrated over the section there must be a resultant force P.

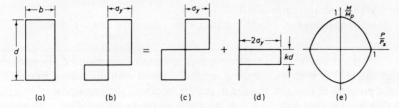

Figure 3.35

This stress distribution can be thought of as being composed of two separate distributions: c showing the fully-plastic stress distribution when $P = 0$ (that is, a moment M_p applied); d showing a stress distribution equivalent to the load P such that $2\sigma_y bkd = P$.

If a purely axial load is applied to the section, the load P_s producing a stress of σ_y over the section is often referred to as the *squash load*. Thus

$$P = 2kP_s$$

and

$$M = M_p - \frac{Pkd}{2} = M_p(1 - 4k^2)$$

Substituting for $k = P/2P_s$

$$\frac{M}{M_p} + \left(\frac{P}{P_s}\right)^2 = 1 \tag{3.45}$$

This equation may be plotted in the form shown in figure 3.35e to give what is called an interaction diagram. Any combination of M and P is safe if it lies within the boundary, and the cross-section is adequate. A point on the boundary indicates that the section is fully plastic and outside the boundary is unsafe.

Problems

3.1 A rope of length l is required to carry a vertical load P at one end. The material is of density ρ and elastic modulus E. The cross-section A is circular but varies such that the axial stress has a constant value σ. Find an expression for A at a distance x above P. What is the extension of the rope when the load is applied?

3.2 A circular brass tube of wall thickness 1 mm fits inside a steel tube of the same wall thickness with a clearance of about 0·1 mm. The mean diameter of the tubes may be taken as 100 mm for calculation purposes. Both tubes are of the same length and are connected together at their ends by heavy flanges. Neglecting any end effects, find the longitudinal stresses for the following cases

(a) a tensile load of 30 kN applied at the ends of the tubes

(b) a temperature rise of 80 °C

(c) a tensile load of 30 kN together with a temperature rise of 80 °C.

3.3 A vessel has a cylindrical centre portion of length l and radius r, the ends are hemispherical. The wall thickness t is small compared with the radius. For an internal pressure P, find the stresses in the cylinder and the hemispheres.

A thin hollow steel sphere of radius r and wall thickness t is filled with water. The temperature of the sphere and contents are raised by $T\,°C$ and water is pumped in until the pressure has risen by P bar. Derive an expression for the volume of water that has to be pumped in. The vessel is then sealed and allowed to cool to its original temperature. What is the final pressure of the water?

3.4 For a case of plane stress find the values of the principal stresses given $\sigma_{xx} = 140$ N/mm^2, $\sigma_{yy} = -50$ N/mm^2, $\tau_{xy} = 80$ N/mm^2. A rosette strain gauge ($0°$, $45°$, $90°$) is fixed to a steel sheet and the same stress system applied. The $90°$ arm is inclined at $40°$ clockwise to the direction of the σ_{xx} stress. Determine the strain in the direction of the gauge arms.

3.5 From a theoretical analysis, the principal stresses at a point on a steel member due to a certain load have been calculated as 65 N/mm^2 tensile and 138 N/mm^2 tensile. To confirm the analysis an electrical resistance strain gauge rosette with arms at $120°$ was placed at the point. When the member was loaded the following changes in the resistance of the arms were noted. $dR_0 = +0{\cdot}122\ \Omega$, $dR_{120} = +0{\cdot}104\ \Omega$, $dR_{240} = +0{\cdot}04\ \Omega$. The nominal resistance of each arm was $120\ \Omega$ and the ratio of electrical strain to mechanical strain or the gauge factor was $2{\cdot}1$.

It is thought that one of the arms of the rosette may not be functioning correctly. Show that the theoretical and experimental results are inconsistent and suggest which arm is incorrect.

3.6 A rectangular beam is to be cut from a circular log of wood 180 mm in diameter and used to span 2·4 m with simple supports at each end. Taking a maximum longitudinal stress of 10 N/mm^2, find the greatest uniformly distributed load that the beam can carry.

3.7 Derive the following relationships used in the bending of beams and state any assumptions on which they are based.

$$\frac{\sigma}{y} = \frac{M}{I} = \frac{E}{R}$$

A rectangular beam of breadth b and depth d is made of a material that has a modulus E_1 in tension and E_2 in compression, $E_1 > E_2$. Find the position of the neutral axis and show that

$$\frac{M}{I} = \frac{E^*}{R}$$

where

$$E^* = \frac{4E_1 E_2}{(E_1^{1/2} + E_2^{1/2})^2}$$

3.8 Explain the terms (a) principal second moment of area; (b) product second moment of area.

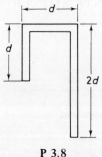

P 3.8

The section shown in P 3.8 is formed from sheet metal of thickness t where $t \ll d$. Find the values of the principal second moments of area. The section is used as a cantilever of length l with the longer side vertical. Find the maximum value of the longitudinal stress when a vertical point load P is applied at the end of the cantilever.

3.9 A short vertical concrete column has a regular hexagonal cross-section of side a. A vertical load P can move along the z axis which lies along a line joining two opposite corners of the hexagon. The maximum allowable compressive stress is σ_m and no tensile stress is permitted. Find the limits of the position of P and the maximum value of P in terms of its distance from the centre. If the load is not restricted to the z axis define the area within which it must be confined if no tension is to develop in the section.

3.10 A beam has a rectangular symmetrical hollow cross-section. The outer dimensions are $2b$ by $2d$ and the inner dimensions are b by d. The z axis is parallel to the b dimension. Find the ratio of the fully-plastic bending moment to the maximum elastic moment for bending about the z axis. The material of the beam is ideal elasto-plastic with a yield stress σ_y.

Find the moment required such that the stress has just reached σ_y at a distance $d/2$ from the neutral axis. What is the residual stress at this position when the moment is removed? Find also the maximum elastic moment of opposite sign that may now be applied.

3.11 Show that the plastic moment of resistance of a mild steel beam of rectangular cross-section of breadth b and depth d is $M_p = \sigma_y bd^2/4$ where σ_y is the yield stress and the material is ideal elasto-plastic.

A prestressing force is applied inwards at the bottom edges of such a beam on the vertical centre-line of its section so that yielding just begins to occur. Show that the effective plastic moment of resistance of the section is now increased from M_p to $\frac{23}{16}M_p$.

4 TORSION AND SHEAR EFFECTS

4.1 Torsion of a circular cross-section

A pure torque is applied to a circular rod of homogeneous isotropic material (figure 4.1a). We shall consider possible deformations that can take place on the section AA. In figure 4.1b it is assumed that the left-hand portion, X, has a convex surface so that the surface of the right-hand portion, Y, must be concave. This is not compatible as may easily be seen by rotating the left-hand portion through 180° about A–A to become Y′ (figure 4.1c). If we now compare b and c, apparently the same torque has produced a concave surface on Y and a convex surface on Y′. Clearly this is impossible. A line that was radial is shown in c assumed deformed into a curve when the torque is applied. A similar argument may be applied to show that this is impossible. The details are left as an exercise for the reader.

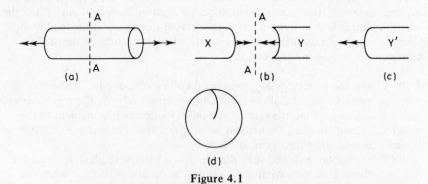

Figure 4.1

It may then be inferred from the above arguments that when a rod of circular cross-section has a pure torque applied

    (1) plane sections remain plane
    (2) a radial line remains straight.

A thin sheet of material has pure shear stresses, τ, applied, and would deform in the manner shown in figure 4.2a. It is possible to bend the sheet round to form

122

the slit tube of radius r (figure 4.2b). The edges could be fixed together in some way. If the ends of the tube have rotated through an angle θ with respect to each other

$$r\theta = l\gamma$$

Note that the shear stress 'flows' round the circular end of the tube. Taking moments about the axis of the tube it can be seen that the same stress pattern would be produced by applying an axial torque of value

$$T = 2\pi r^2 t\tau$$

where t is the thickness of the sheet of material.

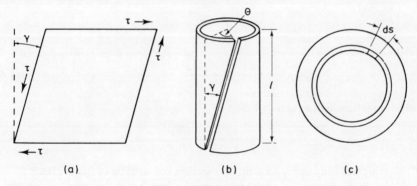

Figure 4.2

A solid rod and indeed a hollow one could be thought of as a large number of cylindrical shells are cemented together. A torque is applied and each shell will rotate through the same angle θ. For a shell at radius r

$$\theta = \frac{l\gamma}{r} \tag{4.1}$$

To derive equation 4.1 no mention has been made of the elastic properties of the material. If the material is elastic

$$\gamma = \frac{\tau}{G} \quad \theta = \frac{\tau l}{Gr}$$

or

$$\frac{\tau}{r} = \frac{G\theta}{l} \tag{4.2}$$

A section through a circular rod of radius R is shown in figure 4.2c, together with a cylindrical shell of radius r and thickness dr. If the shear stress at radius r is

τ then the tangential force on an element of length ds is $\tau\, ds\, dr$ and the moment about the centre is $\tau r\, ds\, dr$. The torque due to the shell is $dT = \tau 2\pi r^2\, dr$ and the total torque on the rod can be written as

$$T = \int_0^R \tau 2\pi r^2\, dr \qquad (4.3)$$

If the rod were hollow the limits of integration would extend between the two radii.

τ is a function of r as given by equation 4.2 for an elastic case, therefore

$$T = \int_0^R \frac{G\theta}{l} 2\pi r^3\, dr = \frac{2\pi G\theta}{l} \frac{R^4}{4}$$

Now the second polar moment of area for a circular rod is J where

$$J = \frac{\pi R^4}{2}$$

thus

$$T = \frac{G\theta J}{l} \qquad (4.4)$$

which together with equation 4.2 gives

$$\frac{\tau}{r} = \frac{T}{J} = \frac{G\theta}{l} \qquad (4.5)$$

This result will only apply to a circular section rod or tube of linear elastic material. When a torque is applied to rods of other shapes the cross-section will no longer remain plane but will warp or deform.

As a simple example of elastic-torsion theory, consider the problem of a rod of circular section built in at each end. One-third of the rod is of radius r and the other two-thirds of radius $2r$. If the shear stress is limited to τ, what is the maximum torque that may be applied at the change in section? Any stress concentrations are to be neglected.

The first point to note is that, at the point of application of the torque, both parts of the rod must have twisted through the same angle θ. Using subscript 1 for the shorter length and 2 for the longer length

$$\frac{\tau_1}{r} = \frac{G\theta}{l/3} \qquad \tau_1 = \frac{3G\theta r}{l}$$

$$\frac{\tau_2}{2r} = \frac{G\theta}{2l/3} \qquad \tau_2 = \frac{3G\theta r}{l}$$

therefore

$$\tau_1 = \tau_2 = \tau$$

$$J_1 = \frac{\pi r^4}{2} \qquad J_2 = \frac{16\pi r^4}{2}$$

therefore

$$T_1 = \frac{\tau}{r}\frac{\pi r^4}{2} \quad T_2 = \frac{\tau}{2r}\frac{16\pi r^4}{2}$$

$$T = T_1 + T_2 = \frac{9\pi r^3 \tau}{2}$$

The case of a rod of one material inside a tube of another material (the two being rigidly connected at the common boundary) will not present any difficulty. The relationship between shear strain and radius will be linear. As there will be a change in the value of the modulus of rigidity at the common boundary, there must be a sudden change in the value of shear stress at this point. The total torque can be found by summing the torque for the rod and the torque for the tube.

4.2 Strain energy due to torsion

Referring back to equation 3.29, the strain energy per unit volume for a case of pure shear would be given by

$$\frac{dU}{dV} = \frac{\tau^2}{2G}$$

Consider a solid circular rod under the action of a torque T. For an annulus at radius r and thickness dr the shear stress τ is given by Tr/J. Therefore the energy stored in the ring of unit length is

$$dU = \frac{T^2 r^2}{J^2 \times 2G} \times 2\pi r\, dr$$

Total energy per unit length equals

$$\frac{T^2}{2GJ^2} \int_0^R 2\pi r^3\, dr = \frac{T^2}{2GJ}$$

So that for a rod of length l

$$U = \frac{1}{2}\frac{T^2 l}{GJ} \tag{4.6}$$

This could have been derived in a simple manner. If θ is the resulting angle of twist when the torque T is applied, the external work done is $T\theta/2$. This must of course be equal to the strain energy stored. Now $\theta = Tl/GJ$ therefore

$$U = \frac{1}{2}\frac{T^2 l}{GJ}$$

4.3 Combined torsion, bending and axial force

A rod of circular section is subjected to an axial force P, a bending moment M about a horizontal diameter and an axial torque T. The question arises, what is the maximum stress that will result?

Taking the x axis coincident with the axis of the tube and the z axis as horizontal, the longitudinal stress produced by the bending moment can be found from simple bending theory. If the rod has a radius r

$$\sigma_{xx} = \frac{4My}{\pi r^4}$$

This can be combined with the stress $\sigma_{xx} = P/\pi r^2$ produced by the axial force, assumed tensile; therefore

$$\sigma_{xx} = \frac{4My}{\pi r^4} + \frac{P}{\pi r^2}$$

The maximum stress will arise when y has a maximum value, that is

$$\sigma_{xx} = \frac{4M}{\pi r^3} + \frac{P}{\pi r^2}$$

$$\sigma_{yy} = \sigma_{zz} = 0$$

The torque will produce a shear stress that will vary from zero at the centre of the rod to $\tau = 2Tr/\pi r^4$ at the outside.

To find the resulting maximum stress when all 'forces' are applied, it will be necessary to perform a principal stress analysis. To simplify the arithmetic it will be assumed that $M = T = Pr$. Thus

$$\sigma_{xx} = \frac{5P}{\pi r^2} \qquad \sigma_{zz} = 0 \qquad \tau_{xz} = \frac{2P}{\pi r^2}$$

$$\sigma_1, \sigma_2 = \frac{P}{\pi r^2} [\tfrac{5}{2} \pm \tfrac{1}{2}(25 + 16)^{1/2}]$$

$$\sigma_1 = \frac{5 \cdot 7P}{\pi r^2} \qquad \sigma_2 = -\frac{0 \cdot 7P}{\pi r^2}$$

$$\tau_{max} = \frac{3 \cdot 2P}{\pi r^2}$$

4.4 Plastic torsion of circular rods

It will be assumed that the shear-stress–shear-strain curve for mild steel, when subjected to pure shear, is similar in form to that assumed for tension and compression in figure 3.27. For a case of pure torsion on a circular rod we shall further assume

(1) the stress remains constant in the plastic range
(2) plane sections remain plane
(3) the material is isotropic and homogeneous.

The torque required to make a section fully plastic can be found assuming that the yield shear stress is τ_y. Considering an annulus at radius r of thickness dr. Moments about the axis give

$$dT = 2\pi r \, dr \times r\tau_y$$

$$T = \tau_y \int_0^R 2\pi r^2 \, dr = \tfrac{2}{3}\pi r^3 \tau_y$$

The fully elastic torque is $\pi r^3 \tau_y/2$, therefore

$$\frac{T_P}{T_E} = \tfrac{4}{3} \tag{4.7}$$

It is not difficult to treat a section that is partially elastic and partially plastic. A specific example will demonstrate the method.

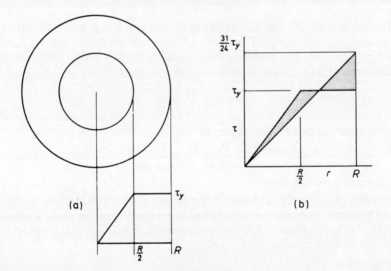

Figure 4.3

A rod of circular cross-section, radius R, is made from an elasto-plastic material with a yield stress τ_y. Find the torque required to produce a plastic stress for one half the radius. The torque is then removed; the residual stresses and the permanent angle of twist per unit length are required.

The stress distribution is shown in figure 4.3a. The torques carried by the two parts of the shaft can be found separately and summed.

$$T_P = \tau_y \int_{R/2}^R 2\pi r^2 \, dr = \tfrac{7}{12}\pi R^3 \tau_y$$

The central elastic portion can be treated as an elastic rod of radius $R/2$

$$T_E = \frac{\pi R^3 \tau_y}{16}$$

$$T = T_P + T_E = \tfrac{31}{48}\pi R^3 \tau_y$$

For the unloading process we shall make exactly the same assumption as for the case of plastic bending, that is, the rod unloads elastically. Effectively this means that we apply a torque of value T in the opposite sense and calculate the stress distribution. This has then to be combined with the initial stress distribution.

$$\tau_{max} = \frac{(31/48)\pi R^3 \tau_y R}{\pi R^4 / 2} = \tfrac{62}{48}\tau_y$$

This stress will vary linearly with r.

The two stress distributions have been superimposed in figure 4.3b, and the residual stress distribution is shown shaded. The two triangular portions of this diagram will have shear stresses of opposite sign such that the resultant torque is zero.

The initial angle of twist θ_1 can be found for the radius at which the stress has just reached τ_y, that is, $R/2$. The value of the shear strain at this point is τ_y/G, so that $\theta_1/l = 2\tau_y/GR$. The change in shear stress at radius R during the elastic unloading process is $(31/24)\tau_y$ so that the change in angle of twist θ_2 will be given by

$$\frac{\theta_2}{l} = \frac{31}{24}\frac{\tau_y}{GR}$$

The permanent angle of twist is given by

$$\frac{\theta_1}{l} - \frac{\theta_2}{l} = \frac{17}{24}\frac{\tau_y}{GR}$$

4.5 Shear stresses in beams

An element of length dx cut from a beam of uniform section is shown in figure 4.4a. The beam is subject to a varying bending moment applied about a principal axis Oz. At the left-hand end A the bending moment is M and at the right-hand end B the bending moment is $M + dM$.

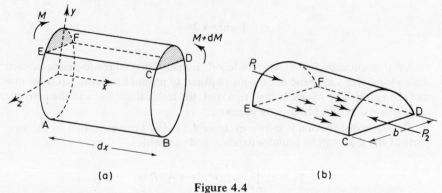

(a)   (b)

Figure 4.4

At a height y above the neutral axis, the stress at section A will be $\sigma_{xx} = -My/I$ and the stress at B will be $\sigma_{xx} + d\sigma_{xx} = -(M + dM)y/I$.

If we consider a small element of area dA at the height y, the force at section A will be $-My\,dA/I$ and the force at section B $-(M + dM)y\,dA/I$. There is then a net difference in longitudinal force on the element of area dA given by $-dMy\,dA/I$. So that there will be a difference in force of $-\int dMy\,dA/I$ on the two areas shown shaded.

The element of the beam above the plane CDEF must be in equilibrium. The only way of providing the necessary longitudinal force is by means of a shear stress acting on the plane CDEF (figure 4.4b). If it is assumed that the shear stress τ_{yx} is uniformly distributed over the width of the beam b

$$\tau_{yx}b\,dx = -\int \frac{dMy\,dA}{I}$$

or

$$\tau_{yx} = -\frac{dM}{dx} \int \frac{y\,dA}{bI}$$

From equation 1.10 $dM/dx = -Q$, therefore

$$\tau_{yx} = \frac{Q\int y\,dA}{bI} \tag{4.8}$$

The horizontal component of shear stress that has been found must be accompanied by a complementary shear τ_{xy}, which of course acts in the vertical direction on the face of the section. We now have a means of calculating the vertical shear stress at any level on a beam.

An alternative method of writing equation 4.8 is

$$\tau_{yx} = \frac{QA\bar{y}}{bI}$$

where A is the cross-sectional area of the beam above the level at which the shear stress is required and $\bar{y}$ is the distance from the neutral axis to the centroid of the area.

Note: The value of I is the second moment of area for the whole section. The assumption has been made that the shear stress is uniformly distributed over the width of the beam. Consequently it can be assumed that the vertical component of shear stress at a particular level is constant.

The shear distribution for a rectangular section will first be found (figure 4.5a). At level CD

$$\tau_{yx} = \tau_{xy} = Q \int_{h}^{d/2} \frac{yb\,dy}{bI} = \frac{Q}{I}\left[\frac{d^2}{8} - \frac{h^2}{2}\right] \tag{4.9}$$

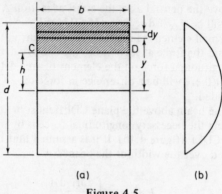

(a)   (b)

Figure 4.5

This implies that the shear stress varies parabolically with a maximum value of $Qd^2/8I$ at the neutral axis. Substituting for I gives

$$\tau_{max} = \frac{3}{2}\frac{Q}{bd}$$

Now Q/bd would be the average value of the shear stress, so that there is an increase of 50 per cent above the average at the neutral axis.

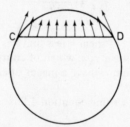

Figure 4.6

The vertical component of shear stress for the circular cross-section (figure 4.6) can be calculated, assuming the shear stress is constant over the breadth of the beam. However, it is not possible for the shear stress to be vertical at C or D as this would infer that the shear stress is crossing an unloaded boundary—an impossibility. It is essential that the shear stress be tangential at C and D and the distribution would appear as shown in the diagram. The maximum stress will still occur at the neutral axis and has a value of 4/3 times the average shear stress.

In the derivation of the simple bending formula $\sigma = My/I$ it was assumed that the bending moment was constant along the length of the beam, and it was perfectly in order to assume that plane sections remained plane. It was further assumed that this latter statement was true even if the bending moment varied.

However, if a shear stress exists over the cross-section of the beam, this must result from the effect of a shear strain. If we consider a cantilever with an end

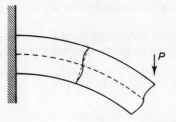

Figure 4.7

load P (figure 4.7), the shear force is constant with a value of $-P$. Due to the effects of shear strain, a section that was plane will no longer be plane but would deform in the manner shown in the figure by the full lines. At the top and bottom surfaces of the beam the shear stress is zero, therefore the line must be normal to the top and bottom surfaces. The angle of inclination to the normal will be given by $\gamma = \tau/G$. Thus for a symmetrical cross-section, the worst angular displacement will occur at the neutral axis. Here we have an example of a section warping, that is, no longer remaining planar.

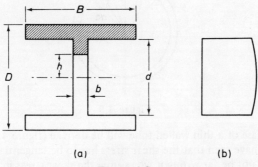

(a)   (b)

Figure 4.8

We shall now deal with two further cases. The first of these is the universal beam shown in figure 4.8a. The shear stress in the web can be found as follows

$$\int y \, dA = \frac{B}{2}\left(\frac{D^2 - d^2}{4}\right) + \frac{b}{2}\left(\frac{d^2}{4} - h^2\right)$$

To obtain the shear stress, the integral would be multiplied by a constant Q/bI. It can be seen that the shear variation is again parabolic with a maximum value at the neutral axis.

$$\tau_{\min} = \frac{QB}{2bI}\left(\frac{D^2 - d^2}{4}\right)$$

$$\tau_{\max} = \frac{Q}{2bI}\left[\frac{BD^2}{4} - \frac{d^2}{4}(B - b)\right]$$

Comparing these values it can be seen that there will only be a small difference between them if $b \ll B$.

If an approximation is made for I, omitting the web completely and also omitting the second moment of area of the flanges about their own centroids

$$I \approx 2B\left(\frac{D-d}{2}\right)\left(\frac{D+d}{4}\right)^2 = \frac{B}{2}\left(\frac{D+d}{2}\right)\left(\frac{D^2-d^2}{4}\right)$$

and the value of τ_{min} becomes

$$\tau_{min} = \frac{2Q}{b(D+d)} \approx \frac{Q}{bd} \tag{4.10}$$

This means that to a reasonable degree of accuracy we may assume that the shear is carried by the web and the shear stress distribution is uniform over the web.

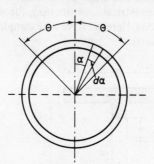

Figure 4.9

Finally the case of a thin walled tube will be derived (figure 4.9). For a solid circular rod we have seen that the shear stress has to be tangential at the boundary. We should then not be far wrong if we assume the shear stress in the case of the tube to be tangential and uniform over the wall thickness t.

$$\tau = \frac{Q\int y\,dA}{bI} = Q\int_{-\theta}^{\theta} \frac{r\cos\alpha rt\,d\alpha}{2tI}$$

note b = 2t

$$= [r^2 t \sin\alpha]_{-\theta}^{\theta}\, \frac{Q}{2tI}$$

Now $I = \pi r^3 t$, therefore

$$\tau = \frac{Q\sin\theta}{\pi rt}$$

τ will be a maximum when $\theta = 90°$, that is, at the neutral axis $\tau = Q/\pi rt$. The average shear stress is $Q/2\pi rt$ so that in this case the maximum is twice the average shear stress.

4.6 Shear-stress distribution in flanges

A similar piece of analysis to that introduced in the previous section can be used to determine the shear-stress distribution in the flanges of the universal beam (figure 4.10a). A small length of beam dx is considered with a bending moment changing from M to $M + dM$. It will be assumed that the flange depth t is small

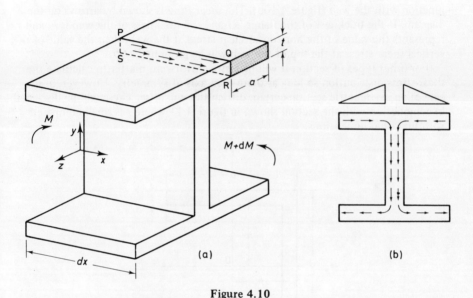

Figure 4.10

compared with the total depth of the beam D, and so it will be a reasonable approximation to assume that for bending about the z axis the stress σ_{xx} in the flanges is constant. It will also be sufficiently accurate to take $y = D/2$ in the simple bending formula, that is, $\sigma_{xx} = -MD/2I$. The total force difference on the two ends of the shaded area can now be written as

$$-\frac{dMDat}{2I}$$

If we isolate the portion of the flange beyond PQRS, it will be seen that to maintain equilibrium it is necessary to have a shear stress acting on the face PQRS. Assuming this to be constant

$$\tau_{zx}t \, dx = -\frac{dMDat}{2I}$$

or

$$\tau_{zx} = \frac{QaD}{2I}$$

This could be written as

$$\tau_{zx} = \frac{QA\bar{y}}{tI}$$

where $A = at$ and $\bar{y} = D/2$, a similar expression to that derived in section 4.5.

This shear stress must be accompanied by a complementary shear stress τ_{xz} which will vary linearly from zero at the outside edge to a maximum value at the junction with the web (figure 4.10b). The *shear flow* is shown by arrows on the diagram. If the thickness of the flange is t and the thickness of the web is b, and τ represents the value of the horizontal shear stress at the web, then the value of the vertical shear stress at the top of the web will be given by $2\tau t/b$.

For other types of section it will be a straightforward matter to calculate the shear-stress distribution so long as there is an axis of symmetry. However, in the case of an asymmetric section certain difficulties will arise. These will be discussed in conjunction with the section shown in figure 4.11 where t is small compared with the other dimensions.

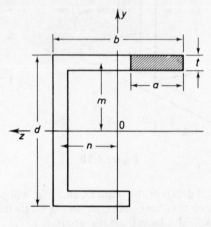

Figure 4.11

In the analysis used so far we have been fortunate in being able to work about principal axes and so the simple bending formula was used. For the case shown we shall assume that the only bending moment applied is about the z axis. Unfortunately it is not known where the principal axis lies. It will therefore be necessary to use the more complicated expression for the horizontal stress given by equation 3.37, which will simplify to

$$\sigma_{xx} = -\frac{M_z(yI_{yy} - zI_{yz})}{I_{yy}I_{zz} - I_{yz}^2}$$

since $M_y = 0$. O is the centroid of the section and the axes are as shown. The total longitudinal force on the shaded area is $\int \sigma_{zz} t \, \mathrm{d}a$. Two integrals are required.

$$\int yt \, \mathrm{d}a = mta \qquad \int zt \, \mathrm{d}a = \left(b - n - \frac{a}{2}\right)at$$

Thus the force difference on two elements dx apart is

$$- \frac{dM_z}{I_{yy}I_{zz} - I_{yz}^2}\left[I_{yy}(mta) - I_{yz}\left(b - n - \frac{a}{2}\right)at\right]$$

thus

$$\tau_{xz} = -\frac{Q}{I_{yy}I_{zz} - I_{yz}^2}\left\{I_{yy}ma - I_{yz}\left[(b - n)a - \frac{a^2}{2}\right]\right\}$$

The shear-stress distribution in the flange is a combination of linear and parabolic relations. Note that the stress is zero when $a = 0$ and also when

$$a = 2\left(b - n - \frac{I_{yy}}{I_{yz}}m\right)$$

The distribution in the other flange can be found in a similar manner. It is left to the reader to show that there will be a parabolic variation of shear stress in the web.

4.7 Shear centre

If the shear-stress distribution is examined it will be seen at once that the vertical components when integrated over the section must be equal to the vertical shear force that is applied.

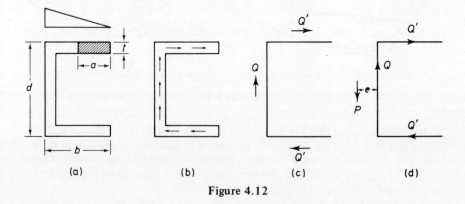

Figure 4.12

For the channel iron (figure 4.12a) the horizontal shear stress in the flange will vary in a linear manner from zero to a maximum value of $Qbd/2I$ where $t \ll b$ or d. Integrating, we shall find that a horizontal force $Q' = Qb^2dt/4I$ is acting in the flange.

Suppose that the channel iron is set up as a cantilever with a point load P at the end. The shear-flow diagram would be as shown in figure 4.12b and the resultant forces as shown in figure 4.12c. Q is in equilibrium with the applied loading P, and horizontal equilibrium is maintained by the two forces Q' acting in opposite directions. Unfortunately they also provide an unbalanced couple of magnitude $Q'd$ which will cause the section to rotate if the applied load P acts through the centroid of the section. However it is possible to prevent this rotation if the applied load is applied at a distance e from the centre line of the web (figure 4.12d). Where

$$Pe = Q'd \quad e = \frac{b^2 d^2 t}{4I} \qquad\qquad (4.11)$$

This particular point is known as the shear centre or the centre of shear for the section.

4.8 Torsion of thin-walled tubes

When the torsion problem was discussed in section 4.1 the point was made that the simple theory derived would only apply to circular sections. It is possible however to derive a theory which will apply to a thin-walled tube of non-circular section.

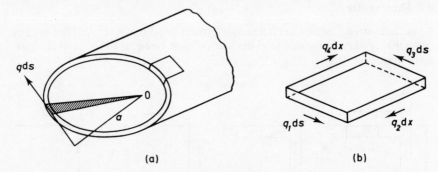

(a)   (b)

Figure 4.13

It will be assumed that the cross-section of the tube does not change with length and that there are not any sharp corners or sudden changes in wall thickness. This, however, does not preclude changes in wall thickness. Instead of referring to shear stress we shall use the term shear flow or shear force per unit length of perimeter, that is, $q = \tau t$.

A length of tube is shown in figure 4.13a. If the equilibrium of the small rectangular element is considered (figure 4.13b) it will be seen at once that $q_1 = q_2 = q_3 = q_4$. From this we may infer that the shear flow is constant round the perimeter of the tube. A small element of the tube perimeter of length ds will have a tangential force $q \, ds$. Taking moments about any point O, $dT = q \, ds a$, where a is the moment arm. Now $a \, ds$ is twice the area shown shaded on the diagram, therefore

$$dT = 2q \, dA$$

or

$$T = 2qA \qquad (4.12)$$

where A is the area enclosed by the mean perimeter of the tube.

We have shown (equation 3.29) that the strain energy per unit volume due to pure shear is $\tau^2/2G$. For a unit length of tube the total energy stored will equal the external work done when a torque T is applied.

$$U = \frac{1}{2G} \int \tau^2 t \, ds = \tfrac{1}{2} T\theta$$

Now

$$\tau t = q = \frac{T}{2A}$$

therefore

$$\tfrac{1}{2} T\theta = \frac{1}{2G} \int \frac{T}{2A} \tau \, ds$$

$$\theta = \frac{1}{2GA} \int \tau \, ds = \frac{q}{2GA} \int \frac{ds}{t} \qquad (4.13)$$

or

$$\theta = \frac{T}{4A^2 G} \int \frac{ds}{t} \qquad (4.14)$$

This theory is readily extended to take account of multi-cellular sections in torsion. A particular example will indicate the procedure used.

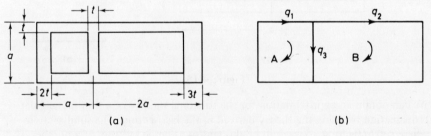

Figure 4.14

A torque T is applied to a cellular structure of shear modulus G. The cross-section is shown in figure 4.14a. The relation between G and θ is required and also the values of the various shear flows. The effects of the sharp corners are to be ignored. $t \ll a$.

It can be seen at once from figure 4.14b that $q_1 = q_2 + q_3$. For the left-hand tube $T_A = 2a^2 q_1$ and for the right-hand $T_B = 4a^2 q_2$. Also for the left-hand tube

$$\theta_A = \frac{1}{2Ga^2}\left[\frac{q_1 a}{2t} + \frac{2q_1 a}{t} + \frac{(q_1 - q_2)a}{t}\right]$$

while for the right-hand tube

$$\theta_B = \frac{1}{4Ga^2}\left[\frac{q_2 a}{3t} + \frac{2q_2 2a}{t} + \frac{(q_2 - q_1)a}{t}\right]$$

It is essential that both tubes rotate by the same amount, that is, $\theta_A = \theta_B$. From this relation

$$q_2 = \tfrac{12}{11}q_1$$

Now

$$T = T_A + T_B = 2a^2 q_1 + 4a^2 q_2$$

Hence

$$q_1 = \frac{11T}{70a^2} \quad q_2 = \frac{12T}{70a^2}$$

Substituting into θ_A gives

$$T = \tfrac{280}{53}Gta^3\theta$$

4.9 Torsion of a thin rectangular section

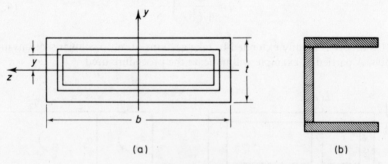

Figure 4.15

We can obtain an approximation for the torsional rigidity of a thin rectangular cross-section by using the theory derived in the last section. The solid section where $t \ll b$ (figure 4.15a) could be thought of as being built up of a number of tubes, one of which is shown at a distance y from the neutral axis, of wall thickness dy. If the wall thickness is constant for the tube τ is constant. The torque in the tube is $2A\,\tau\,dy$ where $A \approx 2yb$. Thus

$$\frac{dT}{dy} = 4by\tau$$

From equation 4.14

$$\theta = \frac{\mathrm{d}T}{4A^2G}\frac{s}{t} \approx \frac{\mathrm{d}T2b}{16y^2b^2G\,\mathrm{d}y}$$

or

$$\frac{\mathrm{d}T}{\mathrm{d}y} = 8by^2G\theta$$

Equating the two values for $\mathrm{d}T/\mathrm{d}y$ gives $\tau = 2yG\theta$. This indicates that the shear-stress distribution is linear over the thickness of the section with a maximum value at $y = \pm t/2$.

$$\mathrm{d}T = 8by^2G\theta\,\mathrm{d}y$$

therefore

$$T = \int_0^{t/2} 8bG\theta y^2\,\mathrm{d}y = \tfrac{1}{3}bt^3G\theta$$

Compare this result with simple torsion theory which only holds for circular sections $T = JG\theta$.

This implies that the equivalent value of the polar second moment of area for the thin rectangle is $bt^3/3$. It is interesting to compare this with

$$J = I_{yy} + I_{zz} = \tfrac{1}{12}(bt^3 + tb^3) = \frac{bt}{12}(b^2 + t^2)$$

This theory can be used to derive approximate torsional constants for other thin-walled sections. The channel iron (figure 4.15b) could be thought of as consisting of three thin rectangles and so the value of J is $\Sigma\,bt^3/3$.

Problems

4.1 A hollow steel shaft with internal diameter one-half the external diameter has to transmit 750 kW at 240 r.p.m. without the shear stress exceeding $110\,\mathrm{N/mm^2}$. Find the dimensions of the cross-section, and show that there is a saving of material of about 22 per cent by making the shaft hollow instead of solid. What will be the angle of twist per metre length of the hollow shaft?

4.2 A particular material has a relation between shear stress and strain

$$\tau = G\gamma - A\gamma^2$$

where G and A are constants. Find the torque required to produce a maximum shear strain of $G/4A$ in a solid circular shaft, length l, radius R. When the torque is removed and shaft unloads elastically with a modulus G. Find the permanent angle of twist in the shaft.

4.3 A solid shaft of radius R is twisted through an angle θ by a torque T. Writing the fully plastic torque as T_p, the angle of twist at first yield as θ_y and the radius at which the yield stress is reached as λR, plot graphs of T/T_p and λ as functions of θ/θ_y.

4.4 A circular steel shaft of length $2l$ is rigidly fixed at both ends. One half of the shaft has a diameter $2d$ and the other half a diameter of d. Assuming that the material behaves in an ideal elasto-plastic manner, with a yield stress of τ_y, find the value of the torque which, when applied at the mid-point, will cause half the cross-sectional area of the shaft with the larger diameter to become plastic. Discuss the residual stress distributions when the torque is removed.

4.5 A torsional assembly consists of a steel rod of radius r and a concentric steel tube of internal radius $1.5r$ and wall thickness $0.1r$. Both are of the same length and the rod and tube are rigidly fixed together at one end. The free ends are fixed together after a clockwise torque T has been applied to the rod. When the torque is released, find the residual torque in the rod and the tube. An anti-clockwise torque $4T$ is now applied to the assembly. Both materials behave in an ideal elasto-plastic manner with a yield stress τ_y. A separate test on the rod indicated that yield first occurred at a torque of $2T$. Assuming that the tube is fully plastic find the depth of yielding in the rod.

4.6 Derive an expression for the shear stress at any point on the cross-section of a beam subject to a shear force. Indicate the types of section to which the result may be applied without serious error. The cross-section of a box beam is square, the side being 120 mm and the thickness 5 mm. A shear force of 8 kN acts along a line through the centre of the square and parallel to two of its sides. Determine the maximum shear stress in the beam.

4.7 Sketch the shear distribution for both the webs and the flanges of the sections shown in P 4.7, when subject to a vertical shear force.

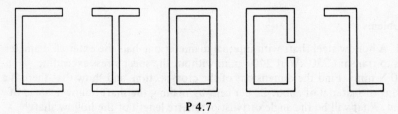

P 4.7

4.8 Find the position of the centre of shear for a circular tube of diameter D and wall thickness t (where $D \gg t$), when the tube is slit along one edge parallel to the axis of the tube.

4.9 A beam has an L-shaped cross-section, the sides of which are of equal length a and constant thickness t ($t \ll a$). A shear force Q acts normally across the beam parallel to one of the section sides. Show that the maximum shear stress is $1.35 Q/at$ and sketch the shear distribution.

4.10 A thin-walled tube is to be formed from a particular strip of sheet metal and welded longitudinally. Compare the torsional rigidity of a circular tube with that of the best possible rectangular tube. Prove any formulae that are used.

4.11 A bar made from a material with shear modulus G has a square cross-section of side a. Assuming that under torsion the stress trajectories form concentric squares, find the torsional stiffness of the bar.

4.12 Determine the torsional rigidity of the tube with the section shown in P 4.12.

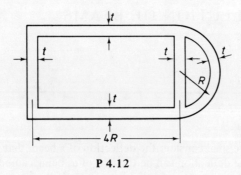

P 4.12

5 DEFLECTION OF BEAMS

5.1 Introduction

In this chapter we shall consider the deflection of a beam that is subject to transverse loading. The deflection can be thought of as being caused by two different effects: the bending moment and the shear force. If the beam is relatively long when compared with its depth, the shear force will only make a small contribution to the deflection and can be neglected. For the time being we shall only consider the deflection due to bending.

In section 3.12 it was shown that when bending takes place about a principal axis coincident with the z axis

$$\frac{M_z}{I_{zz}} = \frac{E}{R_y} \tag{5.1}$$

It is also shown in mathematical books that the equation for the radius of curvature for a plane curve in the x, y plane is

$$\frac{1}{R_y} = \frac{\dfrac{\mathrm{d}^2 v}{\mathrm{d}x^2}}{\left[1 + \left(\dfrac{\mathrm{d}v}{\mathrm{d}x}\right)^2\right]^{3/2}} \tag{5.2}$$

where v is the displacement in the y direction.

In the majority of cases of beams in structures the value of $\mathrm{d}v/\mathrm{d}x$ is small and $(\mathrm{d}v/\mathrm{d}x)^2 \ll 1$. Thus equation 5.2 can be written as

$$\frac{1}{R_y} = \frac{\mathrm{d}^2 v}{\mathrm{d}x^2} \tag{5.3}$$

If equation 5.3 is now substituted into equation 5.1 the following relation is obtained

$$EI_{zz} \frac{\mathrm{d}^2 v}{\mathrm{d}x^2} = M_z \tag{5.4}$$

142

This equation gives a relation between the transverse displacement and the bending moment. Note that a positive bending moment is associated with a positive curvature. From now on we shall confine our attention to bending about the z axis and, for ease of writing, will drop the subscripts in equation 5.4.

It was shown in chapter 1 that $dM/dx = -Q$ and that $d^2M/dx^2 = p$. Thus

$$\frac{d^2}{dx^2}\left(EI\,\frac{d^2v}{dx^2}\right) = p$$

If the beam is of constant cross-section, EI is constant and

$$EI\,\frac{d^4v}{dx^4} = p \tag{5.5}$$

EI is called the flexural rigidity of the beam and is sometimes denoted by B.

5.2 Deflection by direct integration

The deflection curves for several straightforward cases can be found by direct integration of equation 5.4. We shall ignore the dead weight of the beam.

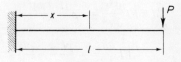

Figure 5.1

A cantilever of uniform section with a point load P at the end is shown in figure 5.1. The bending moment at x from the encastré end is $M = -P(l - x)$. Thus

$$EI\,\frac{d^2v}{dx^2} = -P(l - x)$$

Integrating once

$$EI\,\frac{dv}{dx} = -Plx + \frac{Px^2}{2} + A$$

Integrating again

$$EIv = \frac{-Plx^2}{2} + \frac{Px^3}{6} + Ax + B$$

At the fixed end of the cantilever both the deflection v and the slope dv/dx are zero, enabling the two integration constants A and B to be found. $x = 0$, $dv/dx = 0$ therefore $A = 0$. $x = 0$, $v = 0$ therefore $B = 0$. Hence

$$EIv = -\frac{Px^2}{6}\,(3l - x)$$

At the free end

$$v = -\frac{Pl^3}{3EI}$$

and

$$\frac{dv}{dx} = -\frac{Pl^2}{2EI}$$

The same results would apply for the deflection and slope under the load if it was applied at some other point l' from the fixed end (l' would be substituted for l in the two expressions).

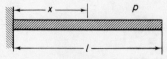

Figure 5.2

For the case of a cantilever with a uniformly distributed load (figure 5.2), the bending moment at x from the fixed end is

$$M = EI\frac{d^2v}{dx^2} = -p\frac{(l-x)^2}{2}$$

$$EI\frac{dv}{dx} = -\frac{pl^2x}{2} + \frac{plx^2}{2} - \frac{px^3}{6} + A$$

$$EIv = -\frac{pl^2x^2}{4} + \frac{plx^3}{6} - \frac{px^4}{24} + Ax + B$$

When $x = 0$, $dv/dx = 0$ therefore $A = 0$; when $x = 0$, $v = 0$ therefore $B = 0$. Thus

$$EIv = -\frac{px^2}{24}(6l^2 - 4lx + x^2)$$

For the particular case $x = l$

$$\frac{dv}{dx} = -\frac{pl^3}{6EI}$$

and

$$v = -\frac{pl^4}{8EI}$$

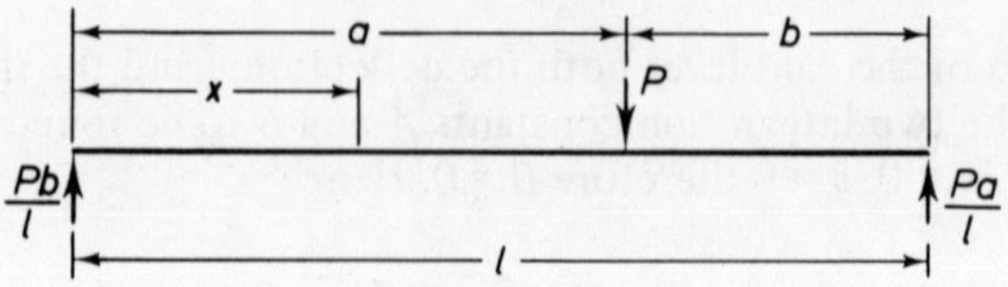

Figure 5.3

The simply supported uniform beam with a concentrated load (figure 5.3) will be somewhat more lengthy to solve. For $x < a$

$$EI \frac{d^2v}{dx^2} = \frac{Pbx}{l}$$

$$EI \frac{dv}{dx} = \frac{Pbx^2}{2l} + A$$

$$EIv = \frac{Pbx^3}{6l} + Ax + B$$

For $x > a$

$$EI \frac{d^2v}{dx^2} = \frac{Pbx}{l} - P(x - a)$$

$$EI \frac{dv}{dx} = \frac{Pbx^2}{2l} - \frac{Px^2}{2} + Pax + C$$

$$EIv = \frac{Pbx^3}{6l} - \frac{Px^3}{6} + \frac{Pax^2}{2} + Cx + D$$

Four constants A, B, C, and D have to be found. Two end conditions are known—when $x = 0$ and l, $v = 0$. Two other conditions are required—these can be found from the fact that when $x = a$, the slope and the deflection must each be the same whether the expressions for $x < a$ or $x > a$ are used.

It is then perfectly possible to obtain expressions for the deflection curve by this means; however it could be extremely tedious. Consider the case of a simply supported beam with three separate concentrated loads applied. This method of approach will not be pursued any further as there is a very much neater method of performing the integration.

5.3 Deflections using singularity functions or the Macaulay method

W. H. Macaulay first suggested that singularity functions could be of considerable use in solving problems on the deflection of beams. Figure 5.4 shows five singularity functions of the set.

$$f_n(x) = [x - a]^n \tag{5.6}$$

The functions have a somewhat unusual behaviour. Dealing first with the two negative indices, the functions are always zero except when $x = a$. Here they are infinite but will integrate as follows

$$\int_0^x [x - a]^{-2} \, dx = [x - a]^{-1}$$

$$\int_0^x [x - a]^{-1} \, dx = [x - a]^0$$

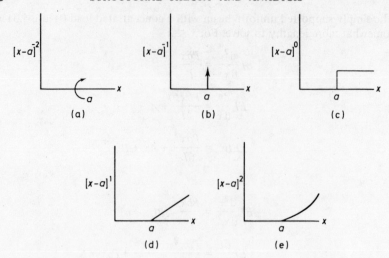

Figure 5.4

The first of these is a unit doublet function or a unit concentrated moment, while the second is a unit impulse function or a unit concentrated load.

For the other functions $n \geqslant 0$. When $x < a$ the value of $f_n(x)$ is zero, but if $x > a$ the value of $f_n(x)$ is $(x - n)^n$. The functions will integrate as follows

$$\int_0^x [x - a]^n \, \mathrm{d}x = \frac{[x - a]^{n+1}}{n + 1} \quad (n \geqslant 0)$$

For $n = 0$ (figure 4.5c), the function is described as a unit step at $x = a$. For $n = 1$ (figure 5.4d) the function is a unit ramp starting at $x = a$.

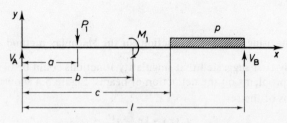

Figure 5.5

We shall now make use of singularity functions to write down an expression for the loading applied to the simply supported beam (figure 5.5). It will be remembered that a positive load is in the positive y direction. Concentrated loads will be considered as figure 5.4b and uniformly distributed loads as step functions (figure 5.4c).

$$p(x) = V_A[x - 0]^{-1} - P_1[x - a]^{-1} + M_1[x - b]^{-2} - p[x - c]^0 + V_B[x - l]^{-1}$$

This may be integrated to find the shear force since $dQ/dx = -p$

$$-Q(x) = V_A[x - 0]^0 - P_1[x - a]^0 + M_1[x - b]^{-1} - p[x - c]^1 + V_B[x - l]^0$$

Integrating again, remembering that $dM/dx = -Q$

$$M = V_A[x - 0]^1 - P_1[x - a]^1 + M_1[x - b]^0 - \frac{p[x - c]^2}{2} + V_B[x - l]^1$$

This expression holds for any positive value of x. To find V_A and V_B take x just greater than l, substitute into the expressions for Q and M which will be zero. It will be seen that the resulting equations are in fact the same as resolving vertically and taking moments about B. It would, of course, be easier to determine V_A by statics.

If we now substitute into equation 5.4, two more integrations can be performed to give both the slope and deflection at any point. Omitting V_B since the beam terminates at $x = l$.

$$EI\frac{dv}{dx} = \frac{V_A x^2}{2} - \frac{P_1[x - a]^2}{2} + M_1[x - b] - \frac{p[x - c]^3}{6} + A$$

$$EIv = \frac{V_A x^3}{6} - \frac{P_1[x - a]^3}{6} + \frac{M_1[x - b]^2}{2} - \frac{p[x - c]^4}{24} + Ax + B$$

A and B can be found by the use of suitable end conditions.

The use of singularity functions has been described here as a method for initially writing down the load applied to a beam. For finding the deflection curve it is usually much simpler to write down an expression for the bending moment straight away using singularity functions. Thus for the beam in figure 5.3

$$EI\frac{d^2v}{dx^2} = \frac{Pbx}{l} - P[x - a]^1$$

$$EI\frac{dv}{dx} = \frac{Pbx^2}{2l} - \frac{P[x - a]^2}{2} + A$$

$$EIv = \frac{Pbx^3}{6l} - \frac{P[x - a]^3}{6} + Ax + B$$

When $x = 0$; $v = 0$ therefore $B = 0$. *Note* $[x - a] = 0$. When $x = l$, $v = 0$ therefore

$$A = \frac{P(l - a)^3}{6l} - \frac{Pbl^2}{6l}$$

The complete deflection curve is given by

$$EIv = \frac{Pbx^3}{6l} - \frac{P[x - a]^3}{6} + \frac{P(l - a)^3 x}{6l} - \frac{Pbl^2 x}{6l}$$

If the deflection is required under the load, put $x = a$

$$v = -\frac{Pa^2b^2}{3EIl}$$

When the load is at the centre of the beam the maximum deflection is obtained, that is $a = b = l/2$ and

$$v = -\frac{Pl^3}{48EI}$$

It can be seen that the singularity-function method is simple to apply. An origin is chosen at one end of the beam and an expression is written down for the bending moment such that every load on the beam is included. Functions such as $[x - a]$ are integrated according to the rules set out at the start of this section. Do not forget that $[x - a]$ cannot be negative. A slight difficulty can arise when a uniformly distributed load extends over part of a beam, this however will be illustrated in the following example.

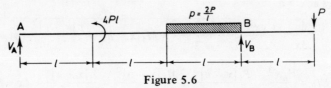

Figure 5.6

The deflection curve is required for the simply supported beam in figure 5.6. The reactions can be found by taking moments: $V_A = 4P/3$; $V_B = 5P/3$.

It should now be possible to write down an expression for the bending moment when x is measured from the left-hand end. If the uniformly distributed load is considered, the bending moment is $-p[x - 2l]^2/2$. This expression however will not suffice once $x > 3l$, as the distributed load terminates when $x = 3l$. A simple way of dealing with this is to continue the load to the right-hand end of the beam and add an equal distributed load acting upwards from B to the free end. The expression for the bending moment is now written

$$-\frac{p[x - 2l]^2}{2} + \frac{p[x - 3l]^2}{2}$$

Thus

$$EI\frac{d^2v}{dx^2} = \frac{4Px}{3} - 4Pl[x - l]^0 - \frac{2P}{l}\frac{[x - 2l]^2}{2} + \frac{2P}{l}\frac{[x - 3l]^2}{2} + \frac{5P}{3}[x - 3l]^1$$

There is no reason to write down the bending moment due to the point load P since the beam terminates when $x = 4l$.

$$EI\frac{dv}{dx} = \frac{4Px^2}{6} - 4Pl[x - l]^1 + \frac{2P}{l}\frac{[x - 2l]^3}{6} + \frac{2P}{l}\frac{[x - 3l]^3}{6} + \frac{5P}{6}[x - 3l]^2 + A$$

$$EIv = \frac{4Px^3}{18} - \frac{4Pl[x - l]^2}{2} + \frac{2P}{l}\frac{[x - 2l]^4}{24} + \frac{2P}{l}\frac{[x - 3l]^4}{24} + \frac{5P}{18}[x - 3l]^3 + Ax + B$$

When $x = 0$; $v = 0$ therefore $B = 0$. When $x = 3l$; $v = 0$ therefore

$$0 = 6Pl^3 - 8Pl^3 - \frac{Pl^3}{12} + 3Al \; ; \; A = \tfrac{25}{36} Pl^2$$

The complete deflection curve has now been determined.

All the deflection problems discussed so far have been statically determinate. However there is no reason why the same methods should not be used for statically

Figure 5.7

indeterminate beam problems. A fairly straightforward case is shown in figure 5.7, where we shall apply beam-deflection theory to determine the fixing moments at the ends of the beam.

$$EI \frac{d^2v}{dx^2} = V_A x - M_A - P[x - a]^1$$

$$EI \frac{dv}{dx} = \frac{V_A x^2}{2} - M_A x - \frac{P[x - a]^2}{2} + A$$

$$EIv = \frac{V_A x^3}{6} - \frac{M_A x^2}{2} - \frac{P[x - a]^3}{6} + Ax + B$$

When $x = 0$, $dv/dx = 0$ and $v = 0$ therefore $A = B = 0$. When $x = l$, $dv/dx = 0$ therefore

$$\frac{V_A l^2}{2} - M_A l - \frac{P(l - a)^2}{2} = 0$$

When $x = l$, $v = 0$ therefore

$$\frac{V_A l^3}{6} - \frac{M_A l^2}{2} - \frac{P(l - a)^3}{6} = 0$$

Solving the two equations gives

$$M_A = \frac{Pab^2}{l^2} \qquad V_A = \frac{Pb^2}{l^3} (3a + b)$$

The moment at B can be found by taking moments $M_B = Pa^2b/l^2$.

Using these integration methods the complete deflection curve for the beam is obtained. This can be a disadvantage if a point deflection is required, since a substitution will have to be made into the final equation for a particular value of x. The method then becomes somewhat tedious and other means exist which will give a more rapid solution.

5.4 Moment-area methods

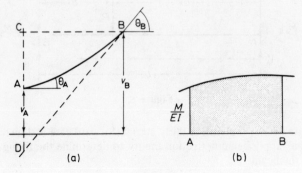

Figure 5.8

There are two theorems associated with moment-area methods and it is essential to have a full understanding of them before attempting an application. Figure 5.8a shows part of the elastic curve of a deflected beam. We shall be concentrating on two points, A and B. Now

$$M = EI \frac{d^2v}{dx^2}$$

or

$$\frac{M}{EI} = \frac{d^2v}{dx^2}$$

$$\int_A^B \frac{M}{EI} dx = \int \frac{d^2v}{dx^2} dx = \left[\frac{dv}{dx} \right]_A^B = \theta_B - \theta_A \tag{5.7}$$

This is the first theorem which states that *the difference in slope between two points on a beam is equal to the area of the M/EI diagram between the two points.* The *M/EI* diagram is shown in figure 5.8b and the shaded area is required.

$$\int_A^B \frac{Mx \, dx}{EI} = \int_A^B \frac{d^2v}{dx^2} x \, dx = \left[x \frac{dv}{dx} \right]_A^B - \int_A^B \frac{dv}{dx} dx$$

$$= x_B \left(\frac{dv}{dx} \right)_B - x_A \left(\frac{dv}{dx} \right)_A - v_B + v_A \tag{5.8}$$

If the origin is now shifted until it is below A

$$\int_A^B \frac{Mx\,dx}{EI} = x_B\theta_B - v_B + v_A \tag{5.9}$$

where $x = 0$ at A . $x_B\theta_B$ is represented by CD in figure 5.8a, and the complete expression is equal to the distance AD.

The second moment-area theorem may now be stated as follows. *The moment about A of the M/EI diagram between points A and B will give the deflection of point A relative to the tangent at point B.*

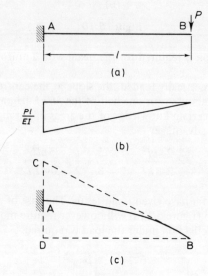

Figure 5.9

We shall first apply the moment-area theorems to find the slope and deflection under the load for the cantilever (figure 5.9a). The M/EI diagram is drawn in figure 5.9b. The difference in slope between A and B is given by the area of the complete M/EI diagram, that is, $Pl^2/2EI$. Now the slope at A is zero, and so the slope at B is $Pl^2/2EI$.

Taking moments about B for the complete area of the M/EI diagram will give the deflection at B, as the tangent at A is horizontal.

$$v_B = \tfrac{2}{3}l \times \frac{Pl^2}{2EI} = \frac{Pl^3}{3EI}$$

If moments had been taken about A, this would give the deflection at A relative to the tangent at B, that is, distance AC (figure 5.9c). The distance AD is in fact required. Now AD = CD − AC, where CD = $l\theta_B$

$$v_B = \frac{Pl^2}{2EI} \times l - \frac{l}{3}\frac{Pl^2}{2EI} = \frac{Pl^3}{3EI}$$

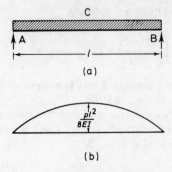

(a)

(b)

Figure 5.10

The central deflection is required for the beam with a uniformly distributed load (figure 5.10a).

As the beam is symmetrically loaded, the slope at the centre of the beam is zero. Taking moments about A for one half of the M/EI diagram will give the required deflection. A certain knowledge of the areas of parabolas and the position of their centroids is an advantage.

$$v_C = \frac{2}{3}\frac{pl^2}{8EI} \times \frac{l}{2} \times \frac{5}{8} \times \frac{l}{2} = \frac{5pl^4}{384EI}$$

The applications so far have given rapid resuls. The case of a point load on a simply supported beam (figure 5.11a) will however require more thought. We shall assume that the deflection under the load is required.

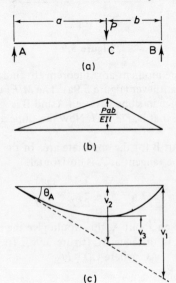

(a)

(b)

(c)

Figure 5.11

The M/EI diagram is drawn in figure 5.11b and a sketch of the deflected form of the beam in c. The first moment-area theorem is not much use here as the slope is not known at any point on the beam.

If moments about B are taken for the complete M/EI diagram the distance v_1 will be obtained. The value of θ_A can then be obtained by dividing this distance by l.

$$v_1 = \frac{Pab^2}{2EIl} \times \tfrac{2}{3}b + \frac{Pa^2b}{2EIl}\left(b + \frac{a}{3}\right) = \frac{Pab}{6EI}(2b + a)$$

therefore

$$\theta_A = \frac{Pab}{6EIl}(2b + a)$$

Now

$$v_2 = a\theta_A = \frac{Pa^2b}{6EIl}(2b + a)$$

v_3 can be obtained by taking moments for the portion AC of the M/EI diagram about C.

$$v_3 = \frac{Pa^2b}{2EIl} \times \frac{a}{3}$$

$$v_C = v_2 - v_3 = \frac{Pa^2b}{6EIl}(2b + a - a) = \frac{Pa^2b^2}{3EIl}$$

Had the beam been built in at both ends (figure 5.12a), the moment-area methods would enable the fixing moments to be determined quite easily. The bending-moment diagram has been sketched (figure 5.12b). If the beam is uniform this can also represent the M/EI diagram.

Applying the first moment-area theorem to the whole of the beam results in the fact that the area of the M/EI diagram is zero, then

$$\frac{Pab}{l} \times \frac{l}{2} - \frac{(M_A + M_B)l}{2} = 0$$

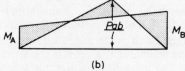

(a)

(b)

Figure 5.12

If the second theorem is applied about end B for the whole of the beam, the first moment is zero.

$$\frac{Pab}{l} \times \frac{a}{2}\left(b + \frac{a}{3}\right) + \frac{Pab}{l} \times \frac{b}{2}\left(\frac{2b}{3}\right) - M_A l \times \frac{l}{2} - (M_B - M_A)\frac{l}{2} \times \frac{l}{3} = 0$$

If these two equations are solved it will be found that

$$M_A = \frac{Pab^2}{l^2} \qquad M_B = \frac{Pa^2 b}{l^2}$$

The deflection at any point can easily be found by applying the second theorem about that point and taking account of the M/EI diagram between the point and one end of the beam.

The moment-area theorems lend themselves equally well to deflection problems where the beam has a variable section, so that the value of I changes. In cases of this kind it is probably best to sketch both the bending-moment and the M/EI diagrams.

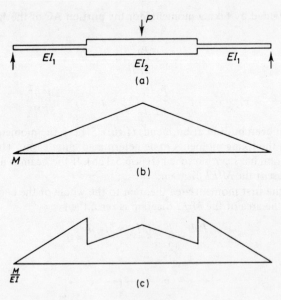

(a)

(b)

(c)

Figure 5.13

The beam in figure 5.13a has a larger cross-section over the central portion of the beam. Sketches of the bending-moment diagram and the M/EI diagram are shown in figures 5.13b and c. No difficulties should arise in the application of the theorems, but the evaluation of the various integrals may be somewhat tedious. In certain cases either graphical or numerical integration could be used to advantage.

An extension of the moment-area theorems is known as the *conjugate-beam method*. With a little thought it can be seen that the slope at any point on a beam is equivalent to finding the 'shear force' at that point when the beam is loaded with the M/EI diagram. The deflection at a point is equivalent to finding the 'bending moment' for the same loading. The equivalent beam is called the 'conjugate' beam and the M/EI diagram is referred to as the 'elastic load'. Slight complications arise over the support conditions of the conjugate beam. At an internal support the deflection is zero and hence the 'bending moment' must be zero. This can be achieved by the insertion of a pin instead of the support. At a built-in end the deflection is zero. This is achieved by making the corresponding point on the conjugate beam into a free end. In a similar manner a free end on the beam becomes a built-in end on the conjugate beam.

5.5 Use of standard cases

Table 5.1

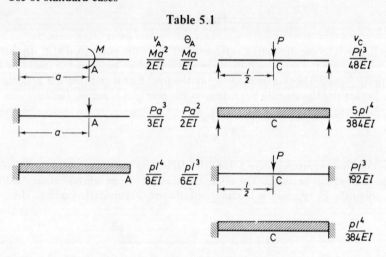

Table 5.1 gives a list of deflections and some slopes for certain loading on beams. The results can sometimes be used in conjunction with the principle of superposition, symmetry and anti-symmetry and Maxwell's reciprocal theorem. We shall have to anticipate Maxwell's theorem which is proved in the next chapter and is now quoted in a somewhat restricted form: *For a beam of linear elastic material the vertical deflection at B caused by a unit vertical load at A is equal to the vertical deflection at A caused by a unit vertical load at B.*

To illustrate the application of this approach the three problems, figures 5.14 to 5.16, will be discussed.

In figure 5.14a the deflection at the end of the cantilever is required when the load is applied at B.

Since the portion BC of the cantilever is unloaded it will remain straight and the deflected form will be as shown in figure 5.14b. The deflection at C is equal to the deflection of B plus the deflection of C relative to B.

$$v_C = v_B + (l - a)\theta_B$$

$$v_C = \frac{Pa^3}{3EI} + \frac{Pa^2}{2EI}(l - a) = \frac{Pa^2}{6EI}(3l - a)$$

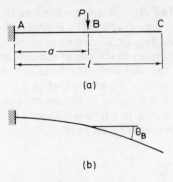

(a)

(b)

Figure 5.14

In figure 5.15 the deflection is required at B for the point load at C. In this case we may use the reciprocal theorem and instead of applying the load P at C and finding the deflection at B, we apply the load P at B and find the deflection at C. In fact this has already been done in the previous example. Hence

$$v_B = \frac{Pa^2}{6EI}(3l - a)$$

In the final example (figure 5.16a) the deflection at the centre of the simply supported beam is required. The effect of this load can be considered as the sum of problems in b—symmetric loading, and in c—anti-symmetric loading. The

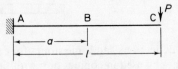

Figure 5.15

deflection at the centre due to the anti-symmetric loading will be zero. This leaves us with the problem in b to solve where the slope at the centre will be zero. This is equivalent to finding the deflection at the end of a cantilever of length $l/2$ loaded as shown in d.

$$v_A = \frac{\frac{P}{2}\left(\frac{l}{2}\right)^3}{3EI} - \frac{\frac{P}{2}\left(\frac{l}{2} - a\right)^3}{3EI} - \frac{\frac{P}{2}\left(\frac{l}{2} - a\right)^2 a}{2EI}$$

This expression will of course simplify.

It would have been possible to approach this last problem in a slightly different manner making use of a reciprocal theorem. The problem is equivalent to applying a central load P to the beam and finding the deflection at a distance a from the support.

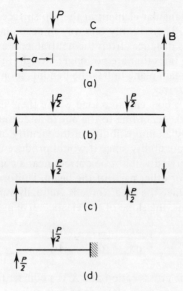

Figure 5.16

5.6 Deflections due to shear

As was mentioned in the introduction to this chapter, there are deflections due to shear forces as well as deflections due to bending moments. Figure 5.17a shows

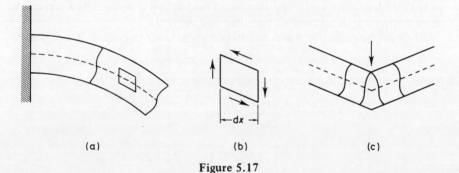

(a) (b) (c)

Figure 5.17

part of a cantilever of rectangular section that has a point load applied at the free end, so that the shear force is constant. The shear stress, as was shown in chapter 4, will vary from zero at the top and bottom edges of the beam to a maximum value at the neutral axis. This will mean that a rectangular element at the neutral axis will deform as shown in figure 5.17b. The slope of the top and bottom edges will be

$$\frac{\mathrm{d}v}{\mathrm{d}x} = \gamma = \frac{\tau_{xy}}{G}$$

However another rectangular element at the top surface will not deform as the shear stress there is zero. There is only one way of accommodating the deformities that take place at various distances from the neutral axis, and that is for the section to warp as shown by the lines that are no longer straight in figure 5.17a. So the plane sections do not remain plane unless the bending moment is constant and hence the shear force is zero.

Difficulties arise near a concentrated load where there is a sudden change in shear force (figure 5.17c). According to the above discussion sections would have to warp as shown—clearly an impossibility in the vicinity of the load. In fact a point force is also an impossibility since it would produce an infinite stress. The strain system we have derived will not be correct near a concentrated load but will be reasonably accurate for other parts of the beam. Uniformly distributed loads produce no difficulties since the shear force is gradually varying.

With these reservations, the slope of the elastic curve may be written as

$$\frac{dv}{dx} = \frac{\tau_{max}}{G} = \frac{KQ}{AG} \tag{5.10}$$

where A is the area of the cross-section and K is a constant to convert the average shear stress to the maximum shear stress for a particular section. For a rectangle the value of K would be $3/2$ while for a universal beam K would be 1 and A would be the area of the web.

To find the deflection due to shear it is only necessary to integrate equation 5.10. This should not present any difficulties and if necessary singularity functions can be used.

We shall now find the ratio of shear deflection to bending deflection at the end of the cantilever with uniformly distributed load shown in figure 5.2. A rectangular cross-section will be assumed of breadth b and depth d.

The bending deflection was found to be $v = -pl^4/8EI$ (section 5.2). For shear

$$Q = -p(l - x)$$

Thus

$$\frac{dv}{dx} = -\frac{Kp}{AG}(l - x)$$

$$v = -\frac{Kp}{AG}\left(lx - \frac{x^2}{2}\right) + C$$

The constant C is zero as $v = 0$ when $x = 0$. Thus the shear deflection at the end of the cantilever is $-Kpl^2/2AG$. The ratio of shear to bending deflection is

$$\frac{4KEI}{AGl^2}$$

Now for the rectangular section $K = 3/2$, $I = bd^3/12$, $A = bd$ also $E/G = 2(1 + v)$. Thus the ratio becomes $(1 + v)d^2/l^2$.

It can now be seen that the statement at the start of the chapter was correct—shear deflection could be ignored when compared with bending deflection unless the span to depth ratio is small.

Problems

5.1 A uniform cantilever of length l has a uniformly distributed load of p per unit length and a load P at the mid-point. Calculate the deflection at the free end. The free end is to be raised, so that it is at the same height as the built-in end, by a pure couple applied at the tip. Find the value of M to do this. What will be the slope at the free end after the couple has been applied?

5.2 A light uniform beam of flexural rigidity EI is simply supported at A and D. A uniform load of p per unit length is applied to AB and a clockwise couple of value $pl^2/8$ is applied at C. AD = l, AB = $l/2$, AC = $3l/4$.

Find the deflection curve for the beam and the value of the deflection and the slope at B. A third support is now introduced at B such that it is on the same level as those at A and D. Find the reactions at the supports when the above loading is applied.

5.3 A horizontal beam of length l, freely supported at the ends, carries equal loads P at distances a and $(l - a)$ from one end. Find the deflection under the loads and show that when the loads divide the beam into three equal parts, the deflection under the loads is

$$\frac{5}{162} \frac{Pl^3}{EI}$$

Find the central deflection when loads of P and $3P$ are applied at third-points of the beam.

5.4 A fixed-ended beam of length l carries a uniformly distributed load p. Its cross-section is optimised such that its bending stiffness is proportional to the magnitude of the bending moment it has to carry. Find the points at which the bending stiffness is zero, assuming that these are not at the ends. Also find the deflection curve assuming that the end bending stiffness of the beam is EI.

5.5 Find the end deflection of a cantilever of length l, constant width b, and height h which increases linearly from zero at the free end to h_0 at the built-in end, when it is subject to a uniformly distributed load of p per unit length.

5.6 A cantilever has an adjustable prop at one end and carries a uniformly distributed load, the maximum allowable bending moment being M. Calculate and compare the load-carrying capacity for the two cases

  (a) the prop at the same level as the built-in end
  (b) the prop adjusted for maximum carrying capacity.

5.7 A Universal beam has a web thickness of $10 \cdot 1$ mm and an effective web depth of 358 mm ($I_{xx} = 26\,900$ cm^4). It is simply supported and carries a uniformly distributed load. Investigate the ratio of shear deflection to total deflection for span to depth ratios of 10, 15 and 20.

6 VIRTUAL WORK AND ENERGY METHODS

6.1 Introduction

Much of the work covered in the last three chapters has been devoted to the behaviour of an individual member when subjected to various types of loading. We shall now investigate framed structures made up of a number of individual members that may either be pinned together at their ends or rigidly jointed. A number of ideas and methods will be presented that will enable the deflections of points on such structures to be found; or the forces may be found for the members in statically indeterminate cases. The chapter also provides an introduction to a later chapter on stiffness and flexibility methods, which are fundamental to a modern approach in structural analysis and can in fact be applied to all types of structures, including framed structures, plates, and shells.

6.2 Strain energy and complementary energy

We have already discussed the strain energy that is stored in a stressed member in section 3.11. Let us again examine the case of a member made from a linear elastic material and subject to a steadily increasing tensile force F. The load–extension curve is shown in figure 6.1a.

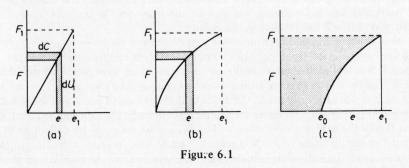

Figure 6.1

As the force increases from F to $F + dF$ let the extension increase from e to $e + de$. The external work done is $(F + dF/2) \, de \approx F \, de$. This will result in strain energy being stored in the member, represented by dU, and equal to the shaded

161

area on the diagram. U is the total strain energy stored in the member and it will be represented by the area under the load–extension curve. That is

$$U = \int_0^{e_1} F \, de \tag{6.1}$$

For a member of constant cross-section A, length l and modulus of elasticity E, $e = Fl/AE$, and on substituting into equation 6.1 the strain energy stored can be expressed as

$$\tfrac{1}{2} F_1 e_1 = \frac{F_1^2 l}{2AE}$$

Now examining the shaded area dC in figure 6.1a

$$dC \approx e \, dF \quad C = \int_0^{F_1} e \, dF \tag{6.2}$$

C is called the complementary energy and is sometimes denoted by U^*.

For a linear elastic material $C = F_1 e_1 /2$, represented by the area above the load–extension curve. It will be seen that the values of the strain and complementary energies are equal.

The load–extension curve for a non-elastic member is shown in figure 6.1b. The same initial arguments will apply such that

$$U = \int_0^{e_1} F \, de \quad C = \int_0^{F_1} e \, dF$$

It will not be possible to integrate these expressions unless the relationship between the load and extension of the member is known. The strain and complementary energies will not be equal unless the relation is linear.

In figure 6.1c the same load–extension curve is shown for the non-elastic member, with an initial extension applied, due for example to a temperature increase before the load is applied. In this case the complementary energy is represented by the shaded area; the strain energy however is the same as for the member without the initial extension. We shall find that complementary energy, even though it has no particular physical meaning, will play an important part in certain energy theorems that will follow later in this chapter.

Now consider a pin-jointed structure that consists of members all made from the same linear elastic material. An external force P_1 is applied to a particular point 1; this will induce forces in the members and strain energy will be stored in them. The total strain energy stored must be equal to the external work done. As the members are linearly elastic the deflection of the loaded point and indeed all other points that are not restrained will increase linearly with the applied load, so that finally the external work done will be $P_1 d_1 /2$ where d_1 is the deflection of point 1 in the direction of P_1. Neither is there any need to restrict the structures to pin-jointed statically determinate cases. The structure can be highly redundant and have rigid joints. The external work done will still be given by $P_1 d_1 /2$ and this must still be equal to the total strain energy stored in the members, which may be due to axial, bending, or torsional effects.

6.3 Flexibility coefficients

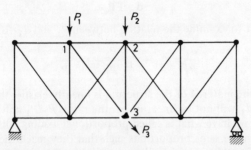

Figure 6.2

The pin-jointed structure in figure 6.2 is constructed from members that are linearly elastic. This will mean that the deflection in any particular direction at a particular point will be proportional to the load applied at any other point.

We shall define flexibility coefficients as follows:

f_{11} is the displacement at point 1 due to a unit load applied at point 1.
f_{21} is the displacement at point 2 due to a unit load applied at point 1.
f_{22}, f_{12}, etc. follow from these definitions.

The first subscript gives the point and direction at which the displacement is to be measured, the second subscript gives the point of application and the direction of the applied unit load. It can be seen that it is perfectly possible to have two or more different subscripts for the same point if the displacement is measured in a different direction to that in which the load is applied. Or indeed we may require the deflection at a point for a moment applied at that point.

Let us refer again to the framework. If P_1 was applied by itself the vertical deflection of point 1 could be written $f_{11}P_1$, the vertical deflection of point 2, $f_{21}P_1$, and the deflection of point 3 in the direction of P_3, $f_{31}P_1$. If the three loads are all applied to the structure, by making use of superposition, the following equations for the deflections of points 1, 2, and 3 can be written down.

$$d_1 = f_{11}P_1 + f_{12}P_2 + f_{13}P_3$$
$$d_2 = f_{21}P_1 + f_{22}P_2 + f_{23}P_3 \qquad (6.3)$$
$$d_3 = f_{31}P_1 + f_{32}P_2 + f_{33}P_3$$

So if we can find the values of the flexibility coefficients, the displacement due to the applied loading can be found for any point.

The three equations 6.3 can be written in the following matrix form

$$\begin{bmatrix} d_1 \\ d_2 \\ d_3 \end{bmatrix} = \begin{bmatrix} f_{11} & f_{12} & f_{13} \\ f_{21} & f_{22} & f_{23} \\ f_{31} & f_{32} & f_{33} \end{bmatrix} \begin{bmatrix} P_1 \\ P_2 \\ P_3 \end{bmatrix} \qquad (6.4)$$

or in a more general form

$$d = FP \qquad (6.5)$$

It is of interest to examine the relation between P_1 and d_1 from equation 6.3

$$\frac{d_1}{P_1} = f_{11} + f_{12}\frac{P_2}{P_1} + f_{13}\frac{P_3}{P_1}$$

Even though the material of the structure is linearly elastic, the relation between P_1 and d_1 will not be linear unless the ratios P_2/P_1 and P_3/P_1 remain constant, that is, we have what is called proportional loading, all the loads are steadily applied in the same proportions such that they reach their maximum values together. For this type of loading the relation between the loads and their corresponding displacements will all be linear and it can be seen at once that the external work done is given by

$$\tfrac{1}{2}P_1 d_1 + \tfrac{1}{2}P_2 d_2 + \tfrac{1}{2}P_3 d_3 \qquad (6.6)$$

The case where the loads are applied in a random manner must be examined in more detail. To simplify the discussion it will be assumed that P_1 is first applied, followed by P_2 and then P_3.

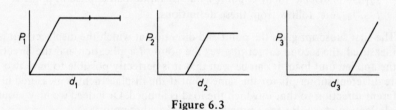

Figure 6.3

The graphs of applied load against deflection for the three loads are shown in figure 6.3. For the case of P_1 there will be a linear portion while P_1 rises to its maximum value, followed by two horizontal portions while P_2 and then P_3 are applied, where the load P_1 does not change but the corresponding deflection increases. The work done will be equal to the area under the curve and will clearly not equal $P_1 d_1/2$.

If we now concentrate our attention on a particular member of the structure, as it is made from linearly elastic material the graph of the force in the member against the extension must be linear whatever the manner of application of the external forces; so the energy stored in the member will always be the same once the loading in the structure is complete. This will apply to all the members of the structure, so that the total strain energy stored must be independent of the loading sequence. This in turns means that the external work done must also be constant and independent of the loading sequence. The final deflection of each point on the structure is also independent of the manner in which the loads are applied.

We have obtained an expression in equation 6.6 for the external work done for proportional loading, and this must represent the work done for any other type of loading.

6.4 Maxwell's reciprocal theorem

In order to determine the deflection of the three points in figure 6.2 it would be necessary to know the values of nine flexibility coefficients, equation 6.3. This however is not quite true as it can be shown that $f_{12} = f_{21}$ and that the matrix is in fact symmetric.

Consider the same structure (figure 6.2) with the case of $P_3 = 0$. Let us obtain an expression for the work done when P_1 and P_2 are applied. It has already been shown that the work done is independent of the way in which the loads are applied.

P_1 is applied first: the deflection at point 1 can be written as $f_{11}P_1$ so that the work done is $f_{11}P_1^2/2$. Now apply P_2, the deflection of point 2 will be $f_{22}P_2$, and the work done $f_{22}P_2^2/2$. When P_2 is applied point 1 will deflect through a further distance of $f_{12}P_2$ and as the load P_1 is displaced through that distance the work done will be $f_{12}P_2P_1$. Consideration of the graphs in figure 6.3 will help to make this point clear.

The total work done or energy stored is

$$\tfrac{1}{2}f_{11}P_1^2 + \tfrac{1}{2}f_{22}P_2^2 + f_{12}P_2P_1 \tag{6.7}$$

The structure is now unloaded and the two loads applied in the reverse order. A similar argument will give the work done as

$$\tfrac{1}{2}f_{22}P_2^2 + \tfrac{1}{2}f_{11}P_1^2 + f_{21}P_1P_2 \tag{6.8}$$

As the work done is independent of the loading sequence the two expressions 6.7 and 6.8 must be equal. If the common terms are removed we are left with

$$f_{12} = f_{21}$$

This can be written in a more general form

$$f_{ij} = f_{ji} \tag{6.9}$$

Expression 6.9 is a form of the Maxwell reciprocal relationship. This states that for a linear elastic system, *the displacement at point i due to a unit load applied at point j is equal to the displacement at point j due to a unit load at point i.* The loads and the displacements must of course correspond, so that when the displacement is measured at *i* it is in the same direction as the unit load applied at *i*. We made advance use of this theorem in chapter 5.

This proof can also be carried out making use of matrix algebra as follows.

Instead of having two loads on the structure, assume that there are *n*, so that the work done or energy stored is given by

$$W = \tfrac{1}{2}P_1 d_1 + \tfrac{1}{2}P_2 d_2 \ldots + \tfrac{1}{2}P_n d_n \tag{6.10}$$

Now both **P** and **d** are column matrices and as the work done is a vector dot product we can express equation 6.10 as

$$W = \tfrac{1}{2}\mathbf{P}^T \mathbf{d}$$

Now $\mathbf{d} = \mathbf{FP}$, (6.5), therefore

$$W = \tfrac{1}{2}\mathbf{P}^T \mathbf{FP} \tag{6.11}$$

If we take the tranpose of both sides of equation 6.5

$$\mathbf{d}^T = \mathbf{P}^T \mathbf{F}^T \tag{6.12}$$

Now as the work done is a scalar the expression for W could also be written

$$W = \tfrac{1}{2} \mathbf{d}^T \mathbf{P}$$

Substituting from equation 6.12 for $\mathbf{d}^T$

$$W = \tfrac{1}{2} \mathbf{P}^T \mathbf{F}^T \mathbf{P} \tag{6.13}$$

Examining the two expressions for W, equations 6.11 and 6.13, it can be seen at once that

$$\mathbf{F} = \mathbf{F}^T \tag{6.14}$$

This implies that the flexibility matrix is symmetrical and that

$$f_{ij} = f_{ji}$$

6.5 The flexibility approach

The elastic beam shown in figure 6.4a is statically indeterminate to the first degree. We require the reactions due to the applied loading. The problem could be solved by making use of singularity functions. If the reaction at B is assumed to be V_B, the other reactions could be found in terms of V_B and the applied loading. A

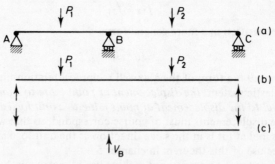

Figure 6.4

bending-moment equation can be written down that will apply for any point on the beam, and this may then be integrated twice to give the deflection equation. There are three end conditions—the deflection is zero at A, B, and C, so that the constants of integration may be determined and the value of V_B can be found.

An alternative approach would be as follows. The problem is first simplified by making it statically determinate. There are several different ways of doing this, any of which would be suitable. However one particular choice can lead to a more rapid solution than another. The choice is really a matter of experience. In this particular problem any one of the supports could be removed, or a pin could be inserted at any point on the beam such that the bending moment at that point becomes zero. We shall make V_B zero by removing the reaction at B (figure 6.4b).

The vertical deflection at B due to applied loads P_1 and P_2 must be calculated, and this is a fairly straightforward matter. The deflection could be found for the two loads applied at the same time or for each load applied separately, and the individual deflections found at B and then summed. Let the total deflection be d_1.

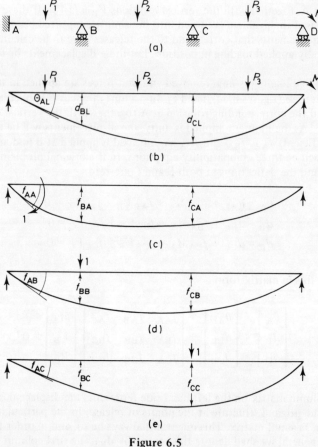

Figure 6.5

The applied loading is next removed and a vertical force equal to the value of the unknown reaction V_B is applied at B (figure 6.4c). Due to this load point B will move vertically upwards and this deflection can be calculated in terms of V_B. Let the deflection be d_2.

In the original problem it is known that the true deflection at B is zero, and this occurs when P_1, P_2, and V_B are all applied together. Superimposing the deflections due to figures 6.4b and c the resulting deflection is zero.

$$d_1 + d_2 = 0 \qquad (6.15)$$

d_1 is known, and d_2 is known in terms of V_B. Thus the value of V_B can be found from equation 6.15. This equation relating the deflections is known as an equation of compatibility or an equation of geometry.

This type of solution is the basis of the flexibility approach, but in order to gain further insight we shall now discuss the problem in figure 6.5.

It is seen that the structure is three times redundant. Instead of choosing the three reactions at B, C, and D as the redundants we shall take M_A the fixing moment at A, together with the vertical reactions V_B and V_C. If these redundants are all released the problem is turned into the simply supported beam of figure 6.5b.

The displacements that correspond to the releases have to be calculated with the externally applied loading in position. Let these displacements be θ_{AL}, d_{BL}, and d_{CL}.

The applied loading is next removed and unit forces are applied in turn to the points of release (figures 6.5c, d and e). In c a unit moment is applied at A, and we can find the corresponding rotation at A together with the vertical deflections at B and C. As the moment applied is unity, the displacements will be the flexibility coefficients f_{AA}, f_{BA} and f_{CA}. A unit load is applied at B in d and at C in e.

There will be three compatibility equations. In the original problem the rotation at A and the deflections at both B and C are zero.

$$\theta_A = \theta_{AL} + f_{AA}M_A + f_{AB}V_B + f_{AC}V_C = 0$$
$$d_B = d_{BL} + f_{BA}M_A + f_{BB}V_B + f_{BC}V_C = 0 \qquad (6.16)$$
$$d_C = d_{CL} + f_{CA}M_A + f_{CB}V_B + f_{CC}V_C = 0$$

or writing this in matrix form

$$\begin{bmatrix} \theta_A \\ d_B \\ d_C \end{bmatrix} = \begin{bmatrix} \theta_{AL} \\ d_{BL} \\ d_{CL} \end{bmatrix} + \begin{bmatrix} f_{AA} & f_{AB} & f_{AC} \\ f_{BA} & f_{BB} & f_{BC} \\ f_{CA} & f_{CB} & f_{CC} \end{bmatrix} \begin{bmatrix} M_A \\ V_B \\ V_C \end{bmatrix} = 0 \qquad (6.17)$$

The column matrix on the left-hand side represents the displacements that occur in the original structure at the points of release. In the particular case discussed this is a null matrix. This might not always be so, and in order to keep the approach general we shall denote this matrix by d_R. The first column on the right-hand side represents the displacements that occur at the points of release with the original loading applied to the released structure, and these will be denoted by d_{RL}. The 3 x 3 matrix is the flexibility matrix F and the last column matrix contains the redundant reactions and is R.

We may now write

$$d_R = d_{RL} + FR \qquad (6.18)$$

Equation 6.18 is the fundamental equation of the flexibility approach when solving redundant structures.

The quantities that are really required are contained in **R**, and these can be found by solving equation 6.18 in the form

$$R = F^{-1}(d_R - d_{RL}) \qquad (6.19)$$

where F^{-1} is the inverse of the flexibility matrix **F**.

6.6 Kinematic indeterminacy

When the stiffness approach is used, the unknowns are chosen to be the joint displacements, as opposed to the flexibility method where the unknowns were taken as the redundant reactions. This will mean that finally we shall need to know the displacement of every joint in the structure. The number of unknowns is equal to what is called the *kinematic indeterminacy*.

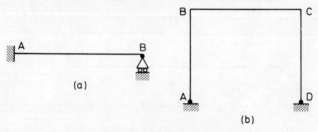

Figure 6.6

Two examples illustrating how the degree of kinematic indeterminacy is obtained are now discussed. The first is the beam in figure 6.6a.

End A is built in so that no deflection or rotation can take place there. At B vertical deflection is prevented, but it is possible for rotation and horizontal displacement to occur. This particular beam would then have a kinematic indeterminacy of 2. Very often axial effects can be neglected when compared with bending, in which case the horizontal movement of B can be omitted, reducing the kinematic indeterminacy to one.

The second example is the portal frame in figure 6.6b. As the feet of the portal are both pinned, rotation can take place at both A and D. At B there are three unknowns, two deflections and a rotation, the same will apply at C. In all, the unknown displacements total 8. However if axial effects are omitted we can reduce this by 3 as B and C will not deflect vertically and the horizontal deflection of B will be the same as that of C. This means that the kinematic indeterminacy is reduced to 5.

6.7 The stiffness approach

As has already been mentioned, when the stiffness approach is used the displacements of the joints are taken as the unknowns. For the beam in figure 6.7a, if axial effects are neglected the kinematic indeterminacy has been shown to be one. The unknown is the rotation at B.

The first step in the analysis is to fix end B such that no rotation can take place. If the external loading is now applied a moment will be developed at B. It is a fairly straightforward matter to determine the value of this moment; at this stage we shall denote it by M_B (figure 6.7b).

The external loading is next removed from the beam and a couple equal and opposite to M_B is applied and the joint allowed to rotate. It is possible to calculate the resulting angle of rotation θ_B (figure 6.7c).

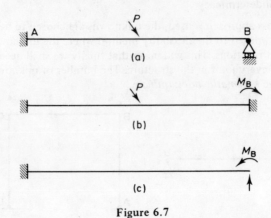

Figure 6.7

These two results can be superimposed and the resulting moment at B will be zero, leaving B pinned as it was in the original problem. θ_B will be the value of the unknown rotation at B. Once this has been found the values of the reactions can be determined.

We shall next attempt to obtain a more general approach using the stiffness method and we shall base the discussion on the beam problem shown in figure 6.8a. This particular problem has already been discussed using a flexibility approach in section 6.5.

The beam is seen to be three times kinematically indeterminate if axial effects are neglected. The displacements θ_B, θ_C, and θ_D corresponding to the joint rotations at B, C, and D will be taken as the unknowns. Proceeding in a similar manner as in the previous example, the beam is built-in at B, C, and D. It is necessary to calculate the values of these restraints when the external loading is applied. Any applied forces corresponding to the unknown displacements are omitted at this stage. In this example the moment M applied at D will be left out, leaving the loads P_1, P_2, and P_3 in position (figure 6.8b).

The moments at B, C, and D for the applied loading have to be found. Thus M_{BF} will be caused by P_1 acting on span AB together with P_2 on span BC. Three moments are required, M_{BF}, M_{CF}, and M_{DF}.

When the individual terms are calculated for a flexibility matrix, a single unit load corresponding to one of the releases is applied and the resulting displacements that correspond to all the releases have to be calculated. These would then form one column of the matrix. The equivalent to this in the case of a stiffness matrix is to apply a unit displacement at one of the restraints that have been introduced,

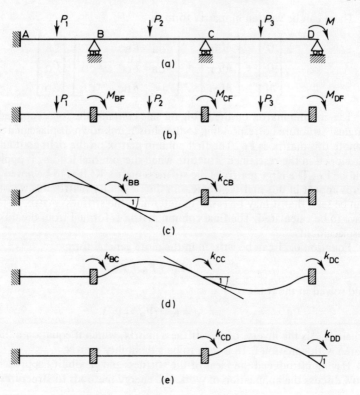

Figure 6.8

and determine the 'forces' that are developed at the positions of all the introduced restraints. One column of the stiffness matrix will result. So that in our problem if a unit rotation is applied at B, with C and D restrained, the moment required to produce this rotation will be the stiffness coefficient k_{BB} and the restraining moments will be k_{CB} and k_{DB}, where the latter is of course zero. Unit rotations are next applied to joints C and D (figures 6.8d and e), and the various stiffness coefficients found.

By the use of superposition we can state that the original forces applied to the structure corresponding to the unknown displacements, will be equal to the forces developed in the restrained structure together with the corresponding forces due to the unit displacements multiplied by the actual displacements. The original force applied at both B and C is zero in each case; at D however there is an externally applied clockwise moment M. The following set of equations will result

$$0 = M_{BF} + k_{BB}\theta_B + k_{BC}\theta_C$$
$$0 = M_{CF} + k_{CB}\theta_B + k_{CC}\theta_C + k_{CD}\theta_D \qquad (6.20)$$
$$M = M_{DF} + k_{DC}\theta_C + k_{DD}\theta_D$$

These can be written in matrix form as

$$
\begin{bmatrix} 0 \\ 0 \\ M \end{bmatrix} = \begin{bmatrix} M_{BF} \\ M_{CF} \\ M_{DF} \end{bmatrix} + \begin{bmatrix} k_{BB} & k_{BC} & 0 \\ k_{CB} & k_{CC} & k_{CD} \\ 0 & k_{DC} & k_{DD} \end{bmatrix} \begin{bmatrix} \theta_B \\ \theta_C \\ \theta_D \end{bmatrix} \tag{6.21}
$$

The column matrix on the left-hand side contains the forces applied to the original structure corresponding to the chosen unknown displacements. We shall denote this matrix as P_R. The first column matrix on the right contains the forces developed in the restrained structure when the external loading is applied. This will be P_F. The large matrix is the stiffness matrix K. It is to be noted that several zeros appear in this matrix, whereas in the flexibility matrix this was not so. It can be seen that it may be easier to determine the stiffness matrix as fewer terms have to be calculated. The final column matrix is formed from the unknown displacements.

Equation 6.21 can be written in the more general form

$$
P_R = P_F + Kd \tag{6.22}
$$

and solved in the form

$$
d = K^{-1}(P_R - P_F) \tag{6.23}
$$

where K^{-1} is the inverse of the stiffness matrix, which if equation 6.23 is compared with equation 6.18 is seen to be a flexibility matrix.

Having introduced the ideas of the stiffness and flexibility approaches we shall now discuss the application of work and energy methods to structures.

6.8 The method of real work

This approach has already been discussed in section 6.2 where it was stated that if a single load is applied to a structure the work done is $P_1 d_1/2$ and this is equal to the energy stored in the structure. So that if the energy stored can be found there will be no difficulty in the determination of d_1.

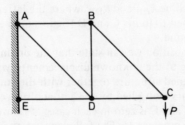

Figure 6.9

The vertical deflection of C is required when the load P is applied to the pin-jointed truss in figure 6.9. The vertical and horizontal members are of length l and the cross-section of all members is A.

To find the energy stored in the members of the truss it will first be necessary to find the forces in the individual members. It is probably best to set out all results in tabular form.

Table 6.1

member	length	area	force	U
AB	l	A	P	k
BD	l	A	$-P$	k
BC	$\sqrt{2}l$	A	$\sqrt{2}P$	$2\sqrt{2}k$
ED	l	A	$-2P$	$4k$
DC	l	A	$-P$	k
AD	$\sqrt{2}l$	A	$\sqrt{2}P$	$2\sqrt{2}k$
				$\Sigma\,(7+4\sqrt{2})k$

The strain energy for a particular member is found from $F^2l/2AE$ and in this case $k = P^2l/2AE$. The total strain energy is found by summing the last column in the table and has a value of

$$\frac{(7+4\sqrt{2})}{2}\,\frac{P^2l}{AE}$$

This must be equal to the external work done, which is $Pd/2$ where d is the vertical deflection of C.

$$\tfrac{1}{2}Pd = \frac{(7+4\sqrt{2})}{2}\,\frac{P^2l}{AE}$$

$$d = (7+4\sqrt{2})\,\frac{Pl}{AE}$$

The beam problem in figure 6.10 has already been solved in chapter 5 where the deflection under the load was required. The strain energy stored in this case

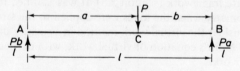

Figure 6.10

will be due to bending action (shear effects are neglected), and was shown in section 3.15 to be given by

$$\int \frac{M^2\,dx}{2EI}$$

If x is measured from support A, $M = Pbx/l$ where $x < a$. The energy stored from A to C is

$$\int_0^a \frac{P^2 b^2 x^2 \, \mathrm{d}x}{2EIl^2} = \frac{P^2 b^2 a^3}{6EIl^2}$$

When $x > a$ the expression for the bending moment is more complicated and it would be much easier to measure x from B. $M = Pax/l$ where $x < b$. It can be seen at once that the energy stored from B to C is $P^2 a^2 b^3 / 6EIl^2$ therefore

$$\tfrac{1}{2}Pd = \frac{P^2 a^2 b^2}{6EIl^2}(a + b)$$

$$d = \frac{Pa^2 b^2}{3EIl}$$

This approach would work equally well if the beam was of variable section; it would only be necessary to include a suitable expression for I in the integration.

The method of real work is seen to be very simple to apply, but unfortunately it has severe limitations in the type of problem that it will solve. If a number of loads are applied to a structure the forces in the members can be found and the strain energy summed, this will be equal to the external work done, which will be of the form $\Sigma \, Pd/2$ but it is impossible to find the deflection of an individual point. Often the deflection is required for a point on a structure at which the load is not applied or the deflection at a particular point in a different direction to the load is wanted. Examples would be the vertical deflection of D or the horizontal deflection of C for the frame in figure 6.9. The method of real work again cannot help in such cases. It is also limited to cases that are statically determinate.

It is therefore necessary to look for an approach that is far more general in its character and the next section discusses such a method.

6.9 Method of virtual work

Virtual work has already been used to determine the forces in the members of a statically determinate framework (section 2.8). It was assumed that there were no large changes in geometry when the loads were applied to the structure.

We shall now consider the virtual-work method in a slightly different form, often referred to as Mohr's equation of virtual work, with particular reference to a pin-jointed framework.

We have already seen in the method of real work that the external work done is equal to the energy stored in the members; we could also equate the external work to the work done *on* the members. If we now consider a member that is in tension with an axial force F_m and the length of the member is changed by a virtual displacement $\mathrm{d}e_m$ considered as an extension, then the work done *on* the member is $F_m \, \mathrm{d}e_m$. (Note that this will not be $F_m \, \mathrm{d}e_m/2$). The same would apply if the member was in compression and the member shortened. If however the member was in tension and the virtual displacement shortened the member, work would be done *by* the member and this would be considered as negative in a work balance equation.

Let us now consider a complete pin-jointed structure under the action of an external set of loads P; the structure is in equilibrium and the force in a particular member is F_m. For the present we shall assume that all members of the structure are inextensible apart from member m, and that a virtual displacement de_m in the form of an extension is applied to this member. As a result of this a number of the joints in the structure will move to new positions and work will be done by the applied loading system. Let $d\Delta_j$ be the corresponding displacement of the load P_j (that is, the displacement in the direction of P_j). Equating the external work done to the work done on the member m

$$\Sigma P_j \, d\Delta_j = F_m \, de_m$$

If all the members of the structure were allowed to extend

$$\Sigma P_j \, d\Delta_j = \Sigma F_m \, de_m \qquad (6.24)$$

So that we may state that the virtual work done by the external loads is equal to the virtual work done on the members.

It will be essential to find a compatible system of deformations $(d\Delta_j, de_m)$ so that the particular information required in a problem can be found easily. Before embarking on this it should be pointed out that the actual loading system applied to the structure has been used together with a compatible system of virtual displacements. The principle would apply equally well if actual displacements and an artificial or virtual set of loads and forces were used; the forces would of course have to satisfy the equilibrium of the system. If actual loads and forces are used together with actual displacements we are of course again using the method of real work.

The derivation of the virtual-work method has been discussed in relation to a pin-jointed structure, but there is no reason why the general principle should not be applied and extended to take account of bending moment, shear force, and torque if these exist in a particular problem.

6.10 Virtual work applied to statically determinate systems

In section 6.9 it was stated that when the equation of virtual work was applied, either virtual displacements and the actual loads on the structure could be used, or the actual displacements and an imaginary or virtual set of loads. The latter will be of more use when dealing with structural problems. The applications that follow are often referred to as unit load methods.

Figure 6.11 shows a Warren truss with external loads P_1 and P_2 applied. The deflection of joint E in the direction of the arrow is required.

So long as there is no large distortion, the equation of virtual work 5.1 can be written

$$\Sigma P\Delta = \Sigma Fe \qquad (6.25)$$

The extensions e of all the members due to the externally applied loads P_1 and P_2 can be found. It is first necessary to find the forces in all the members and from

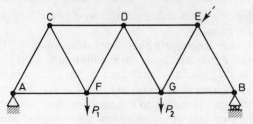

Figure 6.11

a knowledge of the load–extension curve for each member, the individual extensions are found. We shall make use of these extensions together with the actual displacements on the framework in the equation of virtual work as they are known to be a compatible set.

It would be most convenient if we could have only the displacement we are seeking on the left-hand side of equation 6.25. It is quite easy to introduce a set of hypothetical forces that will achieve this; in fact all that is required is a unit load applied in the direction of the required displacement at E. This load will produce forces in the members of the framework, which can be found by resolution and will be denoted by the set F'. The set F' and the unit load are of course in equilibrium.

The left-hand side of equation 6.25 now becomes $1 \times \Delta_E$. The complete virtual-work equation can now be written

$$\Delta_E = F'_{AC}e_{AC} + F'_{CD}e_{CD} + \cdots$$

where the right-hand side is summed for all the members. The value of e for a member is the extension due to the applied loads P_1 and P_2.

If all the members are uniform and linearly elastic the values of e can be found from Fl/AE where F is the force due to the applied loads. Thus

$$\Delta_E = \sum \frac{F'Fl}{AE} \tag{6.26}$$

For the more general case of a curved member of variable section, considering axial effects only, it is necessary to find the extension e of a small element ds where the axial force is F and the axial force due to the unit load is F'. In this case

$$\Delta = \sum \int \frac{F'F\,ds}{AE} \tag{6.27}$$

For bending effects, the bending moment at a particular point is determined with the applied loading in position; the corresponding value of e is the change in angle of the small element ds and is given by $M\,ds/EI$. With the unit load in position the bending moment M' is determined at the same point. Applying the virtual-work principle

$$\Delta = \sum \int \frac{M'M\,ds}{EI} \tag{6.28}$$

where the integral extends over a particular member and the summation is for all the members.

It can be seen that it is possible to extend the above principles to take account of shear force and torsion.

Shear $$\Delta = \sum \int \frac{KQ'Q \, ds}{GA} \qquad (6.29)$$

Torsion $$\Delta = \sum \int \frac{T'T \, ds}{GJ} \qquad (6.30)$$

The constant K is the one introduced in section 5.6 and J is the equivalent polar second moment of area as mentioned in section 4.9.

If a member is subjected to several 'forces' at the same time the virtual-work equation will be formed from the summation of the various terms that appear on the right-hand sides of equations 6.27 to 6.30.

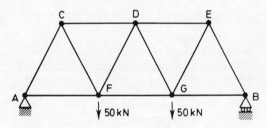

Figure 6.12

As a numerical example of this approach we shall find the vertical deflection of D for the truss in figure 6.12. All members have the same length of 5 m. The cross-sectional areas of the top and bottom chord members are all 2500 mm² and the rest of the members 2000 mm². $E = 210$ kN/mm².

The analysis is best set down in tabular form (table 6.2). The first columns of the table give details of the members. Column 4 gives the force in the members due to the applied loading. Column 5 gives the extension of the members. In column 6 is shown the force in the members due to an applied unit vertical load at D. Column 7 is the product of columns 5 and 6, and the summation of this column will give the vertical deflection of D. This has a value of 2·87 mm.

The working in the table could have been somewhat shorter as both the structure and the applied load are symmetrical and also the deflection at D is required. Thus the contribution from members in similar positions, for example, AF and GB will be the same. This means that the results from one half of the framework could have been doubled and added to those from FG.

Deflections due to temperature change can be treated in a similar manner. The extension e of each member would be the change in length of the member due to a temperature rise, and would be of the form $l\alpha T$ where α is the linear coefficient

Table 6.2

1	2	3	4	5	6	7
Member	Length m	Area mm^2	F kN	e mm	F'	$F'e$
AC	5	2000	−57·8	−0·69	−0·578	0·399
AF	5	2500	28·9	0·276	0·289	0·08
CF	5	2000	57·8	0·69	0·578	0·399
CD	5	2500	−57·8	−0·552	−0·578	0·319
FD	5	2000	0	0	−0·578	0
FG	5	2500	57·8	0·552	0·867	0·478
DE	5	2500	−57·8	−0·552	−0·578	0·319
DG	5	2000	0	0	−0·578	0
EG	5	2000	57·8	0·69	0·578	0·399
EB	5	2000	−57·8	−0·69	−0·578	0·399
GB	5	2500	28·9	0·276	0·289	0·08

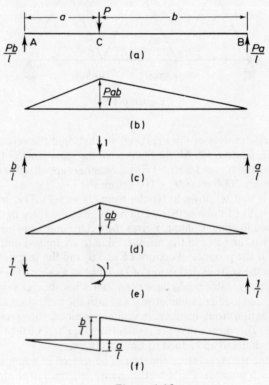

Figure 6.13

of expansion and T the temperature change. Once the extensions have been obtained the working will follow the same steps.

Virtual-work principles will next be applied to find the deflection and slope underneath the load P for the simply supported beam in figure 6.13a. We have already solved this problem by alternative means. The bending-moment diagram M for the applied load system is first drawn. A unit load is next applied at C (figure 6.13c) and the bending-moment diagram M' drawn. The integral $\int (M'M \, ds/EI)$ has to be found. In this case we may put $x = s$.

Measuring x from the left-hand support where $0 < x < a, M' = bx/l$; $M = Pbx/l$ and

$$\int_0^a \frac{Pb^2x^2}{EIl^2} = \frac{Pb^2a^3}{3EIl^2}$$

Measuring x from the right-hand support where $0 < x < b, M' = ax/l$, $M = Pax/l$ and

$$\int_0^b \frac{Pa^2x^2}{EIl^2} = \frac{Pa^2b^3}{3EIl^2}$$

So that

$$1 \times d_C = \frac{Pb^2a^2}{3EIl^2}(a + b) = \frac{Pb^2a^2}{3EIl}$$

Since the value of d_C is positive it will be in the same direction as the applied unit load namely downwards.

To find the slope underneath the load P it will be necessary to apply a unit moment at C (figure 6.13e). The resulting bending-moment diagram is shown at f, when $0 < x < a, M' = -x/l, M = Pbx/l$ and

$$\int_0^a -\frac{Pbx^2}{EIl^2} = -\frac{Pba^3}{3EIl^2}$$

when $0 < x < b, M' = x/l, M = Pax/l$ and

$$\int_0^b \frac{Pax^2}{EIl^2} = \frac{Pab^3}{3EIl^2}$$

so that

$$1 \times \theta_C = \frac{Pab}{3EIl^2}(b^2 - a^2) = \frac{Pab}{3EIl}(b - a)$$

The unit moment was applied in a clockwise direction so that if the expression for θ_C is positive it means that the elastic centre-line of the beam has rotated in a clockwise direction. It can be seen that if $b > l/2$, θ_C is positive and if $b < l/2$, θ_C is negative.

As an example of a problem that is subject to both bending and torsion, consider the case of a steel rod of circular cross-section and radius r bent into a quadrant of radius R, where $R \gg r$. The rod is built in at one end and a load P applied at the free end normal to the plane of the rod (figure 6.14). The deflection of the free end is required in the direction of the load.

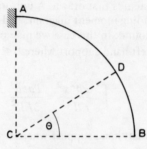

Figure 6.14

A point D on the rod is subject to both bending and torque such that $M = PR \sin \theta$ and $T = PR(1 - \cos \theta)$.

If a unit load is applied at the free end in the direction of the required deflection, $M' = R \sin \theta$, $T' = R(1 - \cos \theta)$, so that

$$d_B = \int \frac{MM' \, ds}{EI} + \int \frac{TT' \, ds}{GJ}$$

where $ds = R \, d\theta$

$$d_B = \int_0^{\pi/2} \frac{PR^3 \sin^2 \theta \, d\theta}{EI} + \int_0^{\pi/2} \frac{PR^3 (1 - \cos \theta)^2 \, d\theta}{GJ}$$

$$= PR^3 \left[\frac{\pi}{4EI} + \frac{(3\pi/4) - 2}{GJ} \right]$$

$$= \frac{PR^3}{Er^4} \left[1 + \frac{2E}{\pi G} \left(\frac{3\pi}{4} - 2 \right) \right]$$

So far all the examples discussed have been for structures made from linearly elastic materials. However this restriction was not applied when the virtual-work applications were discussed; it was only necessary to be able to determine the value of e with the applied loading in position.

The simple framework in figure 6.15 has a vertical load of 200 kN applied and the horizontal deflection that results is required. The relation between the load and extension for member AC is $F = 200e - 20e^2$ while that for BC is $F = 400e$. For both expressions the load is measured in kN and the extension in mm. For AC

$$e = 5 - \left(25 - \frac{F}{20} \right)^{1/2}$$

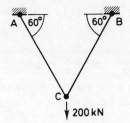

Figure 6.15

For BC

$$e = \frac{F}{400}$$

Due to the 200 kN applied load the forces in AC and BC are each $200/\sqrt{3}$.

If a unit horizontal load is applied to the right at C, $F'_{AC} = 1$, $F'_{BC} = -1$, therefore

$$d_C = 1\left[5 - \left(25 - \frac{200}{20\sqrt{3}}\right)^{1/2}\right] - 1\left[\frac{200}{400\sqrt{3}}\right] = 0.33 \text{ mm}$$

It should be noted that it was necessary to solve a quadratic equation to determine the value of e in terms of the load for member AC. Difficulties of using this approach would be met if the relation had been a cubic equation or higher.

6.11 Energy theorems and their application to statically determinate structures

We shall first derive two energy theorems from basic principles, the first of these is generally known as Castigliano's theorem Part 1.

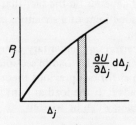

Figure 6.16

Consider a structure that is in equilibrium under a set of applied loads P. No restriction is to be placed on the material of the structure. If the applied loads are increased by dP the points of application of the loads will deflect through $d\Delta$ say.

The external work done at joint j will be $P_j \, \mathrm{d}\Delta_j$, so that the total external work done is $\sum P_j \, \mathrm{d}\Delta_j$, and this must equal the increase in stored energy, that is, strain energy in the structure, so that

$$\mathrm{d}U = \sum P_j \, \mathrm{d}\Delta_j$$

The load–displacement curve for joint j is shown in figure 6.16. The increase in strain energy due to an increase $\mathrm{d}\Delta_j$ could be written as $\mathrm{d}U = (\partial U/\partial\Delta_j)\mathrm{d}\Delta_j$ and the total change as

$$\mathrm{d}U = \sum \frac{\partial U}{\partial \Delta_j} \, \mathrm{d}\Delta_j$$

Equating the two expressions

$$\sum \frac{\partial U}{\partial \Delta_j} \, \mathrm{d}\Delta_j = \sum P_j \, \mathrm{d}\Delta_j; \quad \text{or} \quad \frac{\partial U}{\partial \Delta_j} = P_j \tag{6.31}$$

This relation is known as Castigliano's theorem Part 1.

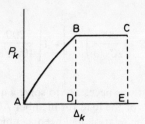

Figure 6.17

So that if the total strain energy is found for a particular structure and differentiated with respect to the deflection at a particular point, the load is obtained at that point. If the point is not subject to a load then equation 6.31 can be written as $\partial U/\partial\Delta_j = 0$. This relation is sometimes referred to as the First Theorem of Minimum Strain Energy, and depends on the fact that the strain energy will have a stationary value if an unloaded joint of a structure is given a small displacement about its equilibrium position.

The second theorem concerns complementary energy which has already been discussed for a member subject to axial load in section 6.2.

Consider the same system as for our previous theorem, that is, a structure in equilibrium with external loads P. If the load at joint j alone increases by $\mathrm{d}P_j$ the increase in complementary energy can be written

$$\mathrm{d}C = \frac{\partial C}{\partial P_j} \, \mathrm{d}P_j \tag{6.32}$$

As a result of this increase the deflection of most of the joints in the structure will increase. Figure 6.17 shows the load–deflection curve for joint k. Note again that no restriction has been placed on the material of the structure. The portion

AB shows the effect as the initial loads are applied. When dP_j is added the value of the load P_k remains constant but the deflection increases by $d\Delta_k$ giving rise to the horizontal part of the graph. Thus the value of the complementary energy does not change. External work will be done however, and this is equal to the area BCDE.

As a result of the increase dP_j there will not be any increase of complementary energy at any of the joints apart from joint j, where the increase can be written as $\Delta_j\,dP_j$.

The two expressions for the complementary energy can now be equated

$$\frac{\partial C}{\partial P_j}\,dP_j = \Delta_j\,dP_j$$

or

$$\frac{\partial C}{\partial P_j} = \Delta_j \tag{6.33}$$

This is known as the First Theorem of Complementary Energy.

It can be seen that if the complementary energy for a particular system can be found and differentiated partially with respect to the load at a particular joint, then the deflection at that joint will be found.

If the structure undergoes a temperature change in addition to the application of an external loading system, the complementary energy for each of the members could be similar to that shown in figure 6.1c. Thus applying the complementary energy theorem to a particular joint in the structure will give the total deflection at that point due to both the temperature change and the external loads.

Basically there is no reason why either of the theorems should not be applied to a redundant structure apart from the fact that it is impossible to calculate either the strain or complementary energy unless the forces are known in the members.

If we restrict the material of the structure so that it is linearly elastic the strain energy will be equal to the complementary energy, that is, $C = U$. It is possible to write equation 6.33 as

$$\frac{\partial U}{\partial P_j} = \Delta_j \tag{6.34}$$

These two derived theorems will next be applied to determine deflections of certain points in different structures.

Let us suppose that the vertical deflection of point F is required in the structure shown in figure 6.11. It would first be necessary to determine the complementary energy for each of the members, and the total complementary energy C would then be known. If the complementary energy theorem is used in the form $\partial C/\partial P_1 = \Delta_F$ the vertical deflection of joint F is obtained.

A slight complication arises if say the deflection of E is required in the direction of the arrow as there is no applied load at E. This difficulty can be overcome by the introduction of a load P_E in the required direction. The complementary energy is found for all the members in terms of the applied loading system and the extra load at E. This can now be differentiated with respect to P_E. The value of P_E can then be set to zero in the resulting expression.

The complementary energy for an axially loaded member is given by $\int_0^{F_1} e\,dF$ where F_1 is the force in the member. For a linearly elastic member the energy will become $F_1 e_1/2$ or $F_1^2 l/2AE$. The total complementary energy will be $\Sigma F^2 l/2AE$. F will be a function of P_1, P_2 and P_E. It is perfectly possible to square the expression for F before differentiating, but this sometimes leads to a rather cumbersome expression. It is often preferable to differentiate in the following manner

$$\frac{\partial}{\partial P_E} \sum \frac{F^2 l}{2AE} = \sum F \frac{\partial F}{\partial P_E} \frac{l}{AE} \tag{6.35}$$

The problem already solved by virtual work will be solved using complementary energy. For convenience the figure has been redrawn in figure 6.18. The vertical deflection of D is required. The calculations are set out in table 6.3. The first three columns are identical with table 6.2. It will be necessary to introduce a further external force P at D. Column 4 gives the total force in each member due to

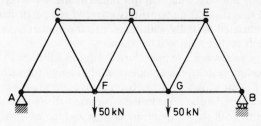

Figure 6.18

the 50 kN applied loads and the vertical load P. In the next column the force is differentiated with respect to P. Column 6 is formed by multiplying together the values in columns 4 and 5, at the same time P is set to zero and the result multiplied by l/AE.

The sum of column 6 gives the required deflection that is, 2·87 mm.

It can be seen that there is very little difference in the calculations that are required when virtual work is compared with complementary energy. In one case a unit load is applied, and in the other a load P which is finally set to zero.

The correspondence between the two methods can be demonstrated in a more theoretical manner as follows.

The force in a member due to a unit load was denoted by F', so that if a load P was applied instead of unity the force in a member would be $F'P$.

The total force in a member can be expressed as $F'' + F'P$ where F'' is the force in the member due to the external loading system, which is of course independent of P.

If C is the complementary energy in the member

$$\frac{\partial C}{\partial P} = \frac{\partial C}{\partial F} \frac{\partial F}{\partial P}$$

Now $F = F'' + F'P$ therefore $\partial F/\partial P = F'$ and $\partial C/\partial F = e$, the extension of the member, therefore

$$\frac{\partial C}{\partial P} = eF'$$

Summing for all the members in the framework where C is now the total complementary energy

$$\frac{\partial C}{\partial P} = \Delta = \sum eF$$

This expression is identical with the virtual-work method of solution.

Table 6.3

1	2	3	4	5	6
Member	Length m	Area mm^2	F kN	$\dfrac{\partial F}{\partial P}$	$F\dfrac{\partial F}{\partial P}\cdot\dfrac{l}{AE}$
AC	5	2000	$-57\cdot8 - 0\cdot578P$	$-0\cdot578$	$0\cdot399$
AF	5	2500	$28\cdot9 + 0\cdot289P$	$0\cdot289$	$0\cdot08$
CF	5	2000	$57\cdot8 + 0\cdot578P$	$0\cdot578$	$0\cdot399$
CD	5	2500	$-57\cdot8 - 0\cdot578P$	$-0\cdot578$	$0\cdot319$
FD	5	2000	$0 \ -0\cdot578P$	$-0\cdot578$	0
FG	5	2500	$57\cdot8 + 0\cdot867P$	$0\cdot867$	$0\cdot478$
DE	5	2500	$-57\cdot8 - 0\cdot578P$	$-0\cdot578$	$0\cdot319$
DG	5	2000	$0 \ -0\cdot578P$	$-0\cdot578$	0
EG	5	2000	$57\cdot8 + 0\cdot578P$	$0\cdot578$	$0\cdot399$
EB	5	2000	$-57\cdot8 - 0\cdot578P$	$-0\cdot578$	$0\cdot399$
GB	5	2500	$28\cdot9 + 0\cdot289P$	$0\cdot289$	$0\cdot08$

$$\Sigma\ 2\cdot872$$

Suppose that the vertical deflection of point F had been required. Using complementary energy the fundamental approach would be to replace the 50 kN load at F with a load P. The calculation could then proceed in the usual manner, and after the differentiation process the value of P could be set to 50 kN.

The truss is symmetrical, as is the loading. It is possible to replace both the 50 kN loads by P and only treat one half of the frame, doubling up the complementary energy for this half and adding in that for member FG. A little care has however to be exercised.

Apply a load P at F and a load P_1 at G where $P = P_1$. Consider two symmetrically placed members, say, AF and GB. For a unit vertical load at F let the force in AF be a and that in GB be b. It follows that for a unit vertical load at G, the force in AF would be b and that in GB would be a.

Thus for the applied loading

$$F_{AF} = Pa + P_1 b \qquad \frac{\partial F}{\partial P} = a$$

$$F_{GB} = P_1 a + Pb \qquad \frac{\partial F}{\partial P} = b$$

Putting $P_1 = P$

$$\sum F \frac{\partial F}{\partial P} = P(a + b)^2 \qquad (6.36)$$

If a load P is placed at F and G

$$F_{AF} = P(a + b) \qquad \frac{\partial F}{\partial P} = a + b$$

$$F_{GB} = P(a + b) \qquad \frac{\partial F}{\partial P} = a + b$$

therefore

$$\sum F \frac{\partial F}{\partial P} = 2P(a + b)^2 \qquad (6.37)$$

It can be seen, comparing equations 6.36 and 6.37 that if both loads are replaced by P, then the resulting deflection will be twice the true value.

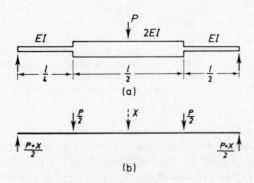

Figure 6.19

Both the complementary energy theorem and the virtual–work method can be applied to find the deflection of points in statically determinate beams. When complementary energy is used it is of course necessary to be able to calculate the complementary energy in the beam. For the case of a linearly elastic material this will be equal to the strain energy stored, that is, $\int M^2 \, ds/2EI$. As an application we shall find the deflection at a quarter-point for a simply supported beam of changing section with a central load (see figure 6.19a).

It is perfectly possible to introduce a load X, say, at the quarter-point and proceed in the usual manner. Unfortunately to find the complementary energy it would be necessary to perform four separate integrations, one for each quarter of the beam. A certain amount of working could be saved. The value of $\partial C/\partial X$ is finally required, and it is possible to differentiate under the integral sign, that is

$$\frac{\partial C}{\partial X} = \int \frac{M(\partial M/\partial X)\,\mathrm{d}x}{EI}$$

this however would still require four slightly simpler integrals. If a little thought is given to the problem much labour is avoided.

We could first apply the reciprocal theorem, change the problem and find the deflection at the centre for a load P applied at the quarter-point. If the load at the quarter-point is now replaced by the sum of two loadings, one symmetrical the other anti-symmetrical, the anti-symmetrical case will result in no deflection at the centre, and we are left with the problem in figure 6.19b to solve. This is symmetrical and it will only be necessary to find the complementary energy for one half of the beam.

Measuring x from the left-hand side

$$0 < x < \frac{l}{4} \quad M = \left(\frac{P + X}{2}\right)x \quad \frac{\partial M}{\partial X} = \frac{x}{2}$$

$$\frac{l}{4} < x < \frac{l}{2} \quad M = \frac{Pl}{8} + \frac{Xx}{2} \quad \frac{\partial M}{\partial X} = \frac{x}{2}$$

$$\int \frac{M\,\partial M/\partial x\,\partial x}{EI} = 2\left\{\int_0^{l/4} \frac{(P + X)x^2\,\mathrm{d}x}{4EI} + \int_{l/4}^{l/2} \frac{(Pl + 4Xx)x\,\mathrm{d}x}{32EI}\right\}$$

There is no point in integrating the terms that have X present since the expression obtained will give the required deflection when X is set equal to zero, so that deflection

$$d = 2\left[\int_0^{l/4} \frac{Px^2\,\mathrm{d}x}{4EI} + \int_{l/4}^{l/2} \frac{Plx\,\mathrm{d}x}{32EI}\right] = \frac{13}{1536}\frac{Pl^3}{EI}$$

The problem of the two pin-jointed members with one member of non-linear material shown in figure 6.15 can be solved by complementary energy. It is of course necessary to calculate the complementary energy in the form $C = \int e\,\mathrm{d}F$. An alternative approach using strain energy and Castigliano Part 1 would be as follows.

It is assumed that when the 200 kN load is applied C displaces to C' with vertical and horizontal components of displacement d_V and d_H (figure 6.20). So that the extension of AC would be $\sqrt{3}d_V/2 + d_H/2 = e_1$ say and that of BC $\sqrt{3}d_V/2 - d_H/2 = e_2$ say.

Now $U = \int F\,\mathrm{d}e$, then for AC

$$F = 200e - 20e^2 \quad U = \int_0^{e_1}(200e - 20e^2)\,\mathrm{d}e$$

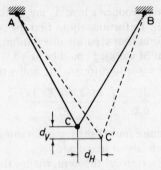

Figure 6.20

for BC

$$F = 400e \qquad U = \int_0^{e_2} 400e \, de$$

The total strain energy is now known in terms of e_1 and e_2 and the substitution for e_1 and e_2 is made, in terms of d_V and d_H.

Arithmetic can be saved by the fact that

$$\frac{\partial U}{\partial \Delta} = \frac{\partial U}{\partial e} \frac{\partial e}{\partial \Delta}$$

for AC

$$\frac{\partial U}{\partial \Delta} = (200 e_1 - 20 e_1^2) \frac{\partial e_1}{\partial \Delta}$$

for BC

$$\frac{\partial U}{\partial \Delta} = 400 e_2 \frac{\partial e_2}{\partial \Delta}$$

Two statements can be made using Castigliano's theorem

$$\frac{\partial U}{\partial d_V} = 200 \text{ kN} \qquad \frac{\partial U}{\partial d_H} = 0$$

$$\left[200 \left(\frac{\sqrt{3}}{2} d_V + \frac{d_H}{2} \right) - 20 \left(\frac{\sqrt{3}}{2} d_V + \frac{d_H}{2} \right)^2 \right] \frac{\sqrt{3}}{2} + \left[400 \left(\frac{\sqrt{3}}{2} d_V - \frac{d_H}{2} \right) \right] \frac{\sqrt{3}}{2}$$

$$= 200$$

or

$$450 d_V - 50\sqrt{3} d_H - 10\sqrt{3} \left(\frac{\sqrt{3}}{2} d_V + \frac{d_H}{2} \right)^2 = 200$$

and

$$\left[200 \left(\frac{\sqrt{3}}{2} d_V + \frac{d_H}{2} \right) - 20 \left(\frac{\sqrt{3}}{2} d_V + \frac{d_H}{2} \right)^2 \right] \frac{1}{2} - \left[400 \left(\frac{\sqrt{3}}{2} d_V - \frac{d_H}{2} \right) \right] \frac{1}{2} = 0$$

or

$$- 50\sqrt{3}d_V + 150\,d_H - 10\left(\frac{\sqrt{3}}{2}\,d_V + \frac{d_H}{2}\right)^2 = 0$$

These equations solve to give $d_V = 0\cdot53$ mm and $d_H = 0\cdot33$ mm. This solution turns out to be more complicated as far as the arithmetic is concerned when compared to the virtual-work solution.

Before proceeding further we ought to consider the type of approach we have been using when applying the various methods that have been discussed. When using virtual work or complementary energy we are in fact using a flexibility approach. The structure is first assumed to be in equilibrium under the action of loads P_1-P_n and the required deflection d_1 at a particular point 1, say, is determined. This could of course be written as

$$d_1 = f_{11}P_1 + f_{12}P_2 + \cdots + f_{1n}P_n$$

The various flexibility coefficients could be determined by virtual work or complementary energy.

With Castigliano's theorem, the displacements are assumed to be unknowns, and using compatibility (that is, the members must fit together both before and after the external loads are applied), the extensions of the various members are found in terms of the displacements. From a knowledge of the properties of the members the individual member forces can be found. Finally equilibrium equations are formed, which can then be solved for the unknown displacements. This is a stiffness approach.

We shall next discuss a graphical approach for determining deflections of pin-jointed statically determinate structures.

6.12 Williot–Mohr displacement diagram

All the methods that we have considered so far give the displacement in a particular direction of one point in the framework for each application of the theorem used. A graphical approach can be used and this will give the displacement of every joint in the framework, the results being displayed on one diagram.

Figure 6.21a shows a basic triangular pin-jointed framework with a vertical load P applied at A. The resulting deflection at A is required.

The forces in all the members can be found, and from a knowledge of the properties of the members the extension of each member can be determined. Let these be e_{BC}, e_{AB} and e_{AC}; the latter two values will of course be negative as the members are in compression.

It is theoretically possible to draw the frame in its displaced position when the external load is applied. As C can only move horizontally its new position can be fixed; it will be at C' a horizontal distance e_{BC} from C. The new position of A will be at the intersection of the arcs BA and CA; the first will have a length $BA - e_{AB}$ and centre B and the second a length $CA - e_{AC}$ and centre C'. The new position of the framework is shown in figure 6.21a. In fact the extensions of the members are small compared with their lengths and it would be completely impossible to attain any accuracy by this method.

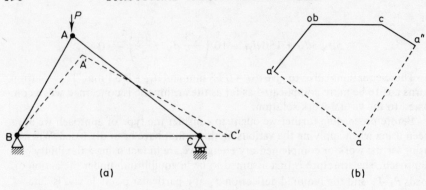

Figure 6.21

Williot decided on a graphical method of construction that overcame these difficulties. The displacement of one end of a member relative to the other end is considered to consist of a displacement in the direction of the member equal to e, the extension due to the force in the member, together with a displacement due to the rotation of the member; since the value of e is small compared to the length of the member it is sufficiently accurate to take the component due to the rotation as being perpendicular to the original direction of the member. The displacement diagram consists only of extension and rotation elements.

A fixed point o is chosen which will also represent any fixed points in the framework; in the given example B is a fixed point (see figure 6.21b). Relative to B, C moves horizontally through a distance e_{BC}; this is drawn as bc on the Williot diagram to a suitable scale, possibly 10 or 20 times full size. Relative to B, A shortens by e_{AB} represented by ba'. The direction of the rotation component is known, perpendicular to AB, but its length is unknown; this direction is shown by a dotted line. The process is repeated considering the displacement of A relative to C, this consists of a length $ca'' = e_{AC}$ and a rotation term. The intersection of the two dotted lines is at the point a. oa will give the displacement of A. The vertical and horizontal components can be measured from the diagram.

A certain amount of care has to be exercised when drawing a Williot diagram. Usually the drawing 'grows' rather rapidly and it is best to make a rough sketch first, enabling a suitable scale to be chosen such that the diagram does not become too large for the size of paper used.

The pin-jointed truss in figure 6.22a is pinned to the wall at A and B and has a vertical load applied at E. The forces in the members would normally have to be found and then the extension of each member calculated. In this example the extensions are given, and are entered on the line drawing.

There are two fixed points A and B and we can start the Williot diagram from these (figure 6.22b). The sequence after fixing the pole, which will also represent A and B, is c, d then e. The only point to note is that since member DC has zero force, and hence no extension, there will only be a rotation component.

It is not always possible to start drawing a Williot diagram unless an assumption is made—the assumption may very well not be true and it will be necessary to correct the diagram when it is completed. A typical assumption in figure 6.23a might be that the displacement of F relative to A is horizontal and has no rotation. With this assumption a horizontal line representing e_{AF} can be drawn, and the rest of the Williot diagram will then follow. When the diagram is complete it will be found that B has both horizontal and vertical components relative to A. A vertical component is of course impossible as the support at B only allows horizontal movement. The frame in its distorted position is shown on figure 6.23b, and it can be seen that it is necessary to rotate the distorted frame about A until B' lies on the horizontal through A.

We shall first consider the effect of rotating the original framework about o through a small angle $d\theta$. The displacement of A will be Ao x $d\theta$ in a direction perpendicular to Ao. This is represented by oa on the displacement diagram figure 6.23c. The displacements of all the other points in the framework can be

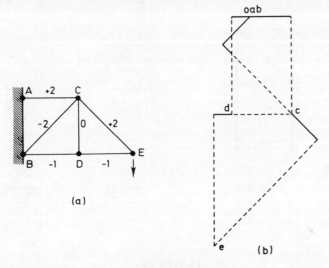

Figure 6.22

drawn. The resulting figure abcdef is similar to the original framework but rotated through 90°. This diagram is known as the Mohr rotation diagram.

Returning to figure 6.23b the correct displacements could be found by rotating the displaced diagram about A through an angle $d\theta$ which would be given by the vertical displacement of B divided by the length AB. An alternative would be to rotate the original frame through the same angle $d\theta$ in the opposite direction. The true displacements would then be given by the displacements of corresponding points measured relative to the distorted framework. This latter method is the one that is used in conjunction with the Williot–Mohr diagram.

The framework in figure 6.23a has the extensions of the members shown on the diagram. The Williot diagram figure 6.23d has been drawn on the assumption that member AF only extends and does not rotate.

It can be seen that point B has a large vertical displacement. To correct this the framework will have to be rotated through an angle bf/AB about A. The Mohr

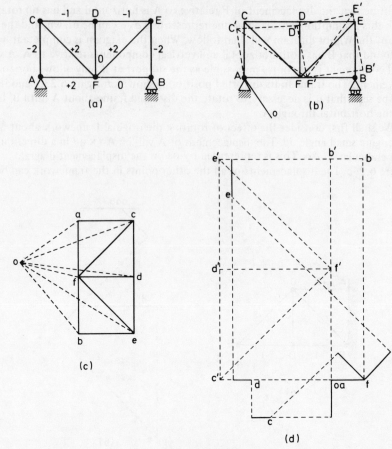

Figure 6.23

diagram is next superimposed on the Williot diagram. This involves drawing a diagram similar to the original framework on the base ab′. The point b′ is easily located since horizontal movement is only possible at B, so that b′ must lie on the horizontal through b and it also has to lie on the perpendicular through a. The Mohr diagram is ab′c′d′e′f′.

The displacements of all the points can now be found. The true displacement of point D for example is given by dd′.

6.13 Virtual work applied to a statically indeterminate truss

The problem in figure 6.24a will first be discussed. The cantilever AB was originally supposed to rest on the spring of flexibility f, the top of which is at the same level as the support at A; unfortunately the cantilever is slightly bent and as a result there is a clearance of λ. A uniformly distributed load p is now applied, end B comes into contact with the spring and compresses it. The force in the spring is required.

The required force will be assumed to be R. The problem can be broken down into a series of steps. If the spring is removed the deflection at B due to the load p will be $pl^4/8EI$ downwards. The force R in the spring acts on the end of the cantilever (figure 6.24b) and by itself would cause a deflection upwards. The force R in the spring will cause a compression of fR. We should now be able to make use of compatibility and determine a relation between the various displacements.

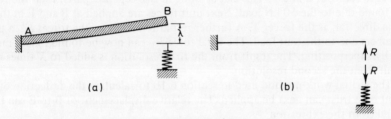

(a)   (b)

Figure 6.24

The downwards deflection of B on the cantilever is

$$\frac{pl^4}{8EI} - \frac{Rl^3}{3EI}$$

which must be equal to the initial clearance plus the compression of the spring. So that

$$\frac{pl^4}{8EI} - \frac{Rl^3}{3EI} = \lambda + fR$$

This solution makes use of the flexibility approach. However it was not necessary to use virtual work since all the deflections involved could be written down. The problem of finding the forces in the truss in figure 6.25 will however not be quite so simple.

All the vertical and horizontal members are of length 2 m; all members have a cross-section of 1000 mm²; $E = 210$ kN/mm².

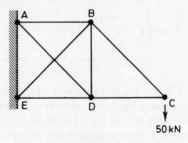

50 kN

Figure 6.25

It will be seen at once that the truss is statically indeterminate to the first degree. Any member can be chosen as the redundant apart from BC or DC, which are excluded since the forces in them are known by resolution. BE is chosen as the redundant and it is assumed that the force is tensile and has a value of X.

If we imagine BE to be cut at each end, in order to maintain equilibrium, forces equal to X must be applied at joints B and E tending to pull them together; while member BE is also subject to tensile forces X. The first step in the analysis will be to find the forces F in all the members due to the applied load and the forces X acting at B and E. This can be done in two separate parts. First the forces are found due to the 50 kN load. Next unit forces are applied at B and E in the same direction as the forces X. A further resolution will give the forces F' in the members due to the unit load. The total force F_T can now be found in any member by superposition. The result from the first resolution is added to X times the result from the second resolution.

The virtual-work method used in section 6.10 to calculate the deflection of a point on a truss can next be applied. The relative displacement of B to E can be found from the expression

$$\Delta = \sum \frac{F'F_T l}{AE}$$

where $F_T = F + F'X$. The summation includes all the members apart from BE.

The extension of BE is given by Xl/AE.

Now the movement of B relative to E must have the same value but be of opposite sign to the extension of BE, hence

$$\sum \frac{F'F_T l}{AE} = - \frac{Xl}{AE} \qquad (6.38)$$

The same result may be obtained by summing the virtual work terms for all the members including BE and equating the result to zero. This is the way in which we shall proceed.

The calculation is best set down in tabular form and this is shown in table 6.4. The column headings are all given and further explanation is not required apart from the fact that $k = AE/l$ where $l = 2$ m.

From equation 6.38 it can be seen that the sum of the elements in column 8 will be zero. Therefore

$$100k\left(1 + \frac{1}{\sqrt{2}}\right) + Xk(\tfrac{3}{2} + 2\sqrt{2}) = 0$$

$$X = -39{\cdot}4 \text{ kN}$$

Once the force in BE is known, the forces in all the members can be found.

The approach here has been flexibility. The structure was first made statically determinate, and displacements were then found due to both the external loads and the redundant force X. Finally compatibility is applied, in that the relative approach of B and E is equal to minus the extension of member BE.

Table 6.4

1	2	3	4	5	6	7	8
Member	Length	Area	50 kN	Unit	Total force	Extension	$(5) \times (7)$
CB	$\sqrt{2}l$	A	$50\sqrt{2}$	0	$50\sqrt{2}$	$100k$	0
CD	l	A	-50	0	-50	$-50k$	0
BD	l	A	-50	$-1/\sqrt{2}$	$-(50 + X/\sqrt{2})$	$-(50 + X/\sqrt{2})k$	$(50/\sqrt{2} + X/2)k$
BA	l	A	50	$-1/\sqrt{2}$	$50 - X/\sqrt{2}$	$(50 - X/\sqrt{2})k$	$(-50/\sqrt{2} + X/2)k$
DE	l	A	-100	$-1/\sqrt{2}$	$-(100 + X/\sqrt{2})$	$-(100 + X/\sqrt{2})k$	$(100/\sqrt{2} + X/2)k$
DA	$\sqrt{2}l$	A	$50\sqrt{2}$	1	$50\sqrt{2} + X$	$(100 + \sqrt{2}X)k$	$(100 + \sqrt{2}X)k$
BE	$\sqrt{2}l$	A	0	1	X	$\sqrt{2}Xk$	$\sqrt{2}Xk$

If the external loading had been accompanied by a temperature rise of T, the forces in the members would be altered. It would not however be difficult to take account of the temperature rise. In fact all that would be altered in the analysis would be the value of e, which means that the values entered in column 7 of table 6.4 would have to be changed and they would now have the form

$$\frac{F_T l}{AE} + \alpha l T$$

At the same time it is easy to take account of any lack of fit. We shall suppose that the truss is being constructed, and that member BE is the last to be placed in position and that it is unfortunately too short by an amount λ. It is however forced into position by straining. The force X in the member due to both the lack of fit and the external loads can be found from the final equation of compatibility, which would state that the relative movement of B and E together with the extension of member BE must be equal to the initial lack of fit.

$$\sum \frac{F' F_T l}{AE} = \lambda \tag{6.39}$$

where the summation of the left-hand side of equation 6.39 includes member BE.

We shall solve a numerical example on the same frame a little later on, but making use of complementary energy.

It should be noted that we can use the virtual-work method of approach for members made from non-linear materials, it is only necessary to have a knowledge of the load–extension characteristic for each member.

6.14 Engesser's theorem of compatibility

It is first necessary to establish this theorem and we shall make use of the redundant pin-jointed truss in figure 6.26a. When the truss was built there was an initial lack of fit of member AC, it was too short by an amount λ. The member was forced into position and the total force F_{AC} in member AC is required.

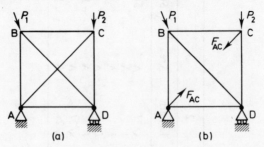

(a)   (b)

Figure 6.26

The complementary energy C_1 can be found for all the members in the truss omitting member AC for the time being. This will arise from two causes, the external loading P_1 and P_2 and the force F_{AC} applied at C (figure 6.26b).

Making use of the first theorem of complementary energy

$$\frac{\partial C_1}{\partial F_{AC}} = \Delta_C$$

where Δ_C is the displacement of joint C in the direction of F_{AC}.

Next consider member AC. The force in the member is F_{AC} and the complementary energy C_{AC} for the member can be found.

Applying the first theorem of complementary energy to this member alone

$$\frac{\partial C_{AC}}{\partial F_{AC}} = e_{AC}$$

where e_{AC} is the extension of member AC due to the force F_{AC}.

Now the displacement of joint C and the extension of AC must together be equal to the original lack of fit of AC

$$\Delta_C + e_{AC} = \lambda$$

or

$$\frac{\partial C_1}{\partial F_{AC}} + \frac{\partial C_{AC}}{\partial F_{AC}} = \lambda$$

Instead of finding the values of C_1 and C_{AC} separately the total complementary energy could be found

$$C = C_1 + C_{AC}$$

and

$$\frac{\partial C}{\partial F_{AC}} = \lambda$$

or in a more general form

$$\frac{\partial C}{\partial P} = \lambda \tag{6.40}$$

This is known as Engesser's theorem of compatibility.

If member AC had been a perfect fit in the first place before the loading P_1, P_2 was applied, the value of λ would be zero; Engesser's theorem could then be written

$$\frac{\partial C}{\partial P} = 0 \tag{6.41}$$

No restriction has been placed on the properties of the load–extension curves of the members. If however the material is linearly elastic the complementary and strain energies will be equal. The statements $\partial U/\partial P = \lambda$, and $\partial U/\partial P = 0$ for no lack of fit, are sometimes referred to as Castigliano's theorem of compatibility.

Both virtual work and Engesser's theorem will give an identical final equation to solve, that is, the compatibility equation is the same. In most cases the analysis is also very similar.

If we consider the same problem again (figure 6.25), the complementary energy in the members can be found from $\Sigma\,F_T^2 l/AE$. However as the expression is to be differentiated with respect to X, this is best carried out under the summation sign, in the form

$$\frac{\partial C}{\partial X} = \Sigma\,F_T\,\frac{\partial F_T}{\partial X}\,\frac{l}{AE} = 0$$

If this is done for one member, say BD

$$F_T = -\left(50 + \frac{X}{\sqrt2}\right)\qquad \frac{\partial F_T}{\partial X} = -\frac{1}{\sqrt2}$$

$$F_T\,\frac{\partial F_T}{\partial X}\,\frac{l}{AE} = \left(\frac{50}{\sqrt2} + \frac{X}{2}\right)\frac{l}{AE}$$

In other words the values so obtained are of course identical to column 8 in table 6.4.

There is really no point in repeating the analysis so we shall consider the same framework with an initial lack of fit.

It will be assumed that when the truss was constructed, member BE was the last member to be put in position and that it was too short, in this case 1·2 mm. Let us find the value of the load that must be applied at C in a vertical direction so that member BE may be fitted. Also, when the member has been fitted the load at C is removed. What is the value of the residual force in BE?

Let the force required at C be P and the force in the member BE be X. The first part of the problem can be solved by making use of the first theorem of complementary energy, and in this case $C = U$. The value of P is required such that there will be a movement of B relative to E of 1.2 mm. We shall introduce a load X in this direction and finally set it to zero.

The forces in the members are found in terms of P and X in table 6.5 column 4. The complementary energy can be found in the form

$$C = \sum \frac{F_T^2 l}{2AE}$$

If this summation is carried out for all the members apart from BE and the result differentiated with respect to X, we can obtain the movement of B relative to E with P applied at C, and X applied at B in the direction BE. However for the first part of the problem X is zero; so that once the differentiation has been carried out X is set to zero and the result equated to the initial lack of fit. It is very much easier to carry out the work in the form

$$\sum_{X=0} F_T\,\frac{\partial F_T}{\partial X}\,\frac{l}{AE} = \Delta$$

This has been set out in the table, columns 5 and 6. l/AE has been put equal to k. Note that member BE has been included in the table as it will be required in the second part of the analysis.

Table 6.5

1	2	3	4	5	6
Member	Length	Area	F_T	$\dfrac{\partial F_T}{\partial X}$	$F_T \dfrac{\partial F_T}{\partial X} \dfrac{l}{AE}$
CB	$\sqrt{2}l$	A	$\sqrt{2}P$	0	0
CD	l	A	$-P$	0	0
BD	l	A	$-(P + X/\sqrt{2})$	$-1/\sqrt{2}$	$(P/\sqrt{2} + X/2)k$
BA	l	A	$P - X/\sqrt{2}$	$-1/\sqrt{2}$	$(-P/\sqrt{2} + X/2)k$
DE	l	A	$-(2P + X/\sqrt{2})$	$-1/\sqrt{2}$	$(2P/\sqrt{2} + X/2)k$
DA	$\sqrt{2}l$	A	$\sqrt{2}P + X$	1	$(2P + \sqrt{2}X)k$
BE	$\sqrt{2}l$	A	X	1	$\sqrt{2}Xk$

Column 6 can be summed and X put equal to zero

$$2P(1 + 1/\sqrt{2})k = 1\cdot2 \times 10^{-3}$$

$$k = 2/(1000 \times 210) \text{ m/kN}$$

therefore

$$P = 36\cdot8 \text{ kN}$$

For the second part of the problem Engesser's theorem of compatibility can be applied with an initial lack of fit. In this case the frame is redundant and the force in the redundant member BE is X while the value of P is zero. There is no need to carry out any further resolution, etc., as the existing column 6 in the table can be used

$$\sum_{P=0} F_T \frac{\partial F_T}{\partial X} \frac{l}{AE} = \lambda$$

where $\lambda = 1\cdot2$ mm.

It is of course necessary this time to take account of the complementary energy stored in member BE. The summation of column 6 with P set to zero will give the required result.

$$X(\tfrac{3}{2} + 2\sqrt{2})k = 1\cdot2 \times 10^{-3}$$

$$X = 29\cdot2 \text{ kN}$$

It is interesting to observe the relatively high load that arises in the member BE for a small initial lack of fit.

6.15 Trusses with several redundants

The methods that have been discussed so far will be equally applicable to a truss that has more than one redundant member. The truss in figure 6.27 has three redundant members; choosing these as BH, CJ, and DK, the forces in these mem-

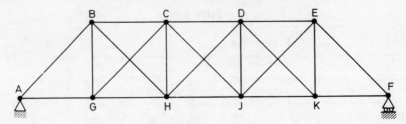

Figure 6.27

bers can be denoted by X_1, X_2 and X_3. The complementary energy can be found in terms of the external loading and these unknown loads. If there is no initial lack of fit Engesser's theorem can be applied in turn to each redundant member. Thus

$$\sum F_T \frac{\partial F_T}{\partial X_1} \frac{l}{AE} = \sum F_T \frac{\partial F_T}{\partial X_2} \frac{l}{AE} = \sum F_T \frac{\partial F_T}{\partial X_3} \frac{l}{AE} = 0 \qquad (6.42)$$

Three simultaneous equations will result and these can be solved for X_1, X_2 and X_3. It can be seen that the method of approach will become rather tedious as the number of redundants increases.

So far we have only considered the cases of trusses with redundant members. Redundant reactions will however not create any special difficulty.

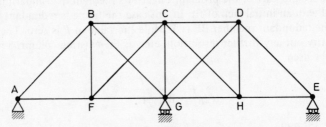

Figure 6.28

Figure 6.28 shows a truss with two redundant members and one redundant reaction. Choosing BG and CH as the redundant members with forces X_1 and X_2 and letting the vertical reaction at G be X_3, the forces in all the members can be found. Engesser's theorem can be applied in the usual manner to BG and CH, to give two simultaneous equations in terms of the three unknowns. The truss rests on the support at G, and assuming that this support does not sink or move when the loads are applied, the first theorem of complementary energy can be applied setting the value of Δ to zero, or

$$\sum F_T \frac{\partial F_T}{\partial X_3} \frac{l}{AE} = 0$$

and hence a third equation will result.

The virtual-work method can of course be applied to the same type of problem.

For figure 6.27 the redundant members are imagined to be cut, to make the truss statically determinate. The forces in all the members can now be found in terms of the external loads. Unit loads will next be applied in the direction of each cut member and the forces found in the rest of the members of the truss. The total force in a member will be of the form

$$F + F_1'X_1 + F_2'X_2 + F_3'X_3 = F_T$$

where F_1', F_2' and F_3' represent the force in the member due to unit loads being applied in the directions of BH, CJ and DK respectively. The extensions of the members can next be calculated and a similar analysis to that used in section 6.4 can be applied.

Thus for BH say

$$\sum (F + F_1'X_1 + F_2'X_2 + F_3'X_3)\frac{F_1'l}{AE} = 0 \qquad (6.43)$$

Two similar equations will result for CJ and DK. It should be emphasised that the equations will be identical to those obtained by applying Engesser's method.

The truss in figure 6.28 can be solved in a similar manner. Two equations result from the redundant members and the third from the statement that the deflection of the support at G is zero. We should perhaps examine equation 6.43 in a little more detail. We shall consider the terms individually.

Once the frame has been made statically determinate the first term $\sum FF_1'l/AE$ gives the deflection of B relative to H. The second term $\sum (F_1')^2X_1l/AE$ gives the deflection of B relative to H with the external loads removed and X_1 only applied. The third term gives the same deflection with only X_2 applied, and the fourth term with only X_3 applied. The complete equation states that the sum of all these terms is zero. We have met this kind of equation before in section 6.5 when discussing flexibility methods. If we compare equation 6.43 with one of those in equation 6.16 the form is identical. We shall rewrite the equation as

$$\sum \frac{FF_1'l}{AE} + X_1 \sum \frac{(F_1')^2 l}{AE} + X_2 \sum \frac{F_1'F_2'l}{AE} + X_3 \sum \frac{F_1'F_3'l}{AE} = 0 \quad (6.44)$$

The summation terms with the Xs outside are flexibility coefficients and the three equations could be written down as a flexibility matrix.

$$\begin{bmatrix} \Sigma FF_1'(l/AE) \\ \Sigma FF_2'(l/AE) \\ \Sigma FF_3'(l/AE) \end{bmatrix}$$

$$+ \begin{bmatrix} \Sigma (F_1')^2(l/AE) & \Sigma F_1'F_2'(l/AE) & \Sigma F_1'F_3'(l/AE) \\ \Sigma F_2'F_1'(l/AE) & \Sigma (F_2')^2(l/AE) & \Sigma F_2'F_3'(l/AE) \\ \Sigma F_3'F_1'(l/AE) & \Sigma F_3'F_2'(l/AE) & \Sigma (F_3')^2(l/AE) \end{bmatrix} \begin{bmatrix} X_1 \\ X_2 \\ X_3 \end{bmatrix} = 0 \qquad (6.45)$$

6.16 The trussed beam

It is not proposed to discuss in detail the application of virtual work and energy to redundant beams, as other methods already described often give a rapid solution in straightforward cases. Suffice it to say that if complementary energy is used, forces X, Y, etc. are chosen as the redundants, which could be moments or reactions. The moment is written down in terms of the applied loading and the redundant forces. For an elastic beam the complementary energy will be given by $\int M^2 \, dx/2EI$. If the supports have not deflected we may make use of the first theorem of complementary energy and state that $\partial C/\partial X = \partial C/\partial Y = 0$. The resulting equations are solved to find the redundants.

It is interesting however to consider the case of a trussed beam which is really a composite structure. It consists of a simple beam that is strengthened by the addition of bracing members.

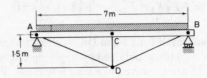

Figure 6.29

A typical example of a trussed beam is shown in figure 6.29. The beam will be subject to both bending and axial force. The stiffening structure is pin-jointed and so the members are only loaded axially.

The beam is formed from two 254 mm × 76 mm steel channels 7 m long. For each channel $I_{zz} = 3367$ cm^4; $A = 3603$ mm^2. The bracing members are made of steel, and pinned at A, C, and B between the two channels that are placed back to back. The cross-section of the tension members is 760 mm^2 and that of the compression member 1600 mm^2. The force in the member DC is required when the beam carries a uniformly distributed load of 20 kN/m.

There are really only two basic approaches to this problem. It can be treated as one structure or it can be split up into a beam and a truss. We shall start by treating it as one structure and will use an energy approach.

The problem is singly redundant and CD will be regarded as the redundant member. Let the compressive force in CD be X.

$$F_{AD} = F_{BD} = 1 \cdot 27X - \text{tensile}$$

The axial force in the beam is $1 \cdot 17X$—compressive.

The vertical reactions at A and B will each be 70 kN. When the bending moment is calculated for any point on the beam allowance must be made for the vertical component of the force in AD or BD.

Measuring x from the left-hand side

$$M_x = \left(70 - \frac{X}{2}\right)x - \frac{20x^2}{2}$$

$0 < x < 3 \cdot 5$ m.

The complementary energy is of the form

$$\int \frac{\dot{M}^2 \, dx}{2EI} + \sum \frac{F^2 l}{2AE}$$

where the first term refers to the bending moment in the beam and the summation in the second term extends to all the members including the beam.

Using Engesser's theorem $\partial C/\partial X = 0$

$$\frac{\partial C}{\partial X} = \int \frac{M(\partial M/\partial X) \, dx}{EI} + \sum F \frac{\partial F}{\partial X} \frac{l}{AE} = 0$$

$$2 \left[\int_0^{3\cdot5} \left(\frac{20x^3}{4} + \frac{Xx^2}{4} - \frac{70x^2}{2} \right) \frac{dx \times 10^8}{6734E} \right]$$

$$+ \frac{1 \cdot 17^2 \times X \times 7 \times 10^6}{7206E} + 2 \left[\frac{1 \cdot 27^2 \times X \times 3 \cdot 8 \times 10^6}{760E} \right] + \frac{X \times 1 \cdot 5 \times 10^6}{1600E} = 0$$

$$X = 74 \cdot 3 \text{ kN}$$

The alternative process is to split the problem up as shown in figure 6.30 into a simply supported beam and a truss. There is an unknown reaction X between them at C. This will also be the force in the member CD. The connecting fact or

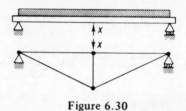

Figure 6.30

compatibility in this case is that at C the deflection of the truss and the deflection of the beam must be the same. The first theorem of complementary energy could be used to find the deflection of the truss at C. The deflection of the beam can easily be found from the list of standard cases in table 5.1. The calculation will be quite straightforward and will include a number of the terms from the previous solution and it will not be repeated here.

6.17 Virtual work and energy methods applied to frames

In this section we shall confine our attention to cases that only have a small number of redundants. Effects due to axial loads will be omitted as in general they will be small when compared with bending effects.

The problem of the portal with pinned feet in figure 6.31 will be solved using energy and virtual work. It is also perfectly possible to make use of moment area theorems, but this is left as an exercise for the reader.

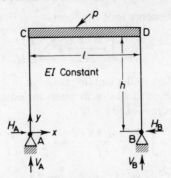

Figure 6.31

In all frames the junction between two or more members will be regarded as rigid, so that the joint as a whole can rotate but there is no relative rotation of the members with respect to each other. This particular problem is only statically indeterminate to one degree, as an equation of condition gives rise to the fact that $V_A = V_B = pl/2$ and $H_A = H_B = H$.

Making use of energy

It will be necessary to obtain an expression for the bending moment at any point on the portal. Choosing an origin at A and adopting the convention that a positive bending moment produces tension on the inside edge of the frame

$$M_{A-C} = M_{B-D} = -Hy \qquad \frac{\partial M}{\partial H} = -y$$

$$M_{C-D} = \frac{plx}{2} - \frac{px^2}{2} - Hh \qquad \frac{\partial M}{\partial H} = -h$$

$$U = C = \int \frac{M^2 \, dx}{2EI}$$

Now $\partial C / \partial H = 0$; or

$$\int \frac{M(\partial M / \partial H) \, dx}{EI} = 0$$

The integration is of course carried out along the length of a member.

$$2 \int_0^h \frac{Hy^2 \, dy}{EI} + \int_0^l \left(-\frac{plhx}{2} + \frac{phx^2}{2} + Hh^2 \right) \frac{dx}{EI} = 0$$

$$\frac{2Hh^3}{3} - \frac{pl^3h}{4} + \frac{pl^3h}{6} + Hlh^2 = 0$$

$$H = \frac{pl^2}{4h(2h + 3l)} \tag{6.46}$$

Once the value of H is known the portal has become statically determinate.

Making use of virtual work

The method developed in section 6.10 for finding the deflection of a point on a beam can be applied

$$\Delta = \int \frac{M'M\,dx}{EI}$$

Let us modify the problem by making the support at A into a roller. The horizontal force is now zero, and CD will be the only member with a bending moment present (figure 6.32a).

$$M_{C-D} = \frac{plx}{2} - \frac{px^2}{2}$$

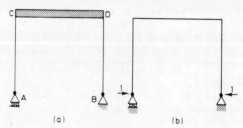

Figure 6.32

A unit load is next applied in the direction of the horizontal reaction with all the applied loading removed (figure 6.32b).

$$M'_{A-C} = M'_{B-D} = -y \quad M'_{C-D} = -h$$

If the applied load is replaced and a force H instead of the unit load is applied

$$M_{A-C} = M_{B-D} = -Hy \quad M_{C-D} = \frac{plx}{2} - \frac{px^2}{2} - Hh$$

In the original problem the horizontal deflection at A is zero.

$$\int \frac{M'M\,dx}{EI} = 0$$

or

$$2\int_0^h \frac{Hy^2\,dy}{EI} + \int_0^l \left(-\frac{plhx}{2} + \frac{phx^2}{2} + Hh^2 \right) \frac{dx}{EI} = 0$$

This equation is of course identical with that obtained by using an energy approach.

As a further example of the application of the unit-load method to a problem where there is more than one redundancy, we shall consider the same portal frame but with both feet built-in. This will introduce a further unknown, namely the moment at the feet (figure 6.33a).

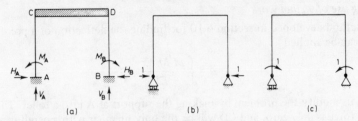

Figure 6.33

From conditions of symmetry $H_A = H_B = H$; $V_A = V_B = pl/2$; $M_A = M_B$.
Choosing the two redundants as H_A and M_A, the bending moment at any point
on the frame could be written in the form

$$M = M_S + M_1'H + M_2'M_A$$

where M_S is the bending moment at the point for the statically determinate frame,
that is, both H_A and M_A are equal to zero. M_1' is the bending moment with no
external load and with $M_A = 0$, $H = 1$. M_2' is the bending moment with no external
load, with $H_A = 0$, $M_A = 1$.

The loading systems for determining M_1' and M_2' are shown in figure 6.33b and
c. The bending moments are set out in table 6.6.

Table 6.6

Member	M_S	M_1'	M_2'
AC, BD	0	$-y$	1
CD	$\dfrac{plx}{2} - \dfrac{px^2}{2}$	$-h$	1

The horizontal deflection at A is zero in the original problem

$$\frac{1}{EI} \int (M_S + M_1'H + M_2'M_A)M_1' \, dx = 0 \tag{6.47}$$

The rotation at A is zero. There is of course no reason why the unit-load tech-
nique should not apply to rotations as well as deflections.

$$\frac{1}{EI} \int (M_S + M_1'H + M_2'M_A)M_2' \, dx = 0 \tag{6.48}$$

Equations 6.47 and 6.48 can of course be written as a flexibility matrix as
follows

$$\begin{bmatrix} \int \dfrac{M_S M_1' \, dx}{EI} \\[2ex] \int \dfrac{M_S M_2' \, dx}{EI} \end{bmatrix} + \begin{bmatrix} \int \dfrac{(M_1')^2 \, dx}{EI} & \int \dfrac{M_1'M_2' \, dx}{EI} \\[2ex] \int \dfrac{M_1'M_2' \, dx}{EI} & \int \dfrac{(M_2')^2 \, dx}{EI} \end{bmatrix} \begin{bmatrix} H \\[2ex] M_A \end{bmatrix} = 0$$

Inserting the values from table 6.6

$$
\begin{bmatrix} \int_0^l \left(\dfrac{plx}{2} - \dfrac{px^2}{2} \right)(-h)\,\mathrm{d}x \\[2ex] \int_0^l \left(\dfrac{plx}{2} - \dfrac{px^2}{2} \right)(1)\,\mathrm{d}x \end{bmatrix}
$$

$$
+ \begin{bmatrix} \int_0^l h^2\,\mathrm{d}x + 2\int_0^h (-y)^2\,\mathrm{d}y & \int_0^l -h\,\mathrm{d}x + 2\int_0^h -y\,\mathrm{d}y \\[2ex] \int_0^l -h\,\mathrm{d}x + 2\int_0^h -y\,\mathrm{d}y & \int_0^l 1^2\,\mathrm{d}x + 2\int_0^h 1^2\,\mathrm{d}x \end{bmatrix} \begin{bmatrix} H \\[2ex] M_A \end{bmatrix} = 0
$$

$$
- \frac{pl^3 h}{12} + \left(h^2 l + \frac{2h^3}{3} \right) H - (hl + h^2)M_A = 0
$$

$$
\frac{pl^3}{12} - (hl + h^2)H + (l + 2h)M_A = 0
$$

Whence

$$
M_A = \frac{pl^3}{12(2l + h)} \qquad H = \frac{pl^3}{4h(2l + h)}
$$

It should be emphasised that this is not necessarily the best method of approach for solving this problem. In actual fact we shall find later that there is a very rapid solution if the method of moment distribution is used.

If point loads are applied to the frame there is no basic change in the approach, using energy or the unit-load methods. All that it will mean is that the expressions for the bending moments in the released structure could be somewhat more complicated.

6.18 Ring and arch problems

Consider a ring loaded by two equal and opposite forces acting at the ends of a diameter. The dimensions of the cross-section are small compared with the radius R of the ring. Axial effects will be neglected in comparison with bending. The maximum bending moment in the ring is required (figure 6.34a).

If the ring is cut through as shown at b, forces must be applied to keep the two halves of the ring in equilibrium. The applied force P must be reacted by $P/2$ at each cut, and there will be bending moments M_1. There also is the possibility that shear forces H can exist. However if both halves of the ring are examined it will be seen that the system of forces shown is not compatible. Each half is identically loaded apart from H which acts in opposite directions on each half of the ring. This is of course impossible and so the value of H must be zero.

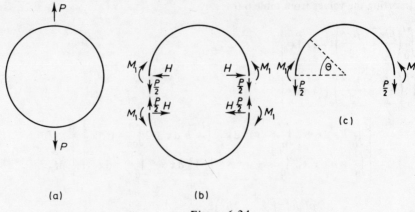

(a)   (b)

Figure 6.34

Consider the upper half of the ring (figure 6.34c). The bending moment at an angle θ is

$$M_\theta = M_1 - \frac{PR}{2}(1 - \cos\theta); \quad 0 < \theta < \frac{\pi}{2}$$

$$C = U = \int \frac{M^2 ds}{2EI} = \int \frac{M^2 R d\theta}{2EI}$$

There will be no need to calculate the value of the energy for more than one quarter of the ring.

There must be no rotation of the ring at the cut. Applying the first theorem of complementary energy

$$\frac{\partial C}{\partial M_1} = 0$$

or

$$\int M \frac{\partial M}{\partial M_1} \frac{R\, d\theta}{EI} = 0$$

$$\frac{\partial M}{\partial M_1} = 1$$

therefore

$$\frac{1}{EI} \int_0^{\pi/2} \left[M_1 - \frac{PR}{2}(1 - \cos\theta) \right] R\, d\theta = 0$$

$$M_1 = PR\left(\frac{1}{2} - \frac{1}{\pi} \right)$$

So that

$$M = PR\left(\frac{1}{2} - \frac{1}{\pi}\right) - \frac{PR}{2}\left(1 - \cos\theta\right)$$

$\theta = 0$

$$M = PR\left(\frac{1}{2} - \frac{1}{\pi}\right)$$

$\theta = \pi/2$

$$M = -\frac{PR}{\pi}$$

The bending moment changes sign and there will be some value of θ where the moment is zero. The maximum bending moment is $-PR/\pi$.

The problem will be taken a stage further to find the change in diameter at right angles to the line of action of P (see figure 6.35).

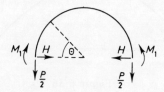

Figure 6.35

A force H will have to be introduced in the direction of the required displacement; H will finally be set to zero.

$$M = M_1 - \frac{PR}{2}\left(1 - \cos\theta\right) - HR\sin\theta$$

A value of M_1 has already been found for the case where H was zero. The question arises whether or not this value of M_1 changes by introducing H. Most certainly M_1 is a function of H, as we should find if we used above expression for M and determined M_1 from $\partial C/\partial M_1 = 0$. But as H is set equal to zero at a later stage of the calculation there is no reason why H should not be made zero when finding the value of M_1. This means that the value of M_1 obtained from the first part of the calculation can be used.

$$M_1 = PR\left(\frac{1}{2} - \frac{1}{\pi}\right)$$

Now

$$\frac{\Delta}{2} = \frac{\partial C}{\partial H} = \int_0^{\pi/2} M \frac{\partial M}{\partial H} \frac{R\,\mathrm{d}\theta}{EI} \; ; \quad \frac{\partial M}{\partial H} = -R\sin\theta$$

therefore

$$\frac{EI\Delta}{2} = \int_0^{\pi/2} \left[PR\left(\frac{1}{2} - \frac{1}{\pi}\right) - \frac{PR}{2}(1 - \cos\theta) - HR\sin\theta \right](-R\sin\theta)R\,d\theta$$

At this stage H may be put equal to zero

$$EI\frac{\Delta}{2} = \int_0^{\pi/2} PR\left(\frac{\cos\theta}{2} - \frac{1}{\pi}\right)(-R\sin\theta)R\,d\theta = PR^3\left(\frac{1}{\pi} - \frac{1}{4}\right)$$

$$\Delta = 0{\cdot}138\,\frac{PR^3}{EI}$$

The basic approach to a two-pin arch is very similar to that of a pin-ended portal frame (section 6.17). We shall use the method of virtual work.

If the arch is released at the right-hand abutment so that horizontal movement can take place, that is, $H = 0$, the bending moment M_s can be written down for

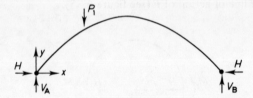

Figure 6.36

any point on the arch due to the external loads. The bending moment M for a unit horizontal load applied at B with the external loading removed can also be found.

The horizontal movement at B due to the external loading is

$$\int \frac{M_s M'\,ds}{EI}$$

If a horizontal load H is applied at B to the unloaded arch the horizontal movement of B is

$$H\int \frac{(M')^2\,ds}{EI}$$

In the original problem no movement is allowed at B. By superposition

$$\int \frac{M_s M'\,ds}{EI} - H\int \frac{(M')^2\,ds}{EI} = 0$$

If the origin of the coordinate system is set up at A (figure 6.36), $M' = y$ and

$$\int \frac{M_s y\,ds}{EI} - H\int \frac{y^2\,ds}{EI} = 0$$

therefore

$$H = \frac{\int M_s y \, ds/EI}{\int y^2 \, ds/EI}$$

The evaluation of these integrals is not always easy. An arch consisting of a segment of a circle with constant EI does not present difficulties.

Another straightforward case is that of an arch of parabolic form but with a secant variation of I. If I_c is the value of the second moment of area at the crown and θ is the slope of the arch at the point considered $I = I_c \sec \theta$, $ds = dx \sec \theta$, therefore

$$\frac{ds}{EI} = \frac{dx}{EI_c}$$

It might seem rather arbitrary to choose a secant variation. Arches in practice do not in general conform to this variation in I, but the value of H obtained by this much easier calculation is often good enough for a first analysis.

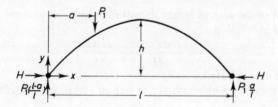

Figure 6.37

As an example we shall find the value of the horizontal thrust for a pinned arch of parabolic form with span l and rise h when a point load is applied at a distance a from the left-hand abutment (figure 6.37). I has a secant variation.

$$M_s = P_1 \frac{(l-a)x}{l} \quad 0 < x < a$$

$$M_s = P_1 \frac{a}{l}(l-x) \quad a < x < l$$

$$\int M_s y \frac{ds}{EI} = \int_0^a \frac{P_1 \frac{(l-a)}{l} xy \, dx}{EI_c} + \int_a^l \frac{P_1 \frac{a}{l}(l-x)y \, dx}{EI_c}$$

For the parabolic arch

$$y = \frac{4hx}{l^2}(l-x)$$

Substituting for y and integrating

$$\int \frac{M_s y \, ds}{EI} = \frac{P_1 h}{3l^3 EI_c} (al^4 - 2a^3 l^2 + a^4 l)$$

$$\int y^2 \, ds = \frac{8}{15} \frac{h^2 l}{EI_c}$$

$$H = \frac{5}{8} \frac{P_1}{h} \left(a - \frac{2a^3}{l^2} + \frac{a^4}{l^3} \right)$$

In this problem the position of P_1 is a variable and the expression that we have derived will give the influence line for H if P_1 is set equal to unity.

The case that we derived in detail presented no difficulties in integration. Cases can arise where the integration cannot be carried out directly and a numerical approach must be used.

While discussing the two-pin arch we ought to consider two other effects: one is the effect of axial forces, often referred to as rib shortening, and the other is the effect of a temperature change.

Consider first the effect of an axial force of value F where the slope of the arch is θ.

An element of the arch of length ds will shorten by an amount $F \, ds/AE$. The change in length dx is

$$\frac{F \, ds \cos \theta}{AE} = \frac{H \, ds}{AE}$$

if shear forces are neglected.

Therefore the change in length of the arch in a horizontal direction is $\int H \, ds/AE$.

The basic equation for a two-pin arch if rib shortening is taken into account will become

$$\int \frac{M_s y \, ds}{EI} - H \int \frac{y^2 \, ds}{EI} = H \int \frac{ds}{AE}$$

The effect of rib shortening will only affect the value of H for the case of a flat arch, and can usually be omitted if the rise to span ratio is greater than $1/10$.

The temperature change in an arch can be treated in a similar manner to a lack-of-fit problem.

If the arch is released at one end the horizontal movement due to a temperature change T would be

$$\int_0^l \alpha T \, dx = \alpha T l$$

where α is the coefficient of linear expansion.

A force H would now have to be applied to the end of the arch to reduce this movement to zero, that is

$$H \int \frac{y^2 \, ds}{EI} = \alpha T l$$

$$H = \frac{\alpha T l}{\int y^2 \, ds/EI}$$

If the arch has external loading and a temperature change takes place the total value of H can be found by superposition.

A number of arches are built-in at the abutments, and can be treated in a similar manner to the two-pin arch. The concept is no more difficult, but the degree of redundancy is higher and the arithmetic becomes tedious. For the two-pin arch we can use the fact that the horizontal displacement is zero; in addition for the built-in arch the rotation at an abutment is zero. There are special methods such as column analogy and elastic centre that help to reduce the arithmetic, but they are beyond the scope of this present work.

6.19 Redundant trusses using the stiffness approach

The redundant structures that have been discussed so far have been solved using a flexibility approach together with virtual work or energy. We shall conclude this chapter by discussing a problem that can be solved more easily if a stiffness approach is used.

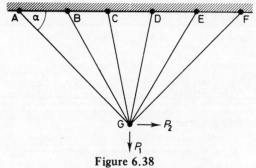

Figure 6.38

The truss in figure 6.38 is redundant to the fourth degree, but it has only two unknown displacements, the vertical and horizontal displacement of G, so that it is kinematically indeterminate to the second degree. This means that on a count of unknowns alone it would probably be better to use stiffness rather than flexibility.

The inclination to the horizontal of a typical member of length l is denoted by α. Let the vertical and horizontal displacements of G be d_1 and d_2 respectively.

Using geometric relations, that is, applying compatibility, the extensions of all the members can be calculated in terms of d_1 and d_2. Since we shall proceed using a virtual-work method, it is probably best to calculate the extensions first of all, in terms of a unit vertical displacement of G and then in terms of a unit displacement to the right of G.

For member AG

extension due to unit vertical displacement = 1 sin α

extension due to unit horizontal displacement = 1 cos α

If now instead of unit displacements d_1 and d_2 are applied together

total extension of AG = $d_1 \sin \alpha + d_2 \cos \alpha$

The force in AG can next be found from the load–extension curve $F = f(e)$. Thus

$$F = f(d_1 \sin \alpha + d_2 \cos \alpha)$$

and in the case of a linear elastic system

$$F = (d_1 \sin \alpha + d_2 \cos \alpha) \frac{AE}{l}$$

The forces in all the members of the truss will be of a similar form

If the horizontal equilibrium of G is considered by resolving the forces in all the members horizontally and vertically, two simultaneous equations will result in terms of d_1 and d_2. These can be solved and if the forces are required in the members, the values of d_1 and d_2 can be substituted into the expression for F. The virtual-work approach differs slightly from this and will be adopted since it is more systematic and will apply to more complicated problems.

Virtual-work equations can be written down in terms of the actual forces in the truss and a virtual compatible set of displacements. Thus the correspondence to the unit-load method will be to use a unit displacement.

Consider first a unit vertical displacement, the extensions due to this have already been found. Applying virtual work

$$P_1 \times 1 = \sum F \sin \alpha$$

For a linear elastic system

$$P_1 = \sum (d_1 \sin^2 \alpha + d_2 \sin \alpha \cos \alpha) \frac{AE}{l}$$

For a unit horizontal displacement

$$P_2 = \sum (d_1 \sin \alpha \cos \alpha + d_2 \cos^2 \alpha) \frac{AE}{l}$$

These equations are solved for d_1 and d_2. The forces in all the members can be found.

The various summation terms on the right-hand side of the equilibrium equations will be stiffness coefficients if d_1 and d_2 are omitted. The equilibrium equations written in the form of a stiffness matrix are as follows

$$
\begin{bmatrix} P_1 \\ P_2 \end{bmatrix} =
\begin{bmatrix} \sum \sin^2 \alpha \dfrac{AE}{l} & \sum \sin \alpha \cos \alpha \dfrac{AE}{l} \\ \sum \sin \alpha \cos \alpha \dfrac{AE}{l} & \sum \cos^2 \alpha \dfrac{AE}{l} \end{bmatrix}
\begin{bmatrix} d_1 \\ d_2 \end{bmatrix}
$$

An alternative approach which would give an identical stiffness matrix, would be to make use of Castigliano's theorem part 1, which was derived in section 6.11. The theorem can be written in the form $\partial U/\partial \Delta = P$.

If we consider the truss that was shown in figure 6.38; under the action of the applied loads the vertical and horizontal displacements of G are d_1 and d_2 respectively. The extensions of all the members meeting at G can be found in terms of d_1 and d_2

$$e_{AG} = d_1 \sin \alpha + d_2 \cos \alpha$$

From a knowledge of the load–extension characteristics of the members the strain energy of all the members can be found.

If the members are linearly elastic

$$U = \sum \frac{1}{2} \frac{AE}{l} e^2 = \sum \frac{1}{2} \frac{AE}{l} (d_1 \sin \alpha + d_2 \cos \alpha)^2$$

$\partial U/\partial d_1 = P_1$, $\partial U/\partial d_2 = P_2$, therefore

$$\sum \frac{AE}{l} (d_1 \sin^2 \alpha + d_2 \sin \alpha \cos \alpha) = P_1$$

$$\sum \frac{AE}{l} (d_1 \sin \alpha \cos \alpha + d_2 \cos^2 \alpha) = P_2$$

These equations are identical to the ones obtained previously by the virtual-work approach.

Problems

6.1 For the pin-jointed frame shown at P 6.1 the cross-section of all tension members is A and that of the compression members $2A$. The length of AB is l. Using the equation of real work show that the vertical deflection of B is approximately $1 \cdot 8 Pl/AE$. Find also the horizontal deflection of D.

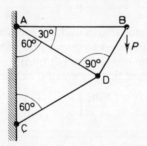

P 6.1

6.2 All the members of the mild steel pin-jointed framework P 6.2 have a length of 5 m and a cross-section of 5000 mm². Find the deflection of the joint E when a 40 kN load is applied as shown.

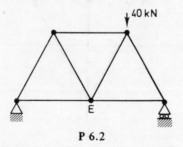

P 6.2

6.3 The pin-jointed truss at P 6.3 has all members equal in length and cross-section. Determine the reaction at C.

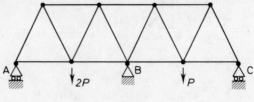

P 6.3

6.4 A thin uniform rod forms a circular arc subtending a right angle at its centre O (P 6.4). The rod is rigidly fixed at A, and is loaded at B by a force P acting in the plane AOB and making an angle ϕ with BO. Find the value of ϕ which will make the deflection of B in the direction of P a maximum.

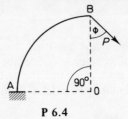

P 6.4

6.5 A ring of radius R is made from a uniform rod whose diameter is small compared with R. The ring is subjected to three equally spaced forces P whose line of action lie in the plane of the ring and pass through its centre. Show that if only bending effects are considered that the greatest bending moment in the ring is

$$\frac{PR}{2}\left(\frac{3}{\pi} - \frac{1}{\sqrt{3}}\right)$$

6.6 The pin-jointed framework P 6.6 has member BD vertical. AD and CD are the same length and inclined to BD at 30°. The relation between the extension and the load in members AD and CD is $e = aF$ and for BD $e = bF^2$, where a and b are constants and $a/b = 3/2$. Find the force in BD when a vertical load P is applied at D.

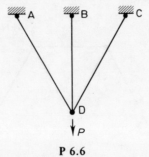

P 6.6

6.7 A circular hoop, made of material whose weight per unit length is w, is suspended from the end of a vertical wire. The dimensions of the cross-section are small compared with the diameter d of the hoop. Find the bending moment at the point of suspension.

6.8 At P 6.8 is shown a structure in which two bars AD and EG are pinned to rigid supports at A, D, E and G. A wire EBFCG is fixed at E and passes over frictionless pulleys at B, F and C to G where it is again anchored. The wire is 10 mm too short before being fitted. Find the tension in the wire when a vertical load of 100 kN acts at F, also find the deflection of F. The second moment of area of the bars is 4000 cm^4, the cross-sectional area of the wire is 60 mm^2, the length l is 1 m. $E = 210$ kN/mm^2.

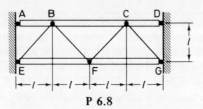

P 6.8

6.9 The plane frame shown at P 6.9 consists of eight identical members pinned to each other. Calculate the forces in the members AB and BC due to the load P by (a) considering compatibility of displacements between C and B, and B and A, and (b) by the use of an energy theorem.

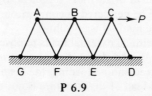

P 6.9

6.10 An open coiled spring with pitch angle α has N complete turns of mean radius R. The wire is of diameter d, where $d \ll R$, Young's modulus E, modulus of rigidity G. Derive an expression for the twist induced by a torque T if the spring is prevented from compressing or extending.

6.11 A portal frame has a beam of length l and vertical legs of length $l/2$ pinned at the feet. A point load P is applied at a point distant $l/3$ from one end of the beam. If EI is constant find the reactions at the feet.

6.12 A semi-circular slender arch of radius R and uniform flexural rigidity EI has fixed supports and carries a vertical load P at its highest point. Calculate the reactions at the supports. Determine the deflection corresponding to P.

7 MOMENT DISTRIBUTION AND SLOPE DEFLECTION

Both moment-distribution and slope-deflection methods can be applied to similar types of problems, namely continuous beams and rigid jointed frameworks. Moment distribution has an arithmetical solution in which a number of successive corrections are applied to an initial set of assumed moments. In the slope-deflection method, equations are written down for the equilibrium of all the joints in the structure; when these equations are solved simultaneously, the moments acting at all the joints can be found.

7.1 Moment distribution

The moment-distribution method was first introduced by Hardy Cross in 1930; it is essentially a displacement or stiffness approach. Deformations due to axial loads and shear forces will be neglected in comparison with those due to bending. The general concept of the method is as follows.

Imagine a framework with all external loads removed and clamps available at all joints so that any joint can be clamped to prevent rotation or left free to rotate. With the clamps fixed and no loading, the bending moments at the joints will all be zero if the dead load of the structure is not considered. If the external loading is now applied to the framework, moments will be developed at each end of the members that are loaded. It is a fairly straighforward job to calculate the values of these moments. The clamp at one joint is now released. If the sum of the moments acting at the joint is not zero, there will be an out of balance moment. This will cause the joint to rotate until equilibrium has been attained.

The rotation will cause moments to be developed at the remote ends of all the members meeting at the joint. This joint is then clamped, and another joint released. The process is repeated for all the joints in the structure and for several cycles until every joint is very nearly in equilibrium and the clamps may be left in the released positions.

When a joint is out of balance we shall need to know what proportion of the out of balance moment is carried by the individual members meeting at the joint. If the joint rotates through an angle θ to reach the equilibrium position, the moments will be distributed among the members according to their stiffnesses. The moment that is developed at the far end of a member due to a rotation of the near end, will be some proportion of the moment at the near end. The ratio of the two moments is called the *carry-over factor*.

218

7.2 Sign convention

Traditionally a different sign convention has been used for moment distribution. Terminal moments will be considered as the moments which act on the ends of the members and not those acting on the joint. All quantities will be considered positive if they are clockwise, whether they be moments, rotations, or the displacement of one end of a member relative to the other.

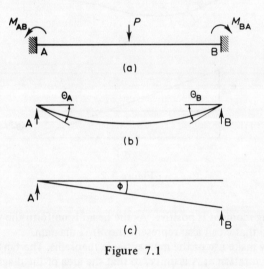

(a)

(b)

(c)

Figure 7.1

Figure 7.1a shows a beam that is built in at both ends. M_{BA} is positive while M_{AB} is negative. In 7.1b rotations have taken place at ends A and B. θ_A is positive and θ_B is negative. A positive displacement is shown at c when the right-hand support has moved downwards relative to the left-hand support.

7.3 Stiffness and carry-over factor

The rotational stiffness or the moment per unit rotation of a member that is pinned at end A and fixed at end B will be written as s_{AB}. Thus s_{AB} is the moment required at end A to produce a unit rotation there. If the member has a uniform cross-section or a cross-section that has a symmetrical variation, then

$$s_{AB} = s_{BA}.$$

When a moment is applied at the pinned end A, a moment will be developed at the fixed end B. c_{AB} is defined as the carry-over factor, that is, the proportion of the moment at end A that is developed at end B. Thus for unit rotation at end A the moment required there will be s_{AB} and the moment developed at end B will be $c_{AB}s_{AB}$.

Figure 7.2a shows a straight uniform member of length l pinned at end A and fixed at end B. A moment s_{AB} is applied at end A, which will cause the member to deflect (b) and a moment $c_{AB}s_{AB}$ is developed at end B. The bending-moment diagram is shown in c, the sign convention adopted for this diagram being that

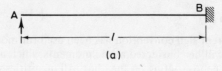

(a)

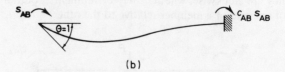

(b)

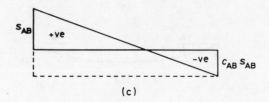

(c)

Figure 7.2

sagging bending moment is positive. As the beam is uniform the value of EI will be constant, so that c can also represent the M/EI diagram.

We can now make use of the moment-area theorems. The tangent at B is horizontal and the rotation at A is unity, so that the area of the diagram between A and B is unity.

$$\frac{1}{EI}\left[\frac{l}{2}(s_{AB} + c_{AB}s_{AB}) - c_{AB}s_{AB}l\right] = 1 \qquad (7.1)$$

The deflection of end A relative to the tangent at B is zero.

$$\frac{1}{EI}\left[\frac{l}{2}(s_{AB} + c_{AB}s_{AB})\frac{l}{3} - c_{AB}s_{AB}l \times \frac{l}{2}\right] = 0 \qquad (7.2)$$

Hence $(1 + c_{AB})/3 = c_{AB}$ or $c_{AB} = 1/2$. Substituting for c_{AB} in equation 7.1.

$$\tfrac{1}{2}(\tfrac{3}{2}s_{AB}) - \tfrac{1}{2}s_{AB} = \frac{EI}{l}$$

or

$$s_{AB} = \frac{4EI}{l} \qquad (7.3)$$

So that the stiffness of a uniform member is $4EI/l$ and the carry-over factor is $1/2$. For non-uniform members it is quite possible to use a similar approach to find the values of stiffness and carry-over factor.

7.4 Distibution factor

Figure 7.3 shows four members that meet at joint A, the far ends of all the members being fixed. A positive moment is applied at joint A and we need to know what proportion of M is carried by each member. If the members are rigidly jointed together at A there will be no relative rotation between the members at the joint. Under the action of M the joint A will rotate through an angle θ say.

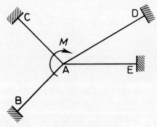

Figure 7.3

The moment developed by member AB will be $s_{AB}\theta$
the moment developed by member AC will be $s_{AC}\theta$, etc.

so that

$$(s_{AB} + s_{AC} + s_{AD} + s_{AE})\theta = M$$

and

$$\frac{M_{AB}}{M} = \frac{s_{AB}}{\Sigma s} = \frac{4E_{AB}I_{AB}/l_{AB}}{\Sigma 4EI/l}$$

This quantity is defined as the distribution factor for member AB.

In general E will be constant for all the members meeting at a joint, therefore

$$\text{D.F.}_{AB} = \frac{I_{AB}/l_{AB}}{\Sigma I/l} \tag{7.4}$$

7.5 Fixed-end moments

These are the moments that will be initially developed at the ends of the members when the external loading is applied and all joints are in the clamped position.

An arbitrary loading system is applied to member AB with both ends fixed (figure 7.4a). The bending-moment diagram is shown at b; this is first drawn for the beam in the simply supported condition and the negative portion is added for the fixed-end moments.

Let the area of the bending-moment diagram for the simply supported beam be A_1. x_A is the distance from end A to the centroid G of the diagram and x_B the distance from end B.

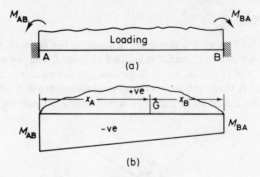

Figure 7.4

The moment of the M/EI diagram about A is zero, and assuming EI is constant

$$Ax_A = M_{BA} \frac{l^2}{2} + (M_{AB} - M_{BA}) \frac{l^2}{6}$$

or

$$\frac{6Ax_A}{l^2} = 2M_{BA} + M_{AB} \qquad (7.5)$$

Similarly the moment of the M/EI diagram about B is zero, hence

$$\frac{6Ax_B}{l^2} = 2M_{AB} + M_{BA} \qquad (7.6)$$

Solving equations 7.5 and 7.6

$$M_{AB} = \frac{2A}{l^2}(2x_B - x_A) \quad M_{BA} = \frac{2A}{l^2}(2x_A - x_B)$$

These two expressions will give the value of the fixed-end moments for any loading on a straight uniform beam. There are however several standard cases that are often required and these will be derived below. It is important to note that if both of the values of M_{AB} and M_{BA} are positive when derived from the above expressions, the implication is that M_{AB} is anti-clockwise and M_{BA} is clockwise. This would mean that M_{AB} would have a negative value when the usual moment-distribution sign convention is used.

Consider a built-in member with a uniformly distributed load p (figure 7.5a). The bending-moment diagram for a simply supported beam would be parabolic with a maximum height at the centre of $pl^2/8$.

$$A = \frac{2}{3} \frac{pl^3}{8} \qquad x_A = x_B = \frac{l}{2}$$

therefore

$$M_{AB} = M_{BA} = \frac{pl^2}{12}$$

Note for moment-distribution sign convention

$$M_{AB} = -\frac{pl^2}{12} \quad M_{BA} = \frac{pl^2}{12} \tag{7.7}$$

For the same member with a concentrated load (figure 7.5b), the bending-moment diagram would be triangular with a maximum height of Pab/l under the load.

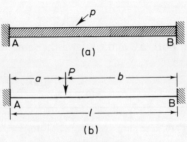

(a)

(b)

Figure 7.5

$A = Pab/2$, $x_A = (2a + b)/3$; $x_B = (a + 2b)/3$, thus

$$2x_A - x_B = a \quad 2x_B - x_A = b$$

Hence

$$M_{AB} = -\frac{Pab^2}{l^2}; \quad M_{BA} = \frac{Pa^2b}{l^2} \tag{7.8}$$

adopting the new sign convention.

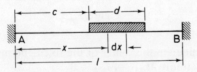

Figure 7.6

The expressions for the fixed-end moments due to a concentrated load are extremely useful for deriving further fixed-end moments in cases where the loading system can be represented as a function of x, the distance along the beam from a support. As an example consider the case of a short uniformly distributed load (figure 7.6). For an element of the load of length dx

$$M_{AB} = -p \, dx \, x \, \frac{(l - x)^2}{l^2}$$

So that for the total load

$$M_{AB} = -\int_c^{d+c} px \, \frac{(l - x)^2}{l^2} \, dx$$

and

$$M_{BA} = \int_{c}^{d+c} px^2 \frac{(l-x)}{l^2}\, dx \qquad (7.9)$$

These expressions can be evaluated quite easily for a specific case.

7.6 Examples

As a simple example, the bending moments at all the supports are required for the beam in figure 7.7a. Both AC and CB have the same values of *EI*.

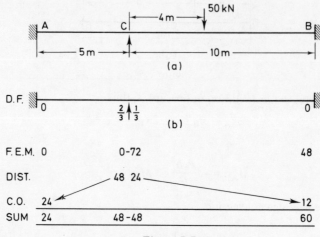

Figure 7.7

It is necessary to calculate the values of the fixed-end moments. AC is unloaded, therefore $M_{AC} = M_{CA} = 0$

$$M_{CB} = \frac{-50 \times 4 \times 6^2}{10^2} = -72 \text{ kNm}$$

$$M_{BC} = \frac{50 \times 4^2 \times 6}{10^2} = 48 \text{ kNm}$$

Next the distribution factors at joint C have to be found. Stiffness of CA = $4EI/5$ and of CB = $4EI/10$, therefore

$$\text{D.F. for CA} = \frac{1/5}{1/5 + 1/10} = \frac{2}{3}$$

and D.F. for CB = $1/3$. The values of the distribution factors have been entered on a line drawing of the beam in figure 7.7b. Next the values of the fixed-end moments are entered at their appropriate points underneath the beam. It is now possible to start the distribution process. As the joints A and B are rigid they will never be unlocked. Examining joint C it can be seen that if it was unlocked it would rotate, since the sum of the moments meeting there is not zero. On unlocking this joint it will rotate until the sum of its moments is zero, i.e. a moment of 72 kNm will have to be distributed at the joint, in the ratio 2/3 to CA and 1/3 to CB, that is, 48 kNm to CA and 24 kNm to CB. These values are entered on the diagram. The distributed moments will cause further moments to be developed at the far ends of the members, the carry-over factor in each case being 1/2. Thus the carry-over moment at A is 24 kNm and at B is 12 kNm. These values are entered on the diagram.

The distribution is now complete for this simple case, since the moments balance at the intermediate support. The final moments are found by summing the values in the diagram, thus

$$M_{AC} = 24 \text{ kNm} \qquad M_{CA} = 48 \text{ kNm}$$

$$M_{CB} = -48 \text{ kNm} \qquad M_{BC} = 60 \text{ kNm}$$

Next consider the same example as before but with a pin joint at B (figure 7.8). Assuming for the time being that joint B is fixed, the fixed-end moments will be the same as before. The distribution factors are the same except for joint B. The new factor is 1, whereas the old one was 0. The reason for having a D.F. of 1 is that each time joint B is unlocked a balancing moment equal and opposite to the out of balance moment has to be applied.

Starting the distribution at joint B, obviously the moment of 48 kNm is incorrect, since the true support there should be pinned, and the resultant moment should be zero. This is achieved by adding -48 kNm at B and will cause a carry-

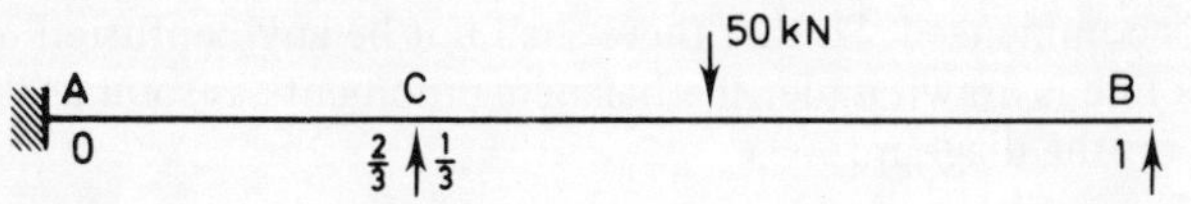

	A	C		B
	0	⅔ ↑ ⅓		1 ↑
F.E.M.	0	0	-72	48
DIST.	0	0	0	-48
C.O.	0	0	-24	0
DIST.	0	64	32	0
C.O.	32	0	0	16
DIST.	0	0	0	-16
C.O.	0	0	-8	0
DIST.	0	5·3	2·7	0
C.O.	2·6	0	0	1·4
DIST.	0	0	0	-1·4
C.O.	0	0	-0·7	0
DIST.	0	0·4	0·2	0
C.O.	0·2	0	0	0·1
DIST.	0	0	0	-0·1
SUM	34·8	69·7	-69·8	0

50 kN

Figure 7.8

over moment of -24 kNm at C. The joint B is now clamped and the one at C unclamped. The total out of balance moment is 96 kNm and this is distributed as 64 kNm to CA and 32 kNm to CB. These moments will in turn cause carry-over moments of 32 and 16 kNm at A and B. Thus joint B is again out of balance, and the distribution is continued until the residuals are small enough to be neglected. The final moments are $M_{AC} = 34 \cdot 8$ kNm, $M_{CA} = 69 \cdot 7$ kNm; $M_{CB} = -69 \cdot 8$ kNm; and $M_{BC} = 0$.

It can be seen that the convergence in this solution has been very slow. If a pinned end has to be treated in this manner the working becomes tedious. Fortunately it is possible to modify the stiffness factor and obtain a rapid convergence. However before considering this, an alternative method of setting out the calculation is shown in figure 7.9.

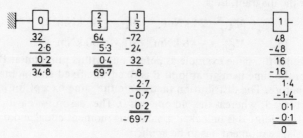

Figure 7.9

The distribution factors are entered in boxes on a line drawing of the beam. The fixed-end moments are written down in their respective places and distribution is commenced. So that there shall not be any confusion over which figures to use, a line is drawn under the balancing moments as soon as they have been entered on the diagram.

7.7 Modified stiffness

Pinned end

The member AB in figure 7.10 is pinned at end B. Consider the following sequence of operations. A unit rotation is applied to A with B clamped. A is then clamped and B released. The moments resulting from this operation are shown on the diagram. Note that in the final state end A has a unit rotation applied and end B has no moment.

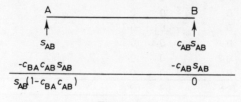

Figure 7.10

Thus if the far end of a member is pinned, the stiffness becomes $s_{AB}(1 - c_{BA}c_{AB})$.

For a uniform member $s_{AB} = s_{BA} = 4EI/l$, $c_{AB} = c_{BA} = 1/2$. Hence the modified stiffness is $3EI/l$.

An alternative method of stating this is to say that the stiffness is reduced to a factor of 3/4 of the normal stiffness.

Symmetry

The member AB (figure 7.11a), is symmetrically loaded. This implies that if the member is uniform then the final moments at each end of the member must be the same but of opposite sign. Thus $\theta_A = -\theta_B$.

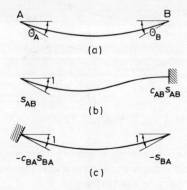

Figure 7.11

Consider the following sequence of operations. Fix end B and apply a unit rotation at end A, and the moments shown at b will result. Fix end A and apply a unit negative rotation at end B, and the moments shown at c will result. Sum the moments that are developed. Note that in the final state end A has a unit positive rotation and end B has a unit negative rotation, thus the total moment must be equal to the stiffness of the member.

$$M_{AB} = s_{AB} - c_{BA}s_{BA}$$
$$M_{BA} = c_{AB}s_{AB} - s_{BA}$$

Again for a uniform member $s_{AB} = s_{BA} = 4EI/l$, $c_{AB} = c_{BA} = 1/2$. Hence

$$M_{AB} = \frac{2EI}{l}$$

and

$$M_{BA} = -\frac{2EI}{l}$$

Thus the modified stiffness is $2EI/l$, and is reduced to a factor of half the normal stiffness.

Antisymmetry

For this type of loading $M_{AB} = M_{BA}$ and for the uniform beam $\theta_A = \theta_B$.

Apply a unit positive rotation at end A with end B fixed (figure 7.12b). Next apply a unit positive rotation to end B with end A fixed (c). The moments for each

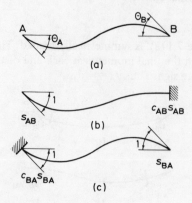

Figure 7.12

case are shown in the diagram and these values can be summed at each joint. Note that in the final state that both A and B have a unit positive rotation, thus the final moments must be equal to the stiffness.

$$M_{AB} = s_{AB} + c_{BA}s_{BA}$$

$$M_{BA} = c_{AB}s_{AB} + s_{BA}$$

Substituting the usual values for s and c will give

$$M_{AB} = M_{BA} = \frac{6EI}{l}$$

Thus the modified stiffness is $6EI/l$, and the stiffness has increased to a factor of 3/2 of the normal stiffness.

Let us reconsider the problem shown in figure 7.8 and this time make use of a modified stiffness to take account of the pin end at B. The values of the fixed-end moments will remain unaltered. There will however be new distribution factors at C. Stiffness of CA = $4EI/5$ and of CB = $3EI/10$, therefore

$$\text{D.F. for CA} = \frac{4/5}{4/5 + 3/10} = \frac{8}{11}$$

and the D.F. for CB is 3/11.

These values, and those of the fixed-end moments are entered on the line drawing (figure 7.13). The distribution should be started at joint B, the pinned end. To reduce the moment there to zero a moment of -48 kNm is required, which results in a carry-over moment of -24 kNm to CB. Once the moment at

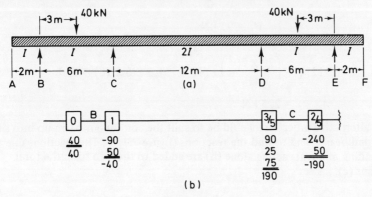

A	8l	C	3l		B
34·9	69·8	−72			48
34·9	69·8	−24			−48
		26·2			0
		−69·8			

Figure 7.13

the pinned end has been reduced to zero no further moments will be developed at this end if a modified stiffness has been used. In other words there is no carry-over moment from CB to BC when joint C is balanced. To complete the analysis of this problem only one further distribution is required, namely at joint C. This will result in a carry-over moment to A but not to B. The distribution is shown completed in figure 7.13. This agrees very well with the previous result. A comparison of the arithmetic required in the two distributions (figures 7.9 and 7.13), shows clearly the advantage of using a modified stiffness.

Figure 7.14

The beam in figure 7.14a carries a uniformly distributed load of 20 kN/m in addition to the two point loads. It can be seen at once that span CD is symmetrically loaded, and thus only half the problem need be treated if a modified value of stiffness is used for CD. In addition to this the beam is pinned at B and E (even though there is an overhang), so a reduced stiffness of $3EI/l$ can be used at C for CB.

The fixed-end moments are first calculated.

$$M_{BA} = \frac{20 \times 2^2}{2} = 40 \text{ kNm}$$

$$M_{BC} = -\left(\frac{20 \times 6^2}{12} + \frac{40 \times 3^3}{6^2}\right) = -90 \text{ kNm}$$

$$M_{CB} = 90 \text{ kNm}$$

$$M_{CD} = -\frac{20 \times 12^2}{12} = -240 \text{ kNm}$$

Distribution factors

$$CB = \frac{3I/6}{3I/6 + 2 \times 2I/12} = \frac{3}{5} \quad CD = \frac{2}{5}$$

The distribution factor for BA must be zero since the end moment BA is 40 kNm and must remain at this value during the distribution.

The distribution is shown in figure 7.14b and is started at joint B. The final moment M_{BA} is 40 kNm, thus the final value of M_{BC} is -40 kNm. To achieve this balance a moment of 50 kNm has to be added to BC, and this causes a carry-over moment of 25 kNm at CB. The joint at B is completely balanced and it is only necessary to perform one further distribution at C. The reactions can be found by taking moments about suitable points.

ΣM_C

$$6V_B = \left(\frac{20 \times 8^2}{2}\right) + 120 - 190$$

$$V_B = 95 \text{ kN}$$

ΣM_D

$$12V_C = \left(\frac{20 \times 20^2}{2}\right) + (40 \times 15) - (95 \times 18) - 190$$

$$V_C = 225 \text{ kN}$$

An alternative procedure would be to split the continuous beam up into simply supported beams, and to find the reactions (figure 7.15a). The reactions due to the bending moments acting alone (b) are added to these to form the total reactions (c).

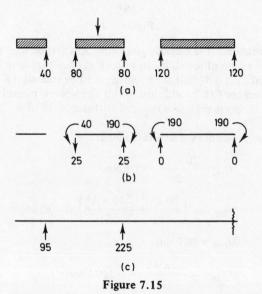

Figure 7.15

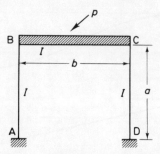

Figure 7.16

As a further example, the bending moments are required at the joints and supports for the portal frame in figure 7.16.

This will turn out to be a very simple calculation if a moment-distribution method is used.

F.E.M.

$$M_{BC} = \frac{-pb^2}{12} \qquad M_{CB} = \frac{pb^2}{12}$$

The member BC is symmetrically loaded.

D.F.

$$BA = \frac{4EI/a}{4EI/a + 2EI/b} = \frac{2b}{a + 2b}$$

$$BC = \frac{a}{a + 2b}$$

Distributing at joint B

$$M_{BA} = \frac{2b}{a + 2b}\left(\frac{pb^2}{12}\right)$$

$$M_{AB} = \frac{b}{a + 2b}\left(\frac{pb^2}{12}\right)$$

The moments at C and D can be written down straight away from these results.

7.8 Deflection of supports and sidesway

The beam problems that have been treated so far have had rigid supports. The question naturally arises as to how the moment-distribution method could be modified to treat problems in which one or more of the supports sinks.

Figure 7.17a shows a member AB that is rigidly built in at each end, and the end B has moved vertically through a distance Δ with respect to A. This is a positive displacement according to the sign convention in section 7.2.

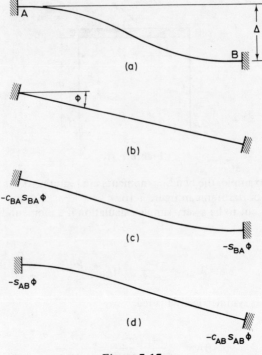

Figure 7.17

Imagine the beam remains straight with the same displacement Δ applied (b). The chord rotates through an angle $\phi = \Delta/l$, and each support rotates clockwise through the same angle such that the bending moment at any point on the beam is zero. To get back to the original configuration as shown at a it is necessary to rotate each end through an angle $-\phi$.

Figures 7.17c and d show the moments developed at each end of the beam when ends B and A are each rotated through an angle of $-\phi$.

Summing the moments will give the total effect due to the rotation of both ends. Thus

$$M_{AB} = -(s_{AB} + c_{BA}s_{BA})\phi$$

$$M_{BA} = -(s_{BA} + c_{AB}s_{AB})\phi$$

With the usual substitution for s and c in the case of a uniform beam,

$$M_{AB} = M_{BA} = -\frac{6EI\phi}{l} \quad \text{or} \quad -\frac{6EI\Delta}{l^2} \qquad (7.10)$$

For the problem in figure 7.18a the deflection at support B is required such that the bending moment at B is zero. $E = 210\ \text{kN/mm}^2$, $I = 41\ 000\ \text{cm}^4$.

It will first of all be necessary to find the bending moment at B without any deflection of the support.

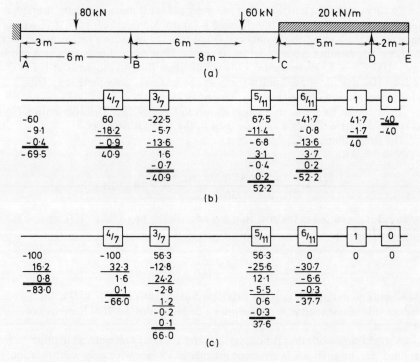

Figure 7.18

Fixed-end moments

$$M_{AB} = -\frac{80 \times 3^3}{6^2} = -60 \text{ kNm} \qquad M_{BA} = 60 \text{ kNm}$$

$$M_{BC} = -\frac{60 \times 6 \times 2^2}{8^2} = -22 \cdot 5 \text{ kNm} \qquad M_{CB} = \frac{60 \times 6^2 \times 2}{8^2} = 67 \cdot 5 \text{ kNm}$$

$$M_{CD} = -\frac{20 \times 5^2}{12} = -41 \cdot 7 \text{ kNm} \qquad M_{DC} = 41 \cdot 7 \text{ kNm} \qquad M_{DE} = -40 \text{ kNm}$$

Distribution factors
Joint B

$$BA = \frac{1/6}{1/6 + 1/8} = \frac{4}{7} \quad BC = \frac{3}{7}$$

Joint C (D is a pin end)

$$CB = \frac{1/8}{1/8 + 3/4 \times 1/5} = \frac{5}{11} \quad CD = \frac{6}{11}$$

The distribution has been carried out in figure 7.18b.

The bending moment at B is 40·9 kNm and this has to be reduced to zero by lowering the support at B.

Consider all the external loading removed and let B move vertically through a distance Δ with all joints in the clamped position. Moments of $-6EI\Delta/l^2$ will be induced at the ends of the members AB and BC. Let us assume (incorrectly) that $6EI\Delta/l^2$ for member AB is equal to 100 units. Thus the end moments at A and B are -100. Since EI is constant for the whole beam the end moments at B and C will be $100 \times 6^2/8^2 = 56\cdot3$ units; these will be positive, since there is an anticlockwise rotation of the chord.

These moments are distributed as shown in figure 7.18c. The distribution factors are of course the same as those used in the first distribution.

The final moment $M_{BC} = 66$ units.

Now for AB

$$\frac{6EI\Delta}{l^2} = \frac{6 \times 210 \times 41\,000 \times 10^{-2}\Delta}{36} = 14\,350\Delta \text{ kNm}$$

where Δ is measured in metres. This was assumed to be equal to 100 units. Thus the true final moment M_{BA} after distribution is

$$-14\,350\Delta \times \frac{66}{100} = -9470\Delta$$

Now with no deflection at B the first distribution gives $M_{BA} = 40\cdot9$ kNm. To reduce this moment to zero will require a deflection of $40\cdot9/9470$ metres or $4\cdot32$ mm.

A rigid framework that is symmetrical and that has symmetrical loading applied, can be analysed in the same manner as a beam with only joint rotation and no displacement taking place; this was illustrated by the example on the portal frame in section 7.7. If however the system is asymmetrical due either to the layout of the members or to the loading, then the joint rotation will be accompanied by translation, or *sidesway*. The joint translations are relatively small although they may have a considerable effect on the values of the bending moments. In the majority of cases there will not be sufficient change in the geometry of the structure for this to be taken into account in the analysis.

The portal frame in figure 7.19a has a lateral load applied. The deformation due to joint translation alone is shown greatly exaggerated in b. This will of course be accompanied by joint rotation. The two column members have deformed in a similar manner to the beam in figure 7.17a, so that an identical analysis can be used to find the resulting moments.

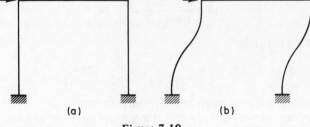

(a)   (b)

Figure 7.19

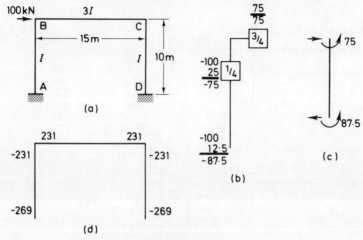

Figure 7.20

Consider first the numerical example in figure 7.20a. If sidesway is allowed to take place with the joints clamped against rotation

$$M_{AB} = M_{BA} = M_{CD} = M_{DC} = -\frac{6EI\Delta}{10^2}$$

Let this be equal to -100 kNm, say. Δ is the horizontal displacement of BC.

When the joints are unclamped to allow rotation to take place at B and C, the member BC will deform antisymmetrically, thus a modified distribution factor can be used in the moment distribution.

D.F.

$$BA = \frac{4I/10}{4I/10 + 6 \times 3I/15} = \frac{1}{4} \quad BC = \frac{3}{4}$$

The distribution has been carried out in figure 7.20b using the arbitrary sidesway values of -100 kNm.

Column AB is shown isolated in figure 7.20c with moments of -75 and -87.5 kNm applied at B and A respectively. To keep the column in equilibrium, shear forces of value $(75 + 87.5)/10 = 16.25$ kN are required. Thus to produce the final moments shown in b a lateral force to the right of $(2 \times 16.25) = 32.5$ kN (for two columns) would have to be applied at B.

In the original problem 100 kN is applied at B, thus the true resulting moments are obtained by multiplying the final moments in the sidesway distribution by a factor of $100/32.5$. The final values are shown at d in kNm units.

The same portal is shown in figure 7.21a but with a vertical load applied to the member BC.

In dealing with a problem of this type it will be necessary to perform two separate distributions. First it is assumed that sidesway is prevented, and a normal distribution is carried out. From the results, the force required to prevent side-

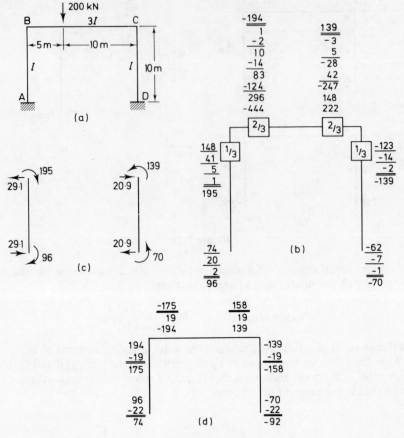

Figure 7.21

sway is found. An arbitrary sidesway is next applied with the original loading removed. After distribution the lateral force required to produce this sidesway can be found. From this result the correct sidesway can be determined to reduce the lateral load from the no-sidesway solution to zero. The true final moments will be the sum of the first distribution and a certain known fraction of the second distribution.

No-sidesway distribution

F.E.M.

$$M_{BC} = -444 \text{ kNm} \quad M_{CB} = 222 \text{ kNm}$$

D.F.

$$BC = CB = \tfrac{2}{3} \quad BA = CD = \tfrac{1}{3}$$

The distribution has been carried out in figure 7.21b. The two columns are shown isolated at c. The total shear is

$$\frac{195 + 96}{10} - \frac{139 + 70}{10} = 29 \cdot 1 - 20 \cdot 9 = 8 \cdot 2 \text{ kN}$$

Thus a force of $8 \cdot 2$ kN to the left is required at B to prevent sidesway.

There is of course no horizontally applied force. If we could find the moments that would result from the application of a force of $8 \cdot 2$ kN to the right at B with no other loads on the frame, we could add this solution to the no sidesway solution and we should have the correct result for the original problem. Unfortunately this cannot be solved directly and the following method is adopted.

For sidesway effects, arbitrary sway moments of -100 are chosen for the columns and the distribution will be identical to that shown in figure 7.20b. This results in a lateral force of $32 \cdot 4$ kN applied to the right at B. To produce a force of $8 \cdot 2$ kN the final moments in figure 7.20b must be multiplied by a factor of $8 \cdot 2/32 \cdot 5$.

The two sets of moments, with their summation are shown in figure 7.21d.

$$M_{AB} = 74 \text{ kNm} \quad M_{BA} = 175 \text{ kNm}$$

$$M_{CD} = -158 \text{ kNm} \quad M_{DC} = -92 \text{ kNm}$$

7.9 Frames with inclined members

For structures of this type the analysis can proceed on somewhat similar lines to the previous analysis. It is however best to determine the shear equation, which will be a relation between the final moments and the external loads. The method of approach will be discussed with reference to the problem in figure 7.22a, where the relative EI/l values of the members are shown in circles.

The simplest method of forming the shear equation is by means of a displacement diagram and virtual work. The method of instantaneous centres can be used instead of a displacement diagram. All joints are considered as pinned and a small clockwise rotation ϕ is applied to AB; the resulting displacement diagram is shown in figure 7.22b.

$$ob = AB \times \phi \quad ox = 9\phi \quad oc = ox \times \frac{CD}{12} = \tfrac{3}{4}\phi CD$$

Thus the rotation of CD is $\tfrac{3}{4}\phi$ clockwise.

$$bx = 3\phi \quad cx = \tfrac{3}{4}\phi \times 4 = 3\phi \quad bc = 6\phi$$

Hence the rotation of BC is ϕ anticlockwise. The horizontal movement of C is 9ϕ. The vertical movement of E is $4\phi - 3\phi = \phi$ upwards.

Making use of the virtual work equation.

$$M_{AB}\phi + M_{BA}\phi - M_{BC}\phi - M_{CB}\phi + M_{CD}\frac{3\phi}{4} + M_{DC}\frac{3\phi}{4} + 20 \times 9\phi - 90 \times \phi = 0$$

Now $M_{BA} + M_{BC} = 0$ and $M_{CB} + M_{CD} = 0$, and the shear equation will simplify to

$$4M_{AB} + 8M_{BA} + 7M_{CD} + 3M_{DC} = -360$$

A no-sidesway distribution is next carried out (figure 7.22c).

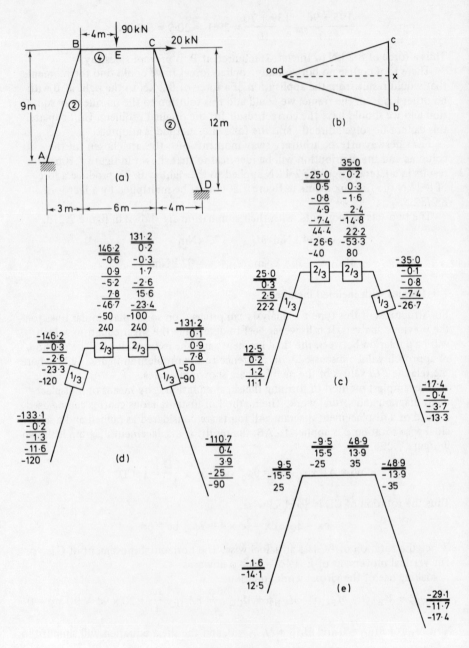

Figure 7.22

F.E.M.

$$M_{BC} = -40 \text{ kNm} \quad M_{CB} = 80 \text{ kNm}$$

D.F.

$$BA = CD = \tfrac{1}{3} \quad BC = CB = \tfrac{2}{3}$$

The moments from this distribution are substituted into the left-hand side of the shear equation. This gives a value of $-47 \cdot 2$ kNm, indicating that sidesway is present. For no sidesway the value would be -360 kNm.

Next a sidesway is applied such that the frame deforms to the right. In the previous analysis the rotations of members were as follows

$$AB = \phi \quad BC = -\phi \quad CD = \tfrac{3}{4}\phi$$

For these rotations the ratio of the sidesway moments $(-6EI\phi/l)$ can be found

$$M_{AB} : M_{BC} : M_{CD} = -6 \times 2\phi : 6 \times 4\phi : -6 \times 2 \times \frac{3\phi}{4}$$

that is

$$-12 : 24 : -9$$

These values multiplied by 10 have been taken as the arbitrary sway moments, and are distributed in figure 7.22d. The final moments from the distribution will be in the correct ratio for the required final sidesway moments, and can be substituted into the left-hand side of the shear equation.

$$-(133 \cdot 1 \times 4) - (146 \cdot 2 \times 8) - (131 \cdot 2 \times 7) - (110 \cdot 7 \times 3) = -2952$$

If k is the multiplying factor for the sidesway solution $-47 \cdot 2 - 2952k = -360$, whence k = 0·106.

The final moments will be given by the sum of the no sway moments and 0·106 times the sidesway moments. These are shown in figure 7.22e.

7.10 Rectangular multi-storey frames

A solution can be obtained for this type of frame using the procedure employed for single-storey structures; unfortunately the process is rather lengthy since a total of $n + 1$ separate distributions are required, where n is the number of storeys.

Basically the procedure is as follows

(1) Carry out distribution with sidesway prevented. Lateral forces will be required at each storey and these can be found from the resulting moments.

(2) Remove all external loading and apply an arbitrary sidesway to the top storey only. Distribute and find the loads at each storey that would cause this sidesway.

(3) Repeat process 2 for each storey in turn. It is then only necessary to combine the solutions in the correct proportion.

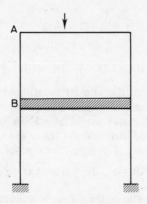

Figure 7.23

For the two-storey frame in figure 7.23. Let the lateral force at A be P_{A1} and B be P_{B1} from the no-sway distribution. Let the lateral force at A be P_{A2} and B be P_{B2} from the top-storey sidesway. Let the lateral force at A be P_{A3} and B be P_{B3} from the bottom-storey sidesway.

Since there are no lateral forces actually applied

$$P_{A1} + k_1 P_{A2} + k_2 P_{A3} = 0$$

and

$$P_{B1} + k_1 P_{B2} + k_2 P_{B3} = 0$$

The two equations can be solved for k_1 and k_2.

The final moments will be given by the sum of the moments from the no-sway solution and k_1 times the moments from the top-storey sidesway and k_2 times the moments for bottom-storey sidesway.

7.11 The slope–deflection equation

This is an alternative method to that of moment distribution, and will solve similar types of problems. Basically a number of simultaneous equations are formed with the unknowns taken as the angular rotations and displacements of each joint. Once these equations have been solved the moments at all the joints may be determined.

The same sign convention as that adopted for moment distribution will be used. It is perfectly possible to derive the slope–deflection equation from results already obtained from moment distribution, but the following derivation will make use of the moment-area theorems.

An arbitrary loading denoted by p is applied to the beam of length l, moments of value M_{AB} and M_{BA} act at the ends (figure 7.24a). It will be assumed that the flexural rigidity has a constant value of EI. The resulting bending-moment diagram

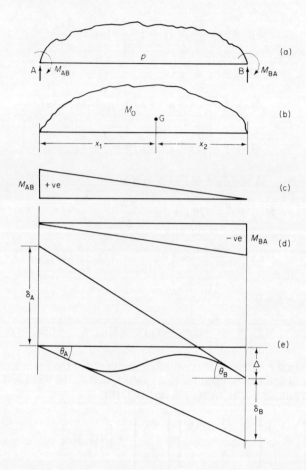

Figure 7.24

can be considered as three separate diagrams, figure 7.24b, to d. M_0 due to the applied loading assuming simple supports, M_{AB} the left-hand support moment, and M_{BA} the right-hand support moment. It will be noted that the sign convention, sagging bending moment positive, has been adopted for these diagrams. A is the area of the M_0 diagram and the position of the centroid has been defined by the distances x_1 and x_2 measured from the left and right-hand supports.

The deflected form of the beam is shown in figure 7.24e. The ends have rotated clockwise through θ_A and θ_B and the right-hand support has deflected through Δ vertically with reference to the left-hand support.

Making use of the second moment-area theorem and taking moments for the $M/(EI)$ diagram about A and then B results in the following equations.

$$-\delta_A = \tfrac{1}{6}\frac{l^2}{EI}M_{AB} - \tfrac{1}{3}\frac{l^2}{EI}M_{BA} + \frac{Ax_1}{EI}$$

$$\delta_B = \tfrac{1}{3}\frac{l^2}{EI}M_{AB} - \tfrac{1}{6}\frac{l^2}{EI}M_{BA} + \frac{Ax_2}{EI}$$

therefore

$$M_{AB} = \frac{2EI}{l^2}(2\delta_B + \delta_A) + \frac{2A}{l^2}(2x_2 - x_1)$$

$$M_{BA} = \frac{2EI}{l^2}(2\delta_A + \delta_B) + \frac{2A}{l^2}(2x_1 - x_2)$$

Now $\delta_A = l\theta_B - \Delta$ and $\delta_B = l\theta_A - \Delta$ so that

$$M_{AB} = \frac{2EI}{l}\left(2\theta_A + \theta_B - \frac{3\Delta}{l}\right) - \frac{2A}{l^2}(2x_2 - x_1)$$

$$M_{BA} = \frac{2EI}{l}\left(2\theta_B + \theta_A - \frac{3\Delta}{l}\right) + \frac{2A}{l^2}(2x_1 - x_2)$$

If $\theta_A = \theta_B = \Delta = 0$

$$M_{AB} = -\frac{2A}{l^2}(2x_2 - x_1) \text{ and } M_{BA} = \frac{2A}{l^2}(2x_1 - x_2)$$

These are the moments that would be developed if the beam were built-in at both ends and the external loading applied, that is, the fixed-end moments.

The equations can be written in matrix form

$$\begin{bmatrix} M_{AB} \\ M_{BA} \end{bmatrix} = \frac{EI}{l}\begin{bmatrix} 4 & 2 & -6 \\ 2 & 4 & -6 \end{bmatrix}\begin{bmatrix} \theta_A \\ \theta_B \\ \dfrac{\Delta}{l} \end{bmatrix} + \begin{bmatrix} \text{F.E.M.} \end{bmatrix}$$

The signs of the fixed-end moments will be the same as those adopted for moment distribution in section 7.5.

As a simple example we will treat the first problem that was solved by moment distribution (figure 7.25).

It is a good idea to list any rotations and relative displacements of the ends that are zero. In this case $\theta_A = \theta_B = 0$ and all Δs are zero.

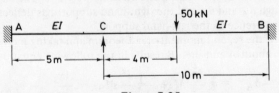

Figure 7.25

The fixed-end moments are next calculated.

$$M_{CB} = -72 \text{ kNm and } M_{BC} = 48 \text{ kNm}$$

The slope–deflection equations are applied to the ends of each member. Writing these in matrix form

$$\begin{bmatrix} M_{AC} \\ M_{CA} \end{bmatrix} = \frac{EI}{5} \begin{bmatrix} 4 & 2 & -6 \\ 2 & 4 & -6 \end{bmatrix} \begin{bmatrix} 0 \\ \theta_C \\ 0 \end{bmatrix} + \begin{bmatrix} 0 \\ 0 \end{bmatrix}$$

$$\begin{bmatrix} M_{CB} \\ M_{BC} \end{bmatrix} = \frac{EI}{10} \begin{bmatrix} 4 & 2 & -6 \\ 2 & 4 & -6 \end{bmatrix} \begin{bmatrix} \theta_C \\ 0 \\ 0 \end{bmatrix} + \begin{bmatrix} -72 \\ 48 \end{bmatrix}$$

We now look for further information that will help to solve the problem. In this case the sum of the moments at the ends of the two members meeting at C must be zero because there is no externally applied bending moment at C.

$$M_{CA} + M_{CB} = 0$$

$$k(4\theta_C) + \frac{k}{2}(4\theta_C) - 72 = 0$$

where $k = EI/5$, therefore

$$k\theta_C = 12$$

This value can now be substituted into the original equations and the values of the moments at the ends of each member may be determined: $M_{AC} = 24$ kNm, $M_{CA} = 48$ kNm, $M_{BC} = 60$ kNm.

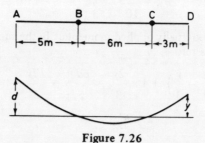

Figure 7.26

A further example is shown in figure 7.26 where the beam is continuous and is supported by pins at B and C, the ends A and D being free. The movement at D is required when A is raised by d.

$$\Delta_{BC} = 0; \quad \Delta_{AB} = d, \quad \theta_C = -y/3 \text{ since CD remains straight.}$$

In addition all the fixed-end moments are zero.

$$\begin{bmatrix} M_{AB} \\ M_{BA} \end{bmatrix} = \frac{EI}{5} \begin{bmatrix} 4 & 2 & -6 \\ 2 & 4 & -6 \end{bmatrix} \begin{bmatrix} \theta_A \\ \theta_B \\ \dfrac{d}{5} \end{bmatrix}$$

$$\begin{bmatrix} M_{BC} \\ M_{CB} \end{bmatrix} = \frac{EI}{6} \begin{bmatrix} 4 & 2 & -6 \\ 2 & 4 & -6 \end{bmatrix} \begin{bmatrix} \theta_B \\ \theta_C \\ 0 \end{bmatrix}$$

Now $M_{CD} = 0$ and since there is no externally applied moment at C, this means that M_{CB} must also be zero. From the last equation substituting for θ_C we find that

$$\theta_B = \frac{2y}{3}$$

and so

$$M_{BC} = \frac{EI}{6} \left(4 \times \frac{2y}{3} - \frac{2y}{3} \right) = \frac{EI_y}{3}$$

Now the bending moment at A is zero, so

$$M_{AB} = \frac{EI}{5} \left(4\theta_A + 2 \times \frac{2y}{3} - \frac{6d}{5} \right) = 0$$

hence

$$\theta_A = \frac{3d}{10} - \frac{y}{3}$$

Because there is no externally applied moment at B, the sum of the end moments at this point is zero.

$$\frac{EI}{5} \left\{ 2 \left(\frac{3d}{10} - \frac{y}{3} \right) + 4 \times \frac{2y}{3} - \frac{6d}{5} \right\} + \frac{EIy}{3} = 0$$

$$y = \frac{9d}{55}$$

As an example of a framework with sidesway present we shall solve the problem shown in figure 7.27 using the slope–deflection method.

It will be seen that there is no rotation at end A, so that $\theta_A = 0$. The sidesway at B must be the same as that at C, $\Delta_{BA} = \Delta_{CD} = \Delta$, say. There is no sidesway for

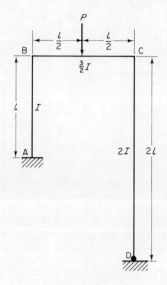

Figure 7.27

BC. The fixed-end moments are $-Pl/8$ at B and $Pl/8$ at C.
The slope–deflection relations may be written down as follows.

$$
\begin{bmatrix} M_{AB} \\ M_{BA} \end{bmatrix} = \frac{EI}{l} \begin{bmatrix} 4 & 2 & -6 \\ 2 & 4 & -6 \end{bmatrix} \begin{bmatrix} 0 \\ \theta_B \\ \dfrac{\Delta}{l} \end{bmatrix} + \begin{bmatrix} 0 \end{bmatrix}
$$

$$
\begin{bmatrix} M_{BC} \\ M_{CB} \end{bmatrix} = \frac{3EI}{2l} \begin{bmatrix} 4 & 2 & -6 \\ 2 & 4 & -6 \end{bmatrix} \begin{bmatrix} \theta_B \\ \theta_C \\ 0 \end{bmatrix} + \begin{bmatrix} -\dfrac{Pl}{8} \\ \dfrac{Pl}{8} \end{bmatrix}
$$

$$
\begin{bmatrix} M_{CD} \\ M_{DC} \end{bmatrix} = \frac{2EI}{2l} \begin{bmatrix} 4 & 2 & -6 \\ 2 & 4 & -6 \end{bmatrix} \begin{bmatrix} \theta_C \\ \theta_D \\ \dfrac{\Delta}{2l} \end{bmatrix} + \begin{bmatrix} 0 \end{bmatrix}
$$

There is a total of four unknown rotations or displacements. Three equations of equilibrium can be obtained, $\Sigma M_B = 0$, $\Sigma M_C = 0$ and $M_D = 0$: three equations relating four unknowns. It is essential that a further equation be found. This is obtained by considering the shear in members AB and CD. Since there is no horizontally applied force the sum of the shears in these members must be zero.

$$\frac{M_{AB} + M_{BA}}{l} + \frac{M_{CD} + M_{DC}}{2l} = 0$$

The four equations may be written down as follows.

$$
\begin{bmatrix}
\Sigma M_B = 0 \\[2mm]
\Sigma M_C = 0 \\[2mm]
M_D = 0 \\[2mm]
\text{Shear} = 0
\end{bmatrix}
= \frac{EI}{l}
\begin{bmatrix}
10 & 3 & 0 & -\dfrac{6}{l} \\[2mm]
3 & 10 & 2 & -\dfrac{3}{l} \\[2mm]
0 & 2 & 4 & -\dfrac{3}{l} \\[2mm]
6 & 3 & 3 & -\dfrac{15}{l}
\end{bmatrix}
\begin{bmatrix}
\theta_B \\[2mm]
\theta_C \\[2mm]
\theta_D \\[2mm]
\Delta
\end{bmatrix}
+
\begin{bmatrix}
-\dfrac{Pl}{8} \\[2mm]
\dfrac{Pl}{8} \\[2mm]
0 \\[2mm]
0
\end{bmatrix}
$$

Solving gives

$$\theta_B = 0.0239 \frac{Pl^2}{EI}, \ \theta_C = -0.0204 \frac{Pl^2}{EI}, \ \theta_D = 0.0168 \frac{Pl^2}{EI}$$

$$\Delta = 0.00886 \frac{Pl^3}{EI}$$

The values of the bending moments can be found by substituting into the original equations.

$$M_{AB} = -0.00536\,Pl, \ M_{BA} = 0.0424\,Pl, \ M_{CB} = 0.0743\,Pl$$

Problems

7.1 Determine the bending-moment diagram and the reactions for the continuous steel beam in P7.1. $I_{AB} = 8000 \text{ cm}^4$; $I_{BC} = 20\,200 \text{ cm}^4$; $I_{CD} = 11\,400 \text{ cm}^4$.

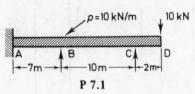

P 7.1

7.2 For the beam in P 7.1 how far would the support at B have to sink to make the bending moment there zero?

7.3 The beam and the columns in P 7.3 have the same cross-section. Find the bending moments at the joints.

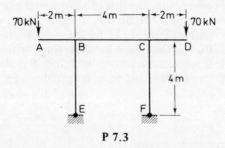

P 7.3

7.4 Find the force exerted on each of a long line of circular pegs by a long mild-steel strip laced through them. The pegs are 10 mm diameter and spaced at intervals of 240 mm with their centres lying on the same straight line. The strip is 20 mm wide and 1·25 mm thick.

7.5 The framework shown at P 7.5 has stiff joints but is pinned at D and H. In addition to the point loads which act at mid-span, the member BF carries a uniformly distributed load of p per unit length, and member CG, $2p$ per unit length, where $pl = P$. Find the moments at the ends of the various members of the framework.

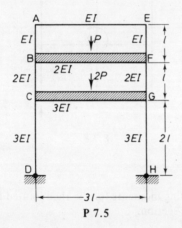

P 7.5

7.6 Find the bending moments for the rigid jointed frame P 7.6. All the members have the same cross-section.

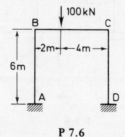

P 7.6

7.7 A single-storey two-bay frame consists of a horizontal beam ABC rigidly connected to vertical stanchions AD, BE, CF, of equal length, which are rigidly fixed at D, E, F. The relative values of EI/l for the members are given by AB = AD = CF = 2, BC = 1, BE = 3. Determine the shear force carried by each stanchion when the frame is subjected to a horizontal force P applied at A in the direction ABC. Discuss how you would solve the problem of a two-storey single-bay portal frame.

7.8 Sketch the framework shown at P 7.8 when it is deformed by the moment M. Make use of the slope–deflection equation to determine the angle of rotation of the centre joint. All members have the same length and the same stiffness K.

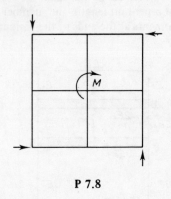

P 7.8

7.9 Determine the bending moments for the rigid jointed frame P 7.9. All members have the same cross-section.

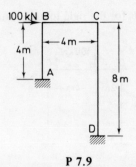

P 7.9

7.10 The mild-steel beam ABC in P 7.10 is encastré at C, supported on a simple support at B, and constrained at A. Vertical movement is prevented at A and the rotation at A is resisted by an elastic bending moment $M(\text{kNm}) = 2000\,\theta$, where θ is the rotation in radians. Determine the moments at A, B and C when the point load is applied at the centre of AB. $I = 2530\ \text{cm}^4$.

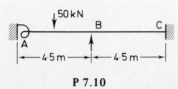

P 7.10

8 STIFFNESS AND FLEXIBILITY METHODS

8.1 Introduction

The basic ideas of the stiffness and flexibility approach to the solution of structural problems were introduced in chapter 6. We shall now proceed to develop these methods in more detail and apply them to specific problems. The solutions to the various problems will be set out in a manner that would be suitable for a computer application so that all the fundamental processes can be demonstrated. As the working will be done by hand it will be necessary to confine our attention to fairly simple structures otherwise the arithmetic will become impossibly tedious. Even so it will be found that the working in most cases will be lengthy and the reader will be able to suggest alternative methods of solution which would give a much more rapid solution. It will appear obvious that once a problem becomes more complex, then a great deal of time and trouble will be saved by making use of a computer together with a stiffness or flexibility approach.

8.2 Formulation of simple stiffness matrices

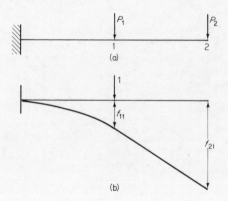

Figure 8.1

250

Consider the relatively straightforward problem shown in figure 8.1a. The deflections under the loads at points 1 and 2 are required. One method of approaching this problem would be to apply a unit load at point 1 and calculate the deflections that would result at points 1 and 2. Assuming that the material used is linear elastic, then these values can be scaled up by P_1 and will give the deflections due to P_1 applied alone. The exercise could be repeated for a unit load applied at 2 and the results scaled up by P_2. The addition of the two sets of results will give the total deflection at 1 and 2 when P_1 and P_2 are applied together. The individual deflections obtained when unit loads are applied are flexibility coefficients; they are illustrated in figure 8.1b. We have already shown in section 6.4 that $f_{12} = f_{21}$. The relation between the displacements and the applied loads can be written in matrix formulation as follows.

$$\begin{bmatrix} d_1 \\ d_2 \end{bmatrix} = \begin{bmatrix} f_{11} & f_{12} \\ f_{21} & f_{22} \end{bmatrix} \begin{bmatrix} P_1 \\ P_2 \end{bmatrix} \tag{8.1}$$

or in the shortened form

$$\mathbf{d} = \mathbf{FP} \tag{8.2}$$

where $\mathbf{d}$ and $\mathbf{P}$ are column vectors and $\mathbf{F}$ is the flexibility matrix.

The relation 8.2 can be rewritten

$$\mathbf{P} = \mathbf{F}^{-1}\mathbf{d}$$

It would, however, be more usual to write this relation as

$$\mathbf{P} = \mathbf{Kd} \tag{8.3}$$

where $\mathbf{K}$ is the stiffness matrix. It can be seen that the inverse of a flexibility matrix is a stiffness matrix, that is

$$\mathbf{K} = \mathbf{F}^{-1}$$

One method for finding the stiffness matrix for the cantilever problem would be to determine the flexibility matrix and then invert it. However it should be possible to determine the stiffness matrix from first principles. We note that in the case of a flexibility matrix a unit load is applied. Presumably a unit displacement has to be applied in order that stiffness coefficients can be determined.

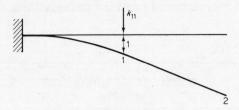

Figure 8.2

As a first attempt try the arrangement shown in figure 8.2. There is obviously something wrong. A unit displacement has been applied at 1 and the force required to produce this displacement has been labelled k_{11}. The question arises, where is

k_{21}? Let us leave this problem for a while and examine an even simpler problem, that of a spring. In this way we shall gain greater insight into the way we should determine a stiffness matrix.

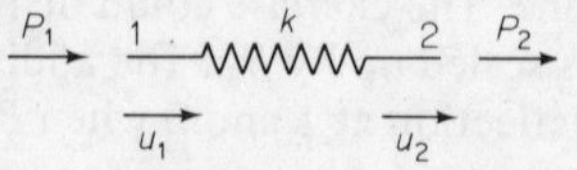

Figure 8.3

The spring in figure 8.3 has a stiffness k, that is to say, if one end of the spring is fixed and a load P is applied, the relation between the load P and the displacement d is given by $P = kd$. The spring as shown is a more general case, as displacements are applied to both the ends of the spring, values u_1 and u_2, the forces required to produce these displacements being P_1 and P_2. The problem can be broken down into two parts.

(1) Apply u_1 and make $u_2 = 0$; hence $P_1 = ku_1$ and $P_2 = -ku_1$.
(2) Apply u_2 and make $u_1 = 0$; hence $P_2 = ku_2$ and $P_1 = -ku_2$.

If u_1 and u_2 are applied together the results can be summed

$$P_1 = ku_1 - ku_2$$
$$P_2 = -ku_1 + ku_2$$

or

$$\begin{bmatrix} P_1 \\ P_2 \end{bmatrix} = \begin{bmatrix} k & -k \\ -k & k \end{bmatrix} \begin{bmatrix} u_1 \\ u_2 \end{bmatrix} \tag{8.4}$$

It would appear that, to find an individual stiffness coefficient, a single unit displacement corresponding to a force is applied and the displacements corresponding to all other forces are set to zero.

As a further example, we shall find the stiffness matrix for the two springs in series, of individual stiffnesses k_1 and k_2, shown in figure 8.4. The nodes or ends of the springs have been numbered 1 to 3 and there is a displacement applied to each node.

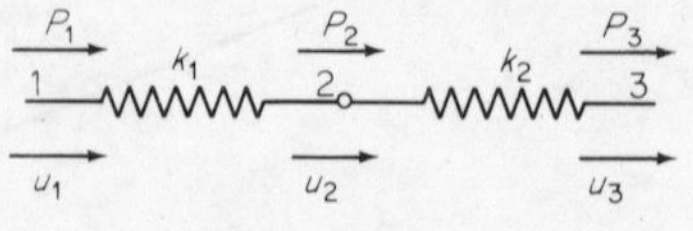

Figure 8.4

(1) Apply u_1 and make $u_2 = u_3 = 0$; hence

$$P_1 = k_1 u_1, P_2 = -k_1 u_1, P_3 = 0$$

(2) Apply u_2 and make $u_1 = u_3 = 0$; hence

$$P_1 = -k_1 u_2, P_2 = (k_1 + k_2) u_2, P_3 = -k_2 u_2$$

(3) Apply u_3 and make $u_1 = u_2 = 0$, hence

$$P_1 = 0, P_2 = -k_2 u_3, P_3 = k_2 u_3$$

If all the displacements are applied together

$$
\begin{bmatrix} P_1 \\ P_2 \\ P_3 \end{bmatrix}
=
\begin{bmatrix} k_1 & -k_1 & 0 \\ -k_1 & k_1 + k_2 & -k_2 \\ 0 & -k_2 & k_2 \end{bmatrix}
\begin{bmatrix} u_1 \\ u_2 \\ u_3 \end{bmatrix}
$$

If the original 2 × 2 stiffness matrices for the individual springs are extended to 3 × 3 matrices by the addition of a row and column of zeros, the two matrices can be added together and will form the stiffness matrix for the combined system.

$$
\begin{bmatrix} P_1 \\ P_2 \\ P_3 \end{bmatrix}
=
\begin{bmatrix} k_1 & -k_1 & 0 \\ -k_1 & k_1 & 0 \\ 0 & 0 & 0 \end{bmatrix}
\begin{bmatrix} u_1 \\ u_2 \\ u_3 \end{bmatrix}
+
\begin{bmatrix} 0 & 0 & 0 \\ 0 & k_2 & -k_2 \\ 0 & -k_2 & k_2 \end{bmatrix}
\begin{bmatrix} u_1 \\ u_2 \\ u_3 \end{bmatrix}
$$

If the determinant of the 3 × 3 matrix is taken it will be found to be zero, indicating that the stiffness matrix is singular. This means that the problem as set cannot be solved. The reason for this is not hard to find. If figure 8.4 is inspected it can be seen that the system is not in equilibrium. It is presumably accelerating to the right under the action of the forces. It is essential to apply an end condition. Consider the case when the first spring is fixed at the left-hand end, that is, $u_1 = 0$.

$$
\begin{bmatrix} P_1 \\ P_2 \\ P_3 \end{bmatrix}
=
\begin{bmatrix} k_1 & -k_1 & 0 \\ -k_1 & k_1 + k_2 & -k_2 \\ 0 & -k_2 & k_2 \end{bmatrix}
\begin{bmatrix} 0 \\ u_2 \\ u_3 \end{bmatrix}
$$

The matrix can be partitioned as shown and will lead to the following, if expanded into two separate matrices.

$$
\begin{bmatrix} P_1 \end{bmatrix} = \begin{bmatrix} -k_1 & 0 \end{bmatrix} \begin{bmatrix} u_2 \\ u_3 \end{bmatrix}
\quad \text{and} \quad
\begin{bmatrix} P_2 \\ P_3 \end{bmatrix} = \begin{bmatrix} k_1 + k_2 & -k_2 \\ -k_2 & k_2 \end{bmatrix} \begin{bmatrix} u_2 \\ u_3 \end{bmatrix}
$$

The second of these when inverted gives

$$
\begin{bmatrix} u_2 \\ u_3 \end{bmatrix}
=
\begin{bmatrix} \dfrac{1}{k_1} & \dfrac{1}{k_1} \\ \dfrac{1}{k_1} & \dfrac{1}{k_1} + \dfrac{1}{k_2} \end{bmatrix}
\begin{bmatrix} P_2 \\ P_3 \end{bmatrix}
$$

so that u_1 and u_2 are known in terms of the applied forces P_2 and P_3. Also

$$\begin{bmatrix} P_1 \end{bmatrix} = \begin{bmatrix} -k_1 & 0 \end{bmatrix} \begin{bmatrix} \dfrac{1}{k_1} & \dfrac{1}{k_1} \\[2ex] \dfrac{1}{k_1} & \dfrac{1}{k_1}+\dfrac{1}{k_2} \end{bmatrix} \begin{bmatrix} P_2 \\[2ex] P_3 \end{bmatrix} = -(P_2+P_3)$$

which is of course correct.

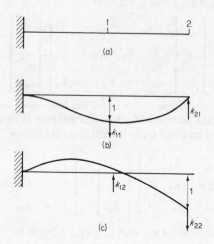

Figure 8.5

We are now in a position to return to our original problem, that of the cantilever, shown again in figure 8.5a. To determine individual stiffness coefficients
only one displacement at a time of unit value is permitted. For a unit displacement
at point 1 no displacement is allowed at point 2. This will mean that a force of
value k_{21} will have to be applied at 2 to prevent any deflection taking place, while
a force of value k_{11} is applied at 1 to produce a deflection of unity there. The
resulting deflection curve is illustrated in figure 8.5b. At c is indicated the deflected
form of the cantilever for finding the other stiffness coefficients. Of course the
stiffness coefficient k_{12} will equal the coefficient k_{21}.

8.3 Pin-jointed plane trusses

A uniform elastic member which has only axial loading applied will behave in
an identical manner to a spring. The relation between the applied loads and the
displacements will be the same as that in equation 8.4 if the value of k is replaced
by EA/l. The two spring problems that we have discussed so far have been the
individual spring and the case of two springs that were collinear. In a truss the

members meeting at a joint are inclined at different angles and we shall want to know how to determine the overall stiffness of the truss in terms of the individual member stiffnesses. The layout of a truss can be specified in terms of one set of axes which we shall refer to as the system axes. At present the stiffness of a single member has been referred to a member set of axes—in fact we took the x axis to lie along the longitudinal axis of the member. We shall now determine the stiffness of a member in terms of the system axes.

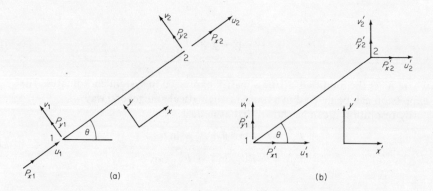

Figure 8.6

A member 12 is shown in figure 8.6a with forces and displacements acting in both the x and y directions; this will make the problem have a more general application. The x and y axes are member axes. The same member is shown again using system coordinates in figure 8.6b, the forces and displacements being distinguished by the addition of a prime. Resolution of the displacements in the system coordinates at the ends of the member into the member coordinates will give

$$u_1 = u'_1 \cos \theta + v'_1 \sin \theta$$
$$u_2 = u'_2 \cos \theta + v'_2 \sin \theta$$

Substituting into equation 8.4

$$\begin{bmatrix} P_{x1} \\ P_{x2} \end{bmatrix} = \frac{EA}{l} \begin{bmatrix} 1 & -1 \\ -1 & 1 \end{bmatrix} \begin{bmatrix} u'_1 \cos \theta + v'_1 \sin \theta \\ u'_2 \cos \theta + v'_2 \sin \theta \end{bmatrix}$$

The member 12 being considered as part of a pin-jointed truss will only have an axial force. Setting P_{y1} and P_{y2} to zero and resolving the forces at the ends of the member will give the following relations

$$P'_{x1} = P_{x1} \cos \theta, \quad P'_{y1} = P_{y1} \sin \theta, \quad P'_{x2} = P_{x2} \cos \theta, \quad P'_{y2} = P_{y2} \sin \theta$$

so that

$$
\begin{bmatrix} P'_{x1} \\ P'_{y1} \\ P'_{x2} \\ P'_{y2} \end{bmatrix} = \frac{EA}{l} \begin{bmatrix} \cos^2 \theta & \sin \theta \cos \theta & -\cos^2 \theta & -\sin \theta \cos \theta \\ \sin \theta \cos \theta & \sin^2 \theta & -\sin \theta \cos \theta & -\sin^2 \theta \\ -\cos^2 \theta & -\sin \theta \cos \theta & \cos^2 \theta & \sin \theta \cos \theta \\ -\sin \theta \cos \theta & -\sin^2 \theta & \sin \theta \cos \theta & \sin^2 \theta \end{bmatrix} \begin{bmatrix} u'_1 \\ v'_1 \\ u'_2 \\ v'_2 \end{bmatrix}
$$

$$(8.5)$$

or

$$\mathbf{P'} = \mathbf{K'd'} \tag{8.6}$$

where K' is the modified stiffness matrix expressed in system coordinates. The expression can be derived in a more mathematical and general way.

By resolution it can be shown that at node 1

$$
\begin{aligned}
P_{x1} &= P'_{x1} \cos \theta + P'_{y1} \sin \theta \\
P_{y1} &= -P'_{x1} \sin \theta + P'_{y1} \cos \theta
\end{aligned}
\tag{8.7}
$$

A similar relation exists at node 2. So that in matrix form we may write

$$
\begin{bmatrix} P_{x1} \\ P_{y1} \\ P_{x2} \\ P_{y2} \end{bmatrix} = \begin{bmatrix} \cos \theta & \sin \theta & 0 & 0 \\ -\sin \theta & \cos \theta & 0 & 0 \\ 0 & 0 & \cos \theta & \sin \theta \\ 0 & 0 & -\sin \theta & \cos \theta \end{bmatrix} \begin{bmatrix} P'_{x1} \\ P'_{y1} \\ P'_{x2} \\ P'_{y2} \end{bmatrix}
$$

This could be written in the shorthand form

$$\mathbf{P} = \mathbf{TP'} \tag{8.8}$$

where $\mathbf{T}$ is a transformation matrix.

This particular matrix may very well be familiar to the reader. In matrix theory for a three-dimensional system, if a positive rotation θ is applied about the z axis, the rotation matrix is given by

$$
\mathbf{R} = \begin{bmatrix} \cos \theta & \sin \theta & 0 \\ -\sin \theta & \cos \theta & 0 \\ 0 & 0 & 1 \end{bmatrix}
\tag{8.9}
$$

This degenerates to the matrix $\mathbf{T}$ in a two-dimensional system.

It can easily be shown that the same relation exists between displacements as that between the forces, thus

$$\mathbf{d} = \mathbf{Td'} \tag{8.10}$$

If we now consider the matrix

$$\begin{bmatrix} \cos\theta & \sin\theta \\ -\sin\theta & \cos\theta \end{bmatrix}$$

it can be seen straight away that the inverse will be given by

$$\begin{bmatrix} \cos\theta & -\sin\theta \\ \sin\theta & \cos\theta \end{bmatrix}$$

and this is equal to the transpose of the original matrix; we have what is termed an orthogonal transformation matrix. Thus

$$\mathbf{T}^{-1} = \mathbf{T}^T$$

therefore

$$\mathbf{P}' = \mathbf{T}^T\mathbf{P} \quad \mathbf{d}' = \mathbf{T}^T\mathbf{d} \tag{8.11}$$

In local coordinates $\mathbf{P} = \mathbf{Kd}$ and in system coordinates $\mathbf{P}' = \mathbf{K'd'}$

$$\mathbf{TP}' = \mathbf{Kd} = \mathbf{KTd}'$$

therefore

$$\mathbf{P}' = \mathbf{T}^{-1}\mathbf{KTd}' = \mathbf{T}^T\mathbf{KTd}'$$

also

$$\mathbf{P}' = \mathbf{K'd'}$$

thus

$$\mathbf{K}' = \mathbf{T}^T\mathbf{KT} \tag{8.12}$$

If this transformation is carried out we shall obtain the same expression as that given by equation 8.5.

If the matrix is expressed in the form

$$\begin{bmatrix} \mathbf{K}'_{11} & \mathbf{K}'_{12} \\ \mathbf{K}'_{21} & \mathbf{K}'_{22} \end{bmatrix}$$

we can see that $\mathbf{K}'_{11} = \mathbf{K}'_{22}$, $\mathbf{K}'_{12} = \mathbf{K}'_{21} = -\mathbf{K}'_{11}$.

The stiffness matrix for each member of a truss can now be found in system coordinates. The stiffness matrix for the complete structure is then assembled. We shall first have a brief discussion on the formation of this matrix.

If the displacements in the x' and y' directions are numbered separately at every joint in the truss, each term in a member stiffness matrix could then be referred to using this numbering system. It should be pointed out that even if a displacement is zero it still must be numbered. As an example, member CD, say, might have displacements 3 and 4 at one end and 7 and 8 at the other end. The stiffness matrix for each member has 16 terms and we can think of the individual stiffness coefficients in terms of the numbered displacements. They will consist of all the combinations of these numbers taken in pairs from k_{33} to k_{88}. Other members may very well have some coefficients with numbers that are common to these. The overall stiffness matrix will be obtained by the summation of all common stiffness coefficients. This can easily be seen if we take the case of a number of members meeting at a particular joint. Let us suppose that the displacement in the x' direction has been numbered 3 with a value d_3. All the members would have a stiffness coefficient that could be numbered k_{33}, the force at 3 in direction 3 due to this displacement alone would be given by $(\Sigma k_{33})d_3$.

Equation 8.6 can now, if desired, be written out in full. The displacements d' will either be zero at a rigid support or have unknown values. The loads P will either be known if they are externally applied or will be unknown if they are reactions acting at the nodes. The case of a point load applied somewhere along the length of a member or a uniformly distributed load present no difficulty. The equivalent resultant forces are applied as point loads to the nodes at the ends of the member.

Equation 8.6 can be rewritten

$$d' = K'^{-1}P'$$

and the displacements found. This would involve inverting the stiffness matrix, a somewhat tedious process particularly if done by hand—it is also rather wasteful of machine storage space. It is often very much easier to solve simultaneous equations. This will be demonstrated in the example that follows.

8.4 Example of a pin-jointed redundant truss

The vertical and horizontal members of the truss shown in figure 8.7a are all of the same length l. All members have the same cross-section and are made from the same material.

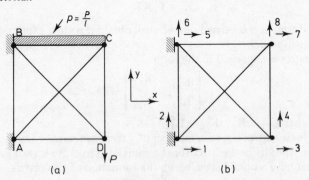

Figure 8.7

It will be seen that the nodes have all been lettered, although it is more usual when using a computer to use numbers. The letters have been used to avoid any possibility of confusion with the displacements, which will be numbered. The members will be referred to in terms of the nodes at their ends. For a particular member the end that has the letter nearer the start of the alphabet will be end 1.

For transferring from a member to a system matrix the angle θ is measured from the x axis such that the rotation is positive with respect to the z axis (right-hand corkscrew).

The problem we have chosen is a very simple one—even so, since there are two unknown displacements at each end of the nodes C and D it will be necessary to solve four simultaneous equations or invert a 4 x 4 matrix.

Details of the various members are set out in table 8.1.

If a computer is used, the coordinates of each node are fed into it, from which it can work out the various rotation matrices.

Table 8.1

Member	Length	Area	θ	$\cos \theta$	$\sin \theta$	$\cos^2 \theta$	$\sin^2 \theta$	$\sin \theta \cos \theta$
AC	$\sqrt{2}l$	A	$45°$	$\dfrac{1}{\sqrt{2}}$	$\dfrac{1}{\sqrt{2}}$	$\frac{1}{2}$	$\frac{1}{2}$	$\frac{1}{2}$
AD	l	A	0	1	0	1	0	0
BC	l	A	0	1	0	1	0	0
BD	$\sqrt{2}l$	A	$315°$	$\dfrac{1}{\sqrt{2}}$	$-\dfrac{1}{\sqrt{2}}$	$\frac{1}{2}$	$\frac{1}{2}$	$-\frac{1}{2}$
CD	l	A	$270°$	0	-1	0	1	0

In figure 8.7b all possible displacements of the nodes have been numbered; the directions are in system coordinates and in the positive x and y directions. The numbering sequence is not the best but has been chosen to illustrate a point further in the development. It would in fact have been better to use displacements 1 to 4 for nodes C and D and 5 to 8 for nodes A and B.

For member AC the angle θ is $45°$ and using equation 8.5 the stiffness matrix in system coordinates is given by

$$
\mathbf{K'_{AC}} = \frac{AE}{\sqrt{2}l}
\begin{array}{cccc}
1 & 2 & 7 & 8 \\
\begin{bmatrix}
\frac{1}{2} & \frac{1}{2} & -\frac{1}{2} & -\frac{1}{2} \\
\frac{1}{2} & \frac{1}{2} & -\frac{1}{2} & -\frac{1}{2} \\
-\frac{1}{2} & -\frac{1}{2} & \frac{1}{2} & \frac{1}{2} \\
-\frac{1}{2} & -\frac{1}{2} & \frac{1}{2} & \frac{1}{2}
\end{bmatrix}
&
\begin{matrix}
1 \\ 2 \\ 7 \\ 8
\end{matrix}
\end{array}
$$

The numbering introduced at the top and side of the matrix corresponds to the numbered displacements at the nodes of the member and will be of assistance when the matrix for the whole structure is assembled. Element 27 (row 2 column 7) would represent the force produced in direction 2 by a unit displacement in direction 7.

The stiffness matrices for all the members can now be written down in system coordinates as follows.

$$\mathbf{K}'_{AC} = \frac{0{\cdot}354EA}{l} \begin{array}{cccc} 1 & 2 & 7 & 8 \\ \left[\begin{array}{cccc} 1 & 1 & -1 & -1 \\ 1 & 1 & -1 & -1 \\ -1 & -1 & 1 & 1 \\ -1 & -1 & 1 & 1 \end{array}\right] & \begin{array}{c} 1 \\ 2 \\ 7 \\ 8 \end{array} \end{array}$$

$$\mathbf{K}'_{AD} = \frac{EA}{l} \begin{array}{cccc} 1 & 2 & 3 & 4 \\ \left[\begin{array}{cccc} 1 & 0 & -1 & 0 \\ 0 & 0 & 0 & 0 \\ -1 & 0 & 1 & 0 \\ 0 & 0 & 0 & 0 \end{array}\right] & \begin{array}{c} 1 \\ 2 \\ 3 \\ 4 \end{array} \end{array}$$

$$\mathbf{K}'_{BD} = \frac{0{\cdot}354EA}{l} \begin{array}{cccc} 5 & 6 & 3 & 4 \\ \left[\begin{array}{cccc} 1 & -1 & -1 & 1 \\ -1 & 1 & 1 & -1 \\ -1 & 1 & 1 & -1 \\ 1 & -1 & -1 & 1 \end{array}\right] & \begin{array}{c} 5 \\ 6 \\ 3 \\ 4 \end{array} \end{array}$$

$$\mathbf{K}'_{BC} = \frac{EA}{l} \begin{array}{cccc} 5 & 6 & 7 & 8 \\ \left[\begin{array}{cccc} 1 & 0 & -1 & 0 \\ 0 & 0 & 0 & 0 \\ -1 & 0 & 1 & 0 \\ 0 & 0 & 0 & 0 \end{array}\right] & \begin{array}{c} 5 \\ 6 \\ 7 \\ 8 \end{array} \end{array}$$

$$\mathbf{K}'_{CD} = \frac{EA}{l} \begin{array}{cccc} 7 & 8 & 3 & 4 \\ \left[\begin{array}{cccc} 0 & 0 & 0 & 0 \\ 0 & 1 & 0 & -1 \\ 0 & 0 & 0 & 0 \\ 0 & -1 & 0 & 1 \end{array}\right] & \begin{array}{c} 7 \\ 8 \\ 3 \\ 4 \end{array} \end{array}$$

The stiffness matrix for the truss can now be assembled and this will be an 8×8 matrix since there are four nodes, each with a possible displacement in two directions. A particular element in the matrix is formed by the addition of any corresponding terms from the individual stiffness matrices. For example element 77 is the sum of $0{\cdot}354$ from $\mathbf{K}'_{AC}$; 1 from $\mathbf{K}'_{BC}$; 0 from $\mathbf{K}'_{CD}$; that is, a total value of $1{\cdot}354$.

There is a connection between nodes A and B formed by the wall, which will have a very high stiffness value. So far as our calculations are concerned it would not matter what value is used for the stiffness as it will be assumed that the displacements at both the nodes are zero. Hence the stiffness can be taken as a null matrix.

$$
K' = \frac{EA}{l}
\begin{bmatrix}
 & 1 & 2 & 3 & 4 & 5 & 6 & 7 & 8 \\
1.354 & 0.354 & -1.0 & 0 & 0 & 0 & -0.354 & -0.354 \\
0.354 & 0.354 & 0 & 0 & 0 & 0 & -0.354 & -0.354 \\
-1.0 & 0 & 1.354 & -0.354 & -0.354 & 0.354 & 0 & 0 \\
0 & 0 & -0.354 & 1.354 & 0.354 & -0.354 & 0 & -1.0 \\
0 & 0 & -0.354 & 0.354 & 1.354 & -0.354 & -1.0 & 0 \\
0 & 0 & 0.354 & -0.354 & -0.354 & 1.354 & 0 & 0 \\
-0.354 & -0.354 & 0 & 0 & -1.0 & 0 & 1.354 & 0.354 \\
-0.354 & -0.354 & 0 & -1.0 & 0 & 0 & 0.354 & 1.354
\end{bmatrix}
$$

Next we shall examine the external loading applied to the truss. The vertical point load at D corresponds to direction 4, and as it is in the negative y direction the value will be $-P$. The only other applied loading is the uniformly distributed load of total value P; this is equivalent to applying a load of $P/2$ downwards at B and C, that is, $-P/2$ at 6 and 8. There will also be reactions at 1, 2, 5 and 6; these are as yet unknown and cannot be determined until the displacements of all nodes have been found. The force vector can be written as

$$
\begin{bmatrix}
\overset{1}{P_1} & \overset{2}{P_2} & \overset{3}{0} & \overset{4}{-P} & \overset{5}{P_5} & \overset{6}{P_6} & \overset{7}{0} & \overset{8}{-\frac{P}{2}}
\end{bmatrix}^T .
$$

Care will have to be taken over the value of P_6 because there is an applied load of value $-P/2$ that will have to be included in the final value of this reaction.

The displacements are zero at 1, 2, 5 and 6 and unknown at 3, 4, 7 and 8. The displacement vector can be written as

$$
\begin{bmatrix}
\overset{1}{0} & \overset{2}{0} & \overset{3}{d_3} & \overset{4}{d_4} & \overset{5}{0} & \overset{6}{0} & \overset{7}{d_7} & \overset{8}{d_8}
\end{bmatrix}^T
$$

The force–displacement relation will next be written out in full, with, however, a change in the order: the points at which the displacements are zero will be grouped together.

$$
\begin{bmatrix}
0 \\ -P/2 \\ 0 \\ -P \\ P_1 \\ P_2 \\ P_5 \\ P_6
\end{bmatrix}
=
\frac{EA}{l}
\begin{bmatrix}
 & 7 & 8 & 3 & 4 & 1 & 2 & 5 & 6 \\
1.354 & 0.354 & 0 & 0 & -0.354 & -0.354 & -1.0 & 0 \\
0.354 & 1.354 & 0 & -1.0 & -0.354 & -0.354 & 0 & 0 \\
0 & 0 & 1.354 & -0.354 & -1.0 & 0 & -0.354 & 0.354 \\
0 & -1 & -0.354 & 1.354 & 0 & 0 & 0.354 & -0.354 \\
-0.354 & -0.354 & -1.0 & 0 & 1.354 & 0.354 & 0 & 0 \\
-0.354 & -0.354 & 0 & 0 & 0.354 & 0.354 & 0 & 0 \\
-1.0 & 0 & -0.354 & 0.354 & 0 & 0 & 1 & 0 \\
0 & 0 & 0.354 & -0.354 & 0 & 0 & 0 & 0
\end{bmatrix}
\begin{bmatrix}
d_7 \\ d_8 \\ d_3 \\ d_4 \\ 0 \\ 0 \\ 0 \\ 0
\end{bmatrix}
$$

It will be noted that the stiffness matrix has been partitioned by the broken lines. The equations for d_7, d_8, d_3 and d_4 can be abstracted and will read

$$\frac{EA}{l}\begin{bmatrix} 1\cdot354 & 0\cdot354 & 0 & 0 \\ 0\cdot354 & 1\cdot354 & 0 & -1\cdot0 \\ 0 & 0 & 1\cdot354 & -0\cdot354 \\ 0 & -1 & -0\cdot354 & 1\cdot354 \end{bmatrix}\begin{bmatrix} d_7 \\ d_8 \\ d_3 \\ d_4 \end{bmatrix} = \begin{bmatrix} 0 \\ -\dfrac{P}{2} \\ 0 \\ -P \end{bmatrix}$$

which when solved gives

$$d_7 = 0\cdot723\,\frac{Pl}{EA} \qquad d_8 = -2\cdot77\,\frac{Pl}{EA}$$

$$d_3 = -0\cdot78\,\frac{Pl}{EA} \qquad d_4 = -2\cdot99\,\frac{Pl}{EA}$$

The reduced matrix relation given above could have been obtained from the original overall stiffness matrix by striking out rows and columns with numbers that correspond to the numbered displacements that are zero. Rows and columns 1, 2, 5 and 6 could be eliminated.

Now that all the displacements are known we can proceed to find the unknown reactions. If we abstract the relations for P_1, P_2, P_5 and P_6 from the partitioned matrix it will be seen at once that we only require the lower left-hand portion of the stiffness matrix. All the displacements are zero that would be multiplied by the corresponding terms in the lower right-hand portion of the stiffness matrix. The following relations will apply.

$$\begin{bmatrix} P_1 \\ P_2 \\ P_5 \\ P_6 \end{bmatrix} = P\;\overset{\displaystyle 7 \quad\;\; 8 \quad\;\; 3 \quad\;\; 4}{\begin{bmatrix} -0\cdot354 & -0\cdot354 & -1\cdot0 & 0 \\ -0\cdot354 & -0\cdot354 & 0 & 0 \\ -1\cdot0 & 0 & -0\cdot354 & 0\cdot354 \\ 0 & 0 & 0\cdot354 & -0\cdot354 \end{bmatrix}}\begin{bmatrix} 0\cdot723 \\ -2\cdot77 \\ -0\cdot78 \\ -2\cdot99 \end{bmatrix} \quad \begin{matrix} \begin{bmatrix} 1\cdot5 \\ 0\cdot72 \\ -1\cdot5 \\ 0\cdot78 \end{bmatrix} & \begin{matrix} 1 \\ 2 \\ 5 \\ 6 \end{matrix} \end{matrix}$$

So that $P_1 = 1\cdot50P, P_2 = 0\cdot72P, P_5 = -1\cdot5P, P_6 = 0\cdot78P$

The value of P_6 will not, however, be correct because it does not include the part of the reaction that is due to the uniformly distributed load on BC, that is, $P/2$. The correct value will be $0\cdot78P + 0\cdot5P = 1\cdot28P$.

A complete solution to the problem will require all the axial forces in the members to be evaluated. Since the values of the end displacements of all members and their stiffnesses are known, this should not be difficult. The values of the end forces in a member are given by equation 8.5 which may be rewritten as follows using $c = \cos\theta$ and $s = \sin\theta$.

$$P'_{x1} = \frac{EA}{l}\left\{ c^2\,(u'_1 - u'_2) + sc\,(v'_1 - v'_2) \right\}$$

$$P'_{y1} = \frac{EA}{l} \left\{ sc\,(u'_1 - u'_2) + s^2\,(v'_1 - v'_2) \right\}$$

The values of P'_{x2} and P'_{y2} are the same as P'_{x1} and P'_{y1} with a sign change.
The axial force in a member is given by equation 8.7.

$$P_{x1} = P'_{x1} \cos \theta + P'_{y1} \sin \theta = P'_{x1}\,c + P'_{y1}\,s$$

Substituting for P'_{x1} and P'_{y1} and determining P_{x2} in similar way will give

$$P_{x1} = \frac{EA}{l} \left\{ c\,(u'_1 - u'_2) + s\,(v'_1 - v'_2) \right\}$$

$$P_{x2} = - \frac{EA}{l} \left\{ c\,(u'_1 - u'_2) + s\,(v'_1 - v'_2) \right\}$$

The values of P_{y1} and P_{y2} will of course be found to be zero. The relations can be written in matrix form as

$$\begin{bmatrix} P_{x1} \\ P_{y1} \\ P_{x2} \\ P_{y2} \end{bmatrix} = \frac{EA}{l} \begin{bmatrix} c & s & -c & -s \\ 0 & 0 & 0 & 0 \\ -c & -s & c & s \\ 0 & 0 & 0 & 0 \end{bmatrix} \begin{bmatrix} u'_1 \\ v'_1 \\ u'_2 \\ v'_2 \end{bmatrix}$$

Using this relation the force in member BD will be given by

$$\begin{bmatrix} P_{x1} \\ P_{y1} \\ P_{x2} \\ P_{y2} \end{bmatrix} = \frac{EA}{\sqrt{(2)}\,l} \times \frac{1}{\sqrt{2}} \begin{bmatrix} 1 & -1 & -1 & 1 \\ 0 & 0 & 0 & 0 \\ -1 & 1 & 1 & -1 \\ 0 & 0 & 0 & 0 \end{bmatrix} \begin{bmatrix} 0 \\ 0 \\ -0\cdot78\,\dfrac{Pl}{AE} \\ -2\cdot99\,\dfrac{Pl}{AE} \end{bmatrix}$$

$$P_{x1} = -1.1P \text{ and } P_{x2} = 1.1P$$

The signs indicate that the member BD is in tension. The axial forces in all the members will be found to be as follows.

Member	Near end x	Far end x
AC	$1\cdot02P$	$-1\cdot02P$
AD	$0\cdot78P$	$-0\cdot78P$
BC	$-0\cdot72P$	$0\cdot72P$
BD	$-1\cdot10P$	$1\cdot10P$
CD	$-0\cdot22P$	$0\cdot22P$

8.5 General outline of the stiffness approach

Before we proceed to derive the stiffness matrix for a member in a space structure we shall set down in a little more detail the various steps that are necessary for the complete solution of a problem using the stiffness approach.

(1) It is necessary to replace any external loading that does not act at a node by equivalent loads at the ends of a member and acting at these nodes. This means that the structure is first analysed with the external loading in position and all the nodes fully restrained. As a result of this a set of fixed-end forces P_F will be obtained. A pin-jointed structure is a special case and as we have seen the ends of the member are not restrained against rotation.

(2) The stiffness matrices for all the individual members are next found. In doing this the orientation of a particular member is ignored and a system of local coordinates is used for each member.

(3) It is next necessary to determine the stiffness matrix for the complete structure. For this it will be necessary to have only one set of coordinates: the system coordinates. We shall require some method such that we can transfer a member stiffness from local to system coordinates. A transformation matrix will be used for this purpose.

(4) It will now be possible to write down the stiffness relations for all the nodes in the structure. In general matrix notation

$$P = P_F + Kd \tag{8.13}$$

This equation states that the external forces P that act at the nodes of the structure, are equal to the forces P_F at each node that would result from the nodes being fully restrained, added to the forces Kd that would develop when those nodes, that are in actual fact not fixed, are released and allowed to rotate and displace through d to their equilibrium position.

The forces P can be divided into two sets, those that are known at the outset of a particular problem P_L and those that are unknown P_R. The displacements corresponding to these forces will also be divided into two sets d_L and d_R. We may now write

$$P = \begin{bmatrix} P_L \\ --- \\ P_R \end{bmatrix} \quad \text{and} \quad d = \begin{bmatrix} d_L \\ --- \\ d_R \end{bmatrix}$$

There is no reason why the forces P_F that arose from the initial restrained condition should not also be partitioned, thus

$$P_F = \begin{bmatrix} P_{FL} \\ ----- \\ P_{FR} \end{bmatrix}$$

The matrix for K may also be partitioned such that equation 8.13 becomes

$$\begin{bmatrix} P_L \\ --- \\ P_R \end{bmatrix} = \begin{bmatrix} P_{FL} \\ ----- \\ P_{FR} \end{bmatrix} + \begin{bmatrix} K_{LL} & \vdots & K_{LR} \\ ----- & \vdots & ----- \\ K_{RL} & \vdots & K_{RR} \end{bmatrix} \begin{bmatrix} d_L \\ --- \\ d_R \end{bmatrix}$$

(5) If the matrix is expanded we may write

$$P_L = P_{FL} + K_{LL}\,d_L + K_{LR}\,d_R$$

and $$P_R = P_{FR} + K_{RL}\,d_L + K_{RR}d_R$$

In the majority of problems that we meet, the displacements d_R corresponding to the unknown forces P_R will be zero. Consider the case of a member built-in at one end: the reactions will all be unknowns whether they be axial forces, shear forces or moments; however, the corresponding deflections and rotations are all zero.

If we take d_R to be zero we may now write

$$P_L = P_{FL} + K_{LL}\,d_L \tag{8.14}$$

and $$P_R = P_{FR} + K_{RL}\,d_L \tag{8.15}$$

These two equations can now be used to determine the values of d_L and P_R. Equation 8.14 would be first solved for d_L; the value of d_L can then be substituted into equation 8.15.

(6) It will be remembered that all the forces and displacments have been written down using system coordinates. The forces at the ends of the members will be required to complete the solution. These will consist of two parts, firstly those arising from the fully constrained condition and secondly those that arise when the structure is allowed to displace and reach its equilibrium under the applied loading. In terms of member coordinates we may write

$$P = P_F + Kd$$

It may very well be easier to determine the end reactions for a member in system coordinates and then rotate these into the member coordinates. This will become apparent in the examples later in the chapter.

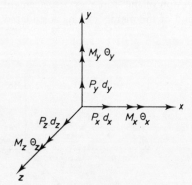

Figure 8.8

The sign convention shown in figure 8.8 will be adopted. A right-handed triad is used for the x, y, and z axes. It will be seen that an axial force or displacement is positive if it acts in the positive sense of the axis concerned. The right-handed screw rule is used to give the positive direction for moments or rotations. This accords with the convention used in chapter 3.

A double arrowhead represents a moment or rotation; the direction indicated is that in which the screw would travel when a clockwise moment is applied.

8.6 Stiffness matrix for a space structure member

We shall first examine the case of a general member in space and determine the stiffness matrix of the member. Once this general case has been obtained the matrix can be simplified and used to treat plane frames and other simpler types of structure.

When discussing the simple spring at the start of this chapter, we found that in order to determine the elements in a stiffness matrix, only one displacement must be applied at a time; the resulting forces to cause this displacement and to suppress other displacements will give the stiffness elements.

It will be assumed that the member is uniform and that the x axis is coincident with the longitudinal axis of the member, while the y and z axes lie along principal axes. We shall only deal with displacements at node 1 of the member since the effect of displacements at node 2 can be written down by inspection, taking care about signs.

The slope–deflection equation will be of considerable help in finding the values of the various elements in the matrix. For the case of zero fixed-end moments

$$M_{12} = \frac{2EI}{l}\left(2\theta_1 + \theta_2 - \frac{3d}{l}\right)$$

The member is shown in figure 8.9a. Each 'force' which acts on the end of a member and displacement of the end of the member will have two subscripts, the first denoting the direction of the 'force' or displacement, and the second, either 1 or 2, denoting the end of the member being considered.

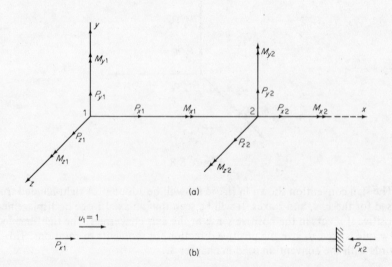

(a)

(b)

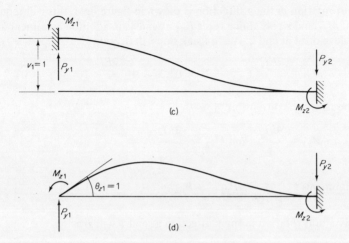

Figure 8.9

For a unit displacement at end 1 in the x direction it can be seen at once from figure 8.9b that $P_{x1} = EA/l$ and $P_{x2} = -EA/l$. All the other restraints will be zero.

Figure 8.9c shows a unit displacement at end 1 in the y direction. There must not be any change of slope at the ends of the member and so moments will be developed there.

From the slope-deflection equation with $\theta_{z1} = \theta_{z2} = 0$ and $\Delta = 1$

$$M_{z1} = \frac{2EI_z}{l}\left(0 + 0 - \frac{3}{l}\right) = -\frac{6EI_z}{l^2} = M_{z2}$$

The negative sign, according to slope-deflection sign convention, indicates that the moments acting at the ends of the beam are anti-clockwise. If we now look at the new sign convention that is being adopted (figure 8.8) an anti-clockwise moment M_z is considered as positive. Thus

$$M_{z1} = M_{z2} = \frac{6EI_z}{l^2}$$

Taking moments about one end of the member

$$P_{y1}l = M_{z1} + M_{z2} = \frac{12EI_z}{l^2} \text{ and } P_{y1} = -P_{y2} = \frac{12EI_z}{l^3}$$

It can be seen at once that a unit displacement at end 1 in the z direction will give

$$P_{z1} = -P_{z2} = \frac{12EI_y}{l^3} \text{ and } M_{y1} = M_{y2} = \frac{6EI_y}{l^2}$$

A unit rotation in the z direction is shown in figure 8.9d; this will require a moment M_{z1} and a restraining force P_{y1} at end 1; in addition a moment and force will be developed at end 2. Again using slope–deflection

$$M_{z1} = \frac{2EI_z}{l} (-2 + 0 + 0), M_{z2} = \frac{2EI_z}{l} (0 - 1 + 0)$$

using the new sign convention

$$M_{z1} = \frac{4EI_z}{l} \text{ and } M_{z2} = \frac{2EI_z}{l}$$

taking moments
$$P_{y1} = -P_{y2} = \frac{6EI_z}{l^2}$$

Similarly a unit rotation in the y direction at end 1 will give

$$M_{y1} = \frac{4EI_y}{l}, M_{y2} = \frac{2EI_y}{l}, P_{z1} = P_{z2} = - \frac{6EI_y}{l^2}$$

The only other displacement that can be applied to end 1 is a rotation about the x axis: a torque is required to produce this and there will be no other restraints.

$$M_{x1} = \frac{GJ_x}{l} \text{ and } M_{x2} = - \frac{GJ_x}{l}$$

As we mentioned earlier the effects of applying displacements at end 2 can now be written down if care is taken with the signs. Alternatively if the terms already found are entered in the matrix, the rest will follow from the fact that the matrix must be symmetrical. The complete matrix is shown in equation 8.16.

This could be expressed as

$$\mathbf{P} = \begin{bmatrix} \mathbf{K}_{11} & \mathbf{K}_{12} \\ \mathbf{K}_{21} & \mathbf{K}_{22} \end{bmatrix} \begin{bmatrix} \mathbf{d} \end{bmatrix}$$

The stiffness matrix that has been derived is 12 x 12, and a matrix of this size would exist for each member of a rigid jointed space frame. It can be seen that it will be well nigh impossible to solve problems of this type by hand, hence access to a computer is essential. It is however possible to reduce the matrix to deal with certain classes of problems. The next sections will be devoted to this end.

$$
\begin{bmatrix}
\dfrac{EA}{l} & 0 & 0 & 0 & 0 & 0 & -\dfrac{EA}{l} & 0 & 0 & 0 & 0 & 0 \\[2mm]
0 & \dfrac{12EI_z}{l^3} & 0 & 0 & 0 & \dfrac{6EI_z}{l^2} & 0 & -\dfrac{12EI_z}{l^3} & 0 & 0 & 0 & \dfrac{6EI_z}{l^2} \\[2mm]
0 & 0 & \dfrac{12EI_y}{l^3} & 0 & -\dfrac{6EI_y}{l^2} & 0 & 0 & 0 & -\dfrac{12EI_y}{l^3} & 0 & -\dfrac{6EI_y}{l^2} & 0 \\[2mm]
0 & 0 & 0 & \dfrac{GJ_x}{l} & 0 & 0 & 0 & 0 & 0 & -\dfrac{GJ_x}{l} & 0 & 0 \\[2mm]
0 & 0 & -\dfrac{6EI_y}{l^2} & 0 & \dfrac{4EI_y}{l} & 0 & 0 & 0 & \dfrac{6EI_y}{l^2} & 0 & \dfrac{2EI_y}{l} & 0 \\[2mm]
0 & \dfrac{6EI_z}{l^2} & 0 & 0 & 0 & \dfrac{4EI_z}{l} & 0 & -\dfrac{6EI_z}{l^2} & 0 & 0 & 0 & \dfrac{2EI_z}{l} \\[2mm]
-\dfrac{EA}{l} & 0 & 0 & 0 & 0 & 0 & \dfrac{EA}{l} & 0 & 0 & 0 & 0 & 0 \\[2mm]
0 & -\dfrac{12EI_z}{l^3} & 0 & 0 & 0 & -\dfrac{6EI_z}{l^2} & 0 & \dfrac{12EI_z}{l^3} & 0 & 0 & 0 & -\dfrac{6EI_z}{l^2} \\[2mm]
0 & 0 & -\dfrac{12EI_y}{l^3} & 0 & \dfrac{6EI_y}{l^2} & 0 & 0 & 0 & \dfrac{12EI_y}{l^3} & 0 & \dfrac{6EI_y}{l^2} & 0 \\[2mm]
0 & 0 & 0 & -\dfrac{GJ_x}{l} & 0 & 0 & 0 & 0 & 0 & \dfrac{GJ_x}{l} & 0 & 0 \\[2mm]
0 & 0 & -\dfrac{6EI_y}{l^2} & 0 & \dfrac{2EI_y}{l} & 0 & 0 & 0 & \dfrac{6EI_y}{l^2} & 0 & \dfrac{4EI_y}{l} & 0 \\[2mm]
0 & \dfrac{6EI_z}{l^2} & 0 & 0 & 0 & \dfrac{2EI_z}{l} & 0 & -\dfrac{6EI_z}{l^2} & 0 & 0 & 0 & \dfrac{4EI_z}{l}
\end{bmatrix}
\begin{Bmatrix}
d_{x1} \\ d_{y1} \\ d_{z1} \\ \theta_{x1} \\ \theta_{y1} \\ \theta_{z1} \\ d_{x2} \\ d_{y2} \\ d_{z2} \\ \theta_{x2} \\ \theta_{y2} \\ \theta_{z2}
\end{Bmatrix}
=
\begin{Bmatrix}
P_{x1} \\ P_{y1} \\ P_{z1} \\ M_{x1} \\ M_{y1} \\ M_{z1} \\ P_{x2} \\ P_{y2} \\ P_{z2} \\ M_{x2} \\ M_{y2} \\ M_{z2}
\end{Bmatrix}
$$

$$(8.16)$$

8.7 The space truss

The relation between the end load and the displacement for a pin-jointed space frame can be obtained from equation 8.5 and has the following form in member axes

$$
\begin{bmatrix} P_{x1} \\ P_{y1} \\ P_{z1} \\ P_{x2} \\ P_{y2} \\ P_{z2} \end{bmatrix}
=
\begin{bmatrix}
1 & 0 & 0 & -1 & 0 & 0 \\
0 & 0 & 0 & 0 & 0 & 0 \\
0 & 0 & 0 & 0 & 0 & 0 \\
-1 & 0 & 0 & 1 & 0 & 0 \\
0 & 0 & 0 & 0 & 0 & 0 \\
0 & 0 & 0 & 0 & 0 & 0
\end{bmatrix}
\begin{bmatrix} d_{x1} \\ d_{y1} \\ d_{z1} \\ d_{x2} \\ d_{y2} \\ d_{z2} \end{bmatrix}
\tag{8.17}
$$

Again it will be necessary to be able to transfer from a member stiffness **K** to a system orientated stiffness **K'**. For this we shall make use of direction cosines (figure 8.10).

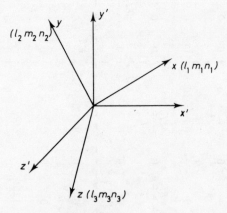

Figure 8.10

l_1 is the cosine of the angle between the x and x' axis
m_1 is the cosine of the angle between the x and y' axis
n_1 is the cosine of the angle between the x and z' axis
l_2 is the cosine of the angle between the y and x' axis

and so on. Then

$$
\begin{aligned}
P_{x1} &= l_1 P'_{x1} + m_1 P'_{y1} + n_1 P'_{z1} \\
P_{y1} &= l_2 P'_{x1} + m_2 P'_{y1} + n_2 P'_{z1} \\
P_{z1} &= l_3 P'_{x1} + m_3 P'_{y1} + n_3 P'_{z1}
\end{aligned}
\tag{8.18}
$$

$$
\begin{bmatrix} P_{x1} \\ P_{y1} \\ P_{z1} \\ P_{x2} \\ P_{y2} \\ P_{z2} \end{bmatrix} = \begin{bmatrix} l_1 & m_1 & n_1 & 0 & 0 & 0 \\ l_2 & m_2 & n_2 & 0 & 0 & 0 \\ l_3 & m_3 & n_3 & 0 & 0 & 0 \\ 0 & 0 & 0 & l_1 & m_1 & n_1 \\ 0 & 0 & 0 & l_2 & m_2 & n_2 \\ 0 & 0 & 0 & l_3 & m_3 & n_3 \end{bmatrix} \begin{bmatrix} P'_{x1} \\ P'_{y1} \\ P'_{z1} \\ P'_{x2} \\ P'_{y2} \\ P'_{z2} \end{bmatrix} \tag{8.19}
$$

The reader is probably familiar with the rotation matrix.

$$
\mathbf{R} = \begin{bmatrix} l_1 & m_1 & n_1 \\ l_2 & m_2 & n_2 \\ l_3 & m_3 & n_3 \end{bmatrix} \tag{8.20}
$$

so that equation 8.18 may be rewritten

$$
\mathbf{P} = \begin{bmatrix} \mathbf{R} & 0 \\ \hline 0 & \mathbf{R} \end{bmatrix} \mathbf{P}' \tag{8.21}
$$

or

$$
\mathbf{P} = \mathbf{T}\mathbf{P}'
$$

$\mathbf{K}$ can be transformed into $\mathbf{K}'$ in exactly the same way as for a plane truss, that is

$$
\mathbf{K}' = \mathbf{T}^T \mathbf{K} \mathbf{T}
$$

If the manipulations are carried out the following expression will be obtained

$$
\mathbf{K}' = \frac{AE}{l} \begin{bmatrix} l_1^2 & l_1 m_1 & l_1 n_1 & -l_1^2 & -l_1 m_1 & -l_1 n_1 \\ l_1 m_1 & m_1^2 & m_1 n_1 & -l_1 m_1 & -m_1^2 & -m_1 n_1 \\ l_1 n_1 & m_1 n_1 & n_1^2 & -l_1 n_1 & -m_1 n_1 & -n_1^2 \\ -l_1^2 & -l_1 m_1 & -l_1 n_1 & l_1^2 & l_1 m_1 & l_1 n_1 \\ -l_1 m_1 & -m_1^2 & -m_1 n_1 & l_1 m_1 & m_1^2 & m_1 n_1 \\ -l_1 n_1 & -m_1 n_1 & -n_1^2 & l_1 n_1 & m_1 n_1 & n_1^2 \end{bmatrix} \tag{8.22}
$$

It will be noticed that only the direction cosines of the x axis relative to the x' axis are required.

The values of l_1, m_1, and n_1 can be found quite easily if the coordinates of all the nodes are known.

The solution of a problem will follow along similar lines to that of a plane truss, and it is not proposed to solve an actual example here.

8.8 Continuous beams

In this type of problem it is assumed that the bending action is confined to one plane and that there are no axial or torsional effects. This means that the general stiffness matrix will reduce to equation 8.23. It will not be necessary to consider member and system coordinates in this case since the two axes can be made coincident.

$$
\begin{bmatrix} P_{y1} \\ M_{z1} \\ P_{y2} \\ M_{z2} \end{bmatrix} = \begin{bmatrix} \dfrac{12EI_z}{l^3} & \dfrac{6EI_z}{l^2} & -\dfrac{12EI_z}{l^3} & \dfrac{6EI_z}{l^2} \\ \dfrac{6EI_z}{l^2} & \dfrac{4EI_z}{l} & -\dfrac{6EI_z}{l^2} & \dfrac{2EI_z}{l} \\ -\dfrac{12EI_z}{l^3} & -\dfrac{6EI_z}{l^2} & \dfrac{12EI_z}{l^3} & -\dfrac{6EI_z}{l^2} \\ \dfrac{6EI_z}{l^2} & \dfrac{2EI_z}{l} & -\dfrac{6EI_z}{l^2} & \dfrac{4EI_z}{l} \end{bmatrix} \begin{bmatrix} d_{y1} \\ \theta_{z1} \\ d_{y2} \\ \theta_{z2} \end{bmatrix} \tag{8.23}
$$

The solution of a continuous-beam type of problem can be illustrated by the fairly straightforward example shown in figure 8.11a. This has been chosen with fixed ends so that the number of unknown displacements is small and hence the

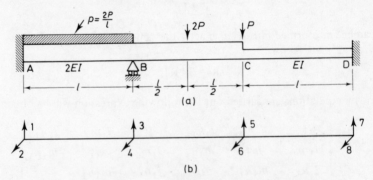

Figure 8.11

size of matrix to be inverted is also small. It would appear at first sight that there is only one unknown, the rotation at node B. There is however a complication to the problem since there is a sudden change of section at C. One method of dealing with this is to introduce another node at C, making a total of three beams. Proceeding in this way we shall also be able to find the displacements at all supports and also that at C.

The beam is restrained and all possible displacements are numbered (figure 8.11b). It will be noted that a sequence of numbering has been adopted such that the nodes have been taken in order along the beam, and two displacement numbers allocated to each node. The **K** matrix for each member can be written down, and each of these will have the form of equation 8.23.

$$\mathbf{K_{AB}} = \frac{EI}{l^3} \begin{array}{c} \\ \left[\begin{array}{cccc} 24 & 12l & -24 & 12l \\ 12l & 8l^2 & -12l & 4l^2 \\ -24 & -12l & 24 & -12l \\ 12l & 4l^2 & -12l & 8l^2 \end{array}\right] \begin{array}{c} 1 \\ 2 \\ 3 \\ 4 \end{array} \end{array}$$

$$\begin{array}{cccc} 1 & 2 & 3 & 4 \end{array}$$

$$\mathbf{K_{BC}} = \frac{EI}{l^3} \left[\begin{array}{cccc} 24 & 12l & -24 & 12l \\ 12l & 8l^2 & -12l & 4l^2 \\ -24 & -12l & 24 & -12l \\ 12l & 4l^2 & -12l & 8l^2 \end{array}\right] \begin{array}{c} 3 \\ 4 \\ 5 \\ 6 \end{array}$$

$$\begin{array}{cccc} 3 & 4 & 5 & 6 \end{array}$$

$$\mathbf{K_{CD}} = \frac{EI}{l^3} \left[\begin{array}{cccc} 12 & 6l & -12 & 6l \\ 6l & 4l^2 & -6l & 2l^2 \\ -12 & -6l & 12 & -6l \\ 6l & 2l^2 & -6l & 4l^2 \end{array}\right] \begin{array}{c} 5 \\ 6 \\ 7 \\ 8 \end{array}$$

$$\begin{array}{cccc} 5 & 6 & 7 & 8 \end{array}$$

It will be noted that the three unknown displacements are 4, 5, and 6. Time will be saved if the system stiffness matrix is formed at once with the partitioned term K_{LL} at the top left-hand corner. The terms will then not need regrouping.

$$\mathbf{K} = \frac{EI}{l^3} \left[\begin{array}{ccc|ccc|cc} 36 & -6l & -12l & -24 & 0 & 0 & -12 & 6l \\ -6l & 12l^2 & 4l^2 & 12l & 0 & 0 & -6l & 2l^2 \\ -12l & 4l^2 & 16l^2 & 0 & 12l & 4l^2 & 0 & 0 \\ \hline -24 & 12l & 0 & 48 & -24 & -12l & 0 & 0 \\ 0 & 0 & 12l & -24 & 24 & 12l & 0 & 0 \\ 0 & 0 & 4l^2 & -12l & -12l & 8l^2 & 0 & 0 \\ -12 & -6l & 0 & 0 & 0 & 0 & 12 & -6l \\ 6l & 2l^2 & 0 & 0 & 0 & 0 & -6l & 4l^2 \end{array}\right] \begin{array}{c} 5 \\ 6 \\ 4 \\ 3 \\ 1 \\ 2 \\ 7 \\ 8 \end{array}$$

$$\begin{array}{cccccccc} 5 & 6 & 4 & 3 & 1 & 2 & 7 & 8 \end{array}$$

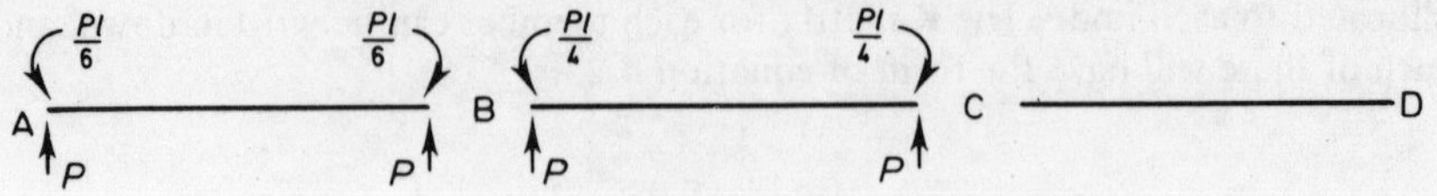

Figure 8.12

The matrix of loads applied at the nodes is known and that due to the imposed restraints can be found (figure 8.12). Using the same order of terms as in the overall stiffness matrix

$$
\mathbf{P} - \mathbf{P_F} =
\begin{bmatrix}
-P \\
0 \\
0 \\
0 \\
0 \\
0 \\
0 \\
0
\end{bmatrix}
-
\begin{bmatrix}
P \\
-\dfrac{Pl}{4} \\
\dfrac{Pl}{12} \\
2P \\
P \\
\dfrac{Pl}{6} \\
0 \\
0
\end{bmatrix}
=
\begin{bmatrix}
-2P \\
\dfrac{Pl}{4} \\
-\dfrac{Pl}{12} \\
-2P \\
-P \\
-\dfrac{Pl}{6} \\
0 \\
0
\end{bmatrix}
\begin{matrix}
5 \\
6 \\
4 \\
3 \\
1 \\
2 \\
7 \\
8
\end{matrix}
$$

Now $\mathbf{K_{LL}d_L} = \mathbf{P_L} - \mathbf{P_{FL}}$

$$
\frac{EI}{l^3}
\begin{bmatrix}
36 & -6l & -12l \\
-6l & 12l^2 & 4l^2 \\
-12l & 4l^2 & 16l^2
\end{bmatrix}
\begin{bmatrix}
d_5 \\
d_6 \\
d_4
\end{bmatrix}
=
\begin{bmatrix}
-2P \\
\dfrac{Pl}{4} \\
-\dfrac{Pl}{12}
\end{bmatrix}
$$

Solving these equations gives

$$
d_5 = -\frac{350}{4608}\frac{Pl^3}{EI}
$$

$$
d_6 = \frac{18Pl^2}{4608EI}
$$

$$
d_4 = -\frac{291}{4608}\frac{Pl^2}{EI}
$$

So that all the displacements are known.

Now $\mathbf{P_R} = \mathbf{P_{FR}} + \mathbf{K_{RL}d_L}$

$$\mathbf{P_R} = \begin{bmatrix} 2P \\ P \\ \dfrac{Pl}{6} \\ 0 \\ 0 \end{bmatrix} + \dfrac{P}{4608l} \begin{bmatrix} -24 & 12l & 0 \\ 0 & 0 & 12l \\ 0 & 0 & 4l^2 \\ -12 & -6l & 0 \\ 6l & 2l^2 & 0 \end{bmatrix} \begin{bmatrix} -350l \\ 18 \\ -291 \end{bmatrix}$$

Hence $P_3 = 3 \cdot 87P$, $P_1 = 0 \cdot 245P$, $P_2 = -0 \cdot 086Pl$, $P_7 = 0 \cdot 885P$, $P_8 = 0 \cdot 445Pl$. A check is obtained since $P_3 + P_1 + P_7 = 5P$.

A calculation of the end forces for member AB is set out below: the subscript to each force P again refers to a force in the direction of a particular displacement.

$$\begin{bmatrix} P_1 \\ P_2 \\ P_3 \\ P_4 \end{bmatrix} = \begin{bmatrix} P \\ \dfrac{Pl}{6} \\ P \\ -\dfrac{Pl}{6} \end{bmatrix} + \dfrac{EI}{l^3} \begin{bmatrix} 24 & 12l & -24 & 12l \\ 12l & 8l^2 & -12l & 4l^2 \\ -24 & -12l & 24 & -12l \\ 12l & 4l^2 & -12l & 8l^2 \end{bmatrix} \dfrac{Pl^2}{4608EI} \begin{bmatrix} 0 \\ 0 \\ 0 \\ -291 \end{bmatrix}$$

therefore $P_1 = 0 \cdot 245P$, $P_2 = -0 \cdot 086Pl$, $P_3 = 1 \cdot 755P$, $P_4 = -0 \cdot 672Pl$.

8.9 The plane frame

The bending action in this case is confined to one plane, there are no torsional effects, and axial forces are taken into account. This means that the matrix will be similar to that shown in equation 8.23, but allowance must be made for axial effects. Thus the matrix becomes

$$\begin{bmatrix} P_{x1} \\ P_{y1} \\ M_{z1} \\ P_{x2} \\ P_{y2} \\ M_{z2} \end{bmatrix} = \begin{bmatrix} \dfrac{EA}{l} & 0 & 0 & -\dfrac{EA}{l} & 0 & 0 \\ 0 & \dfrac{12EI_z}{l^3} & \dfrac{6EI_z}{l^2} & 0 & -\dfrac{12EI_z}{l^3} & \dfrac{6EI_z}{l^2} \\ 0 & \dfrac{6EI_z}{l^3} & \dfrac{4EI_z}{l} & 0 & -\dfrac{6EI_z}{l^2} & \dfrac{2EI_z}{l} \\ -\dfrac{EA}{l} & 0 & 0 & \dfrac{EA}{l} & 0 & 0 \\ 0 & -\dfrac{12EI_z}{l^3} & -\dfrac{6EI_z}{l^2} & 0 & \dfrac{12EI_z}{l^3} & -\dfrac{6EI_z}{l^2} \\ 0 & \dfrac{6EI_z}{l^2} & \dfrac{2EI_z}{l} & 0 & -\dfrac{6EI_z}{l^2} & \dfrac{4EI_z}{l} \end{bmatrix} \begin{bmatrix} d_{x1} \\ d_{y1} \\ \theta_{z1} \\ d_{x2} \\ d_{y2} \\ \theta_{z2} \end{bmatrix}$$

$$(8.24)$$

Once again we must transfer member coordinates to system coordinates. Equation 8.9 gives the rotation matrix $\mathbf{R}$ for a rotation θ about the z axis. The required matrix can be expressed as

$$T = \begin{bmatrix} \mathbf{R} & \vdots & \mathbf{0} \\ \cdots & + & \cdots \\ \mathbf{0} & \vdots & \mathbf{R} \end{bmatrix} \tag{8.25}$$

As in equation 8.12

$$\mathbf{K}' = \mathbf{T}^T \mathbf{K} \mathbf{T}$$

Making the substitution $\lambda = \cos\theta$ and $\mu = \sin\theta$ the rotation matrix becomes

$$\mathbf{R} = \begin{bmatrix} \lambda & \mu & 0 \\ -\mu & \lambda & 0 \\ 0 & 0 & 1 \end{bmatrix} \tag{8.26}$$

Writing $I = I_z$

$$K' = \frac{E}{l} \left[\begin{array}{ccc:ccc} A\lambda^2 + \frac{12I}{l^2}\mu^2 & A\lambda\mu - \frac{12I}{l^2}\lambda\mu & -\frac{6I\mu}{l} & -A\lambda^2 - \frac{12I}{l^2}\mu^2 & -A\lambda\mu + \frac{12I}{l^2}\lambda\mu & -\frac{6I\mu}{l^2} \\ A\lambda\mu - \frac{12I}{l^2}\lambda\mu & A\mu^2 + \frac{12I}{l^2}\lambda^2 & \frac{6I\lambda}{l} & -A\lambda\mu + \frac{12I}{l^2}\lambda\mu & -A\mu^2 - \frac{12I}{l^2}\lambda^2 & \frac{6I\lambda}{l} \\ -\frac{6I\mu}{l} & \frac{6I\lambda}{l} & 4I & \frac{6I}{l}\mu & -\frac{6I\lambda}{l} & 2I \\ \hdashline -A\lambda^2 - \frac{12I}{l^2}\mu^2 & -A\lambda\mu + \frac{12I}{l^2}\lambda\mu & \frac{6I\mu}{l} & A\lambda^2 + \frac{12I}{l^2}\mu^2 & A\lambda\mu - \frac{12I\lambda\mu}{l^2} & \frac{6I\mu}{l} \\ -A\lambda\mu + \frac{12I}{l^2}\lambda\mu & -A\mu^2 - \frac{12I}{l^2}\lambda^2 & -\frac{6I\lambda}{l} & A\lambda\mu - \frac{12I}{l^2}\lambda\mu & A\mu^2 + \frac{12I\lambda^2}{l^2} & -\frac{6I\lambda}{l} \\ -\frac{6I\mu}{l^2} & \frac{6I\lambda}{l} & 2I & \frac{6I\mu}{l} & -\frac{6I\lambda}{l} & 4I \end{array} \right] \tag{8.27}$$

It should be noted that if the matrix is partitioned as shown by the dashed lines and expressed as

$$\begin{bmatrix} \mathbf{K}'_{11} & \mathbf{K}'_{12} \\ \mathbf{K}'_{21} & \mathbf{K}'_{22} \end{bmatrix}$$

then $\mathbf{K}'_{11} = \mathbf{K}'_{22}$ apart from the terms $6I/l$ where the signs are changed. Also $\mathbf{K}'_{12} = \mathbf{K}'^T_{21}$.

As an example we shall find the displacements and the forces at the nodes for the problem in figure 8.13a. The two members are identical and have the following properties: $A = 10^4$ mm^2; $I = 500 \times 10^6$ mm^4; $E = 210$ kN/mm^2. The uniformly distributed load has a total value of $3P$.

The following quantities will be useful when setting up the stiffness matrix: $I/l = 10^5$ mm^3; $I/l^2 = 20$ mm^2.

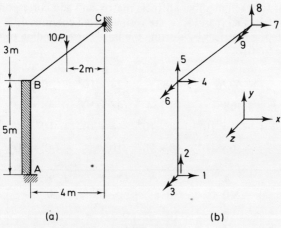

Figure 8.13

For member AB: $\lambda = 0$; $\mu = 1$.

$$
\mathbf{K}'_{AB} = \frac{E}{l}
\begin{array}{c}
\begin{array}{cccccc}
1 & \quad 2 & \quad 3 & \quad 4 & \quad 5 & \quad 6
\end{array} \\
\left[
\begin{array}{cccccc}
240 & 0 & -6 \times 10^5 & -240 & 0 & -6 \times 10^5 \\
0 & 10^4 & 0 & 0 & -10^4 & 0 \\
-6 \times 10^5 & 0 & 2 \times 10^9 & 6 \times 10^5 & 0 & 10^9 \\
-240 & 0 & 6 \times 10^5 & 240 & 0 & 6 \times 10^5 \\
0 & -10^4 & 0 & 0 & 10^4 & 0 \\
-6 \times 10^5 & 0 & 10^9 & 6 \times 10^5 & 0 & 2 \times 10^9
\end{array}
\right]
\begin{array}{c}
1 \\ 2 \\ 3 \\ 4 \\ 5 \\ 6
\end{array}
\end{array}
$$

For member BC: $\lambda = 4/5$; $\mu = 3/5$

$$
\mathbf{K}'_{BC} = \frac{E}{l}
\begin{array}{c}
\begin{array}{cccccc}
4 & \quad 5 & \quad 6 & \quad 7 & \quad 8 & \quad 9
\end{array} \\
\left[
\begin{array}{cccccc}
6486 & 4685 & -36 \times 10^4 & -6486 & -4685 & -36 \times 10^4 \\
4685 & 3754 & 48 \times 10^4 & -4685 & -3754 & 48 \times 10^4 \\
-36 \times 10^4 & 48 \times 10^4 & 2 \times 10^9 & 36 \times 10^4 & -48 \times 10^4 & 10^9 \\
-6486 & -4685 & 36 \times 10^4 & 6486 & 4685 & 36 \times 10^4 \\
-4685 & -3754 & -48 \times 10^4 & 4685 & 3754 & -48 \times 10^4 \\
-36 \times 10^4 & 48 \times 10^4 & 10^9 & 36 \times 10^4 & -48 \times 10^4 & 2 \times 10^9
\end{array}
\right]
\begin{array}{c}
4 \\ 5 \\ 6 \\ 7 \\ 8 \\ 9
\end{array}
\end{array}
$$

It would now be possible to assemble the complete stiffness matrix, but as we have seen in the previous examples we shall only require $\mathbf{K}'_{LL}$ and $\mathbf{K}'_{RL}$ to complete the calculations. Time and effort is saved if these sub-matrices are written down.

From figure 8.13b it is seen that the unknown displacements are composed of the horizontal and vertical displacements and the rotation at B, and also the rotation at C where there is a pin joint. The matrix $\mathbf{K}'_{LL}$ will then be a 4 x 4 matrix with terms involving the displacements at 4, 5, 6, and 9: so that we need to abstract terms for each member stiffness matrix and add corresponding terms.

In fact the matrices K'_{LL} and K'_{AL} are formed from columns 4, 5, 6, and 9. We can then separate the matrices by shifting the lines corresponding to 4, 5, 6, and 9 to the top of the matrix.

$$K'_{LL} = \frac{E}{l} \begin{bmatrix} \overset{4}{6726} & \overset{5}{4685} & \overset{6}{24 \times 10^4} & \overset{9}{-36 \times 10^4} \\ 4685 & 13\,754 & 48 \times 10^4 & 48 \times 10^4 \\ 24 \times 10^4 & 48 \times 10^4 & 4 \times 10^9 & 10^9 \\ -36 \times 10^4 & 48 \times 10^4 & 10^9 & 2 \times 10^9 \end{bmatrix} \begin{matrix} 4 \\ 5 \\ 6 \\ 9 \end{matrix}$$

$$K'_{RL} = \frac{E}{l} \begin{bmatrix} \overset{4}{-240} & \overset{5}{0} & \overset{6}{-6 \times 10^5} & \overset{9}{0} \\ 0 & -10^4 & 0 & 0 \\ 6 \times 10^5 & 0 & 10^9 & 0 \\ -6486 & -4685 & 36 \times 10^4 & 36 \times 10^4 \\ -4685 & -3754 & -48 \times 10^4 & -48 \times 10^4 \end{bmatrix} \begin{matrix} 1 \\ 2 \\ 3 \\ 7 \\ 8 \end{matrix}$$

The load matrix P_L is a null matrix since none of the applied loading acts at the nodes. We shall next have to find the column matrix P_{FL}. The fixed-end forces for each member are given below.

$$\begin{matrix} & 1 & 2 & 3 & 4 & 5 & 6 \\ AB & \{-1{\cdot}5P & 0 & 1250P & -1{\cdot}5P & 0 & -1250P\} \end{matrix}$$

$$\begin{matrix} & 4 & 5 & 6 & 7 & 8 & 9 \\ BC & \{0 & 5P & 5000P & 0 & 5P & -5000P\} \end{matrix}$$

therefore

$$\begin{matrix} & 1 & 2 & 3 & 4 & 5 & 6 & 7 & 8 & 9 \\ P_L - P_{FL} = \{1{\cdot}5P & 0 & -1250P & 1{\cdot}5P & -5P & -3750P & 0 & -5P & 5000P\} \end{matrix}$$

Applying equation 8.3

$$\frac{E}{l} \begin{bmatrix} 6726 & 4685 & 24 \times 10^4 & -36 \times 10^4 \\ 4685 & 13\,754 & 48 \times 10^4 & 48 \times 10^4 \\ 24 \times 10^4 & 48 \times 10^4 & 4 \times 10^9 & 10^9 \\ -36 \times 10^4 & 48 \times 10^4 & 10^9 & 2 \times 10^9 \end{bmatrix} \begin{bmatrix} d_4 \\ d_5 \\ d_6 \\ d_9 \end{bmatrix} = \begin{bmatrix} 1{\cdot}5P \\ -5P \\ -3750P \\ 5000P \end{bmatrix}$$

Solving these equations gives

$$d_4 = 104 \times 10^{-5} \frac{Pl}{E}$$

$$d_5 = -78 \cdot 6 \times 10^{-5} \frac{Pl}{E}$$

$$d_6 = -186 \times 10^{-8} \frac{Pl}{E}$$

$$d_9 = 380 \times 10^{-8} \frac{Pl}{E}$$

Substituting $l = 5000$ mm; $E = 210$ kN/mm^2 gives the following values for displacements, where P is in kN

$d_4 = 0 \cdot 0248P$ mm $\qquad\qquad d_5 = -0 \cdot 0187P$ mm

$d_6 = -4 \cdot 421 \times 10^{-5}P$ rad $\qquad d_7 = 9 \cdot 058 \times 10^{-5}P$ rad

The values of the reactions at the supports can be found from

$$P_R = P_{FR} + K'_{RL} d_L$$

The matrix K'_{RL} has already been formed and d_L is now known, and we shall find that

$$P_1 = -1 \cdot 5P + 0 \cdot 864P = -0 \cdot 636P$$
$$P_2 = \quad 0 \quad + 7 \cdot 858P = \quad 7 \cdot 858P$$
$$P_3 = 1250P - 1233P = 17P$$
$$P_7 = \quad 0 \quad - 2 \cdot 363P = -2 \cdot 363P$$
$$P_8 = \quad 5P \quad - 2 \cdot 858P = 2 \cdot 142P$$

To find the individual member forces

$$P' = P'_F + K'd$$

is applied for each member, where K' is the member stiffness matrix in system coordinates.

For member AB

$P'_{Ax} = -1 \cdot 5P + 0 \cdot 864P = -0 \cdot 636P \qquad P'_{Bx} = -1 \cdot 5P - 0 \cdot 864P = -2 \cdot 364P$

$P'_{Ay} = 0 + 7 \cdot 858P = 7 \cdot 858P \qquad\qquad P'_{By} = 0 - 7 \cdot 858P = -7 \cdot 858P$

$P'_{Az} = 1250P - 1233P = 17P \qquad\quad P'_{Bz} = -1250P - 3089P = -4339P$

For member BC

$P'_{Bx} = 0 + 2 \cdot 364P = 2 \cdot 364P \qquad\quad P'_{Cx} = 0 - 2 \cdot 364P = -2 \cdot 364P$

$P'_{By} = 5P + 2 \cdot 858P = 7 \cdot 858P \qquad P'_{Cy} = 5P - 2 \cdot 858P = 2 \cdot 142P$

$P'_{Bz} = 5000P - 660P = 4340P \qquad P'_{Cz} = -5000P + 5000P = 0$

These results can be transferred to member oriented axes by making use of the rotation matrix (equation 8.27).

8.10 Grillages

In this type of structure the members all lie in a single plane but the loading is applied in a direction that is normal to this plane (figure 8.14a). A typical example of this would be a system of beams that forms the floor of a building. In general not a great deal of accuracy will be sacrificed if axial forces and M_y are neglected. This will mean that for a member whose longitudinal axis is in the x direction, we shall only concern ourselves with the moments and forces in figure 8.14b and their corresponding displacements. Note that torsional effects are introduced.

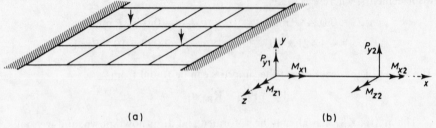

(a)   (b)

Figure 8.14

The stiffness matrix can be found by omitting terms from equation 8.16, and becomes equation 8.28

$$
\begin{bmatrix} P_{y1} \\ M_{x1} \\ M_{z1} \\ P_{y2} \\ M_{x2} \\ M_{z2} \end{bmatrix} =
\begin{bmatrix}
\dfrac{12EI_z}{l^3} & 0 & \dfrac{6EI_z}{l^2} & -\dfrac{12EI_z}{l^3} & 0 & \dfrac{6EI_z}{l^2} \\[2mm]
0 & \dfrac{GJ_x}{l} & 0 & 0 & -\dfrac{GJ_x}{l} & 0 \\[2mm]
\dfrac{6EI_z}{l^2} & 0 & \dfrac{4EI_z}{l} & -\dfrac{6EI_z}{l^2} & 0 & \dfrac{2EI_z}{l} \\[2mm]
-\dfrac{12EI_z}{l^3} & 0 & -\dfrac{6EI_z}{l^2} & \dfrac{12EI_z}{l^3} & 0 & -\dfrac{6EI_z}{l^2} \\[2mm]
0 & -\dfrac{GJ_x}{l} & 0 & 0 & \dfrac{GJ_x}{l} & 0 \\[2mm]
\dfrac{6EI_z}{l^2} & 0 & \dfrac{2EI_z}{l} & -\dfrac{6EI_z}{l^2} & 0 & \dfrac{4EI_z}{l^2}
\end{bmatrix}
\begin{bmatrix} d_{y1} \\ \theta_{x1} \\ \theta_{z1} \\ d_{y2} \\ \theta_{x2} \\ \theta_{z2} \end{bmatrix}
$$

(8.28)

The rotation matrix **R** is the same as that given in equation 8.26. The combination of sine and cosine terms will be similar to that for a plane frame when transferring from member to system coordinates, equation 8.27. It is not proposed to write out the **K'** matrix; this can be obtained from other more detailed works on stiffness and flexibility matrix methods. It is also not necessary to work an example since the procedure is the same as that used for plane frames.

8.11 Special cases

A pinned support in a rigid jointed structure presents no special difficulty—this was seen in the example that was solved. In effect all that results is a further statement that the value of the moment at the support is zero, and the rotation at the support becomes an unknown displacement.

If symmetry is present the order of the stiffness matrix can be considerably reduced, it will only be necessary to treat one half of the structure. It will be possible to set to zero certain displacements in the plane of symmetry.

The effects due to temperature change can be included very readily. If a member is free to expand when the temperature rises the extension due to a temperature rise of T would be $l\alpha T$. If the ends of the members are fixed, a compression force of value $El\alpha T$ would be developed in the member. This effect can be included in the initial state when we consider the structure to be built in at the nodes and determine the fixed-end effects. It is only necessary to modify equation 8.1 slightly.

$$\mathbf{P} = \mathbf{P}_F + \mathbf{P}_T + \mathbf{K}\mathbf{d} \qquad (8.29)$$

where $\mathbf{P}_T$ is a column matrix representing the forces induced by the temperature change. The matrix $\mathbf{P}_T$ can of course be partitioned into $\mathbf{P}_{TL}$ and $\mathbf{P}_{TR}$, and the various equations introduced at the start of the chapter may be suitably modified.

8.12 Outline of the flexibility method

A brief description of the flexibility approach was given in chapter 6. Basically a redundant structure was first made statically determinate by removing a suitable number of restraints or redundancies. There can of course be a number of different forces that could be considered as the redundants in a particular problem. With the structure now statically determinate we have what is sometimes referred to as the primary structure. Displacements corresponding to the redundants are obtained for the primary structure.

When the external loading is removed we have what is referred to as the secondary structure; unit loads are applied in the directions of the redundants and the displacements in these directions are determined. We have shown that

$$\mathbf{d}_R = \mathbf{d}_{RL} + \mathbf{F}\mathbf{R} \qquad (8.30)$$

where L referred to an applied load and R to a redundant.

It is possible to make the solution of a particular problem more general than this, but equation 8.30 will have to be written in a more specific way. A matrix $\mathbf{F}_{RL}$ can be introduced which will give the displacements corresponding to the redundants in the primary structure when unit loads are applied in the same direction as the applied loading $\mathbf{L}$. The flexibility matrix $\mathbf{F}$ in equation 8.30 should be more specific and would be better written $\mathbf{F}_{RR}$ as it refers to the displacements corresponding to the redundants when unit loads are applied in the direction of the redundants. Hence

$$\mathbf{d}_R = \mathbf{F}_{RL}\mathbf{L} + \mathbf{F}_{RR}\mathbf{R} \qquad (8.31)$$

We stated that we were trying to make the application more general. This might mean that we require the displacements of various points on the structure. It will be seen that the displacements corresponding to applied loads L can be written

$$d_L = F_{LL}L + F_{LR}R \qquad (8.32)$$

Equations 8.31 and 8.32 can be combined and written as a partitioned matrix as follows

$$d_S = \begin{bmatrix} d_L \\ \hline d_R \end{bmatrix} = \begin{bmatrix} F_{LL} & \vdots & F_{LR} \\ \hline F_{RL} & \vdots & F_{RR} \end{bmatrix} \begin{bmatrix} L \\ \hline R \end{bmatrix} \qquad (8.33)$$

It will be noted that the displacement matrix d has a subscript S. The reason for this is to differentiate between the system S and a member M. We shall use member and system flexibilities in the same way as we did in dealing with the stiffness approach. It is easier here however to make use of the letters S and M. The partitioned matrix in 8.33 should be F_S.

We shall need to transfer from a member flexibility matrix to a system flexibility matrix. This can be demonstrated by using an energy approach.

The external work done when a set of loads P are applied to a structure can be written as

$$\tfrac{1}{2}P^T F_S P \qquad (8.34)$$

In this case P will consist of external loads L and the redundants R. It is obvious that the system flexibility matrix must be used for this purpose.

If the flexibility matrix were determined for an individual member, the energy stored in the member when the external loads are applied could be written down in terms of the force in the member P_M and the flexibility matrix. Summing the energies for the individual members will give the total energy stored. This can be written as

$$\tfrac{1}{2}P_M^T F_M P_M \qquad (8.35)$$

Now the external work done must equal the total energy stored.

$$P^T F_S P = P_M^T F_M P_M \qquad (8.36)$$

It is quite possible to make use of a transformation matrix T_{MS} to find the values of the forces in the members in terms of P

$$P_M = T_{MS}P \qquad (8.37)$$

Hence from equation 8.36

$$P^T F_S P = P^T T_{MS}^T F_M T_{MS} P$$

or

$$F_S = T_{MS}^T F_M T_{MS} \qquad (8.38)$$

Now the transformation matrix can be partitioned in terms of two sub-matrices, one containing the terms involving L and the other with terms involving R.

Thus

$$F_S = \begin{bmatrix} T^T_{ML} \\ \hline T^T_{MR} \end{bmatrix} F_M [T_{ML} \mid T_{MR}] = \begin{bmatrix} F_{LL} & F_{LR} \\ \hline F_{RL} & F_{RR} \end{bmatrix}$$

It follows that

$$
\begin{aligned}
F_{LL} &= T^T_{ML} F_M T_{ML} \\
F_{LR} &= T^T_{ML} F_M T_{MR} = F^T_{RL} \\
F_{RL} &= T^T_{MR} F_M T_{ML} = F^T_{LR} \\
F_{RR} &= T^T_{MR} F_M T_{MR}
\end{aligned}
\tag{8.39}
$$

Very often the displacements of various joints in the structure will be required. We shall denote these by the column matrix d_J.

If the end forces in the members were known it would be quite easy to use a transformation matrix to transform the member end actions P_M to system orientated P_J.

$$P_M = T_{MJ} P_J \tag{8.40}$$

T_{MJ} would consist of member end forces due to unit forces applied in the direction of system oriented loads or displacements.

According to equation 8.11 and using a slightly different notation since we have several different transformation matrices

$$d_J = T^T_{MJ} d_M$$

Now

$$d_M = F_M P_M$$

Thus

$$d_J = T^T_{MJ} F_M P_M$$

We may also write

$$P_M = T_{ML} L + T_{MR} R$$

therefore

$$d_J = T^T_{MJ} F_M (T_{ML} L + T_{MR} R) \tag{8.41}$$

Use can be made of other transformation matrices, such that when unit loads are applied in the directions of the loads L or redundants R the values of the reactions or constraints at the supports C are

$$P_{CL} = T_{CL} L \quad P_{CR} = T_{CR} R$$

The support reactions or constraints can be found from

$$P_C = P_{CL} + P_{CR} = T_{CL} L + T_{CR} R \tag{8.42}$$

We are now in a position to set down the various steps that have to be followed when using the flexibility method together with a member approach. It will be assumed that a complete solution is required, and the following may have to be found: forces at the ends of members, support reactions, displacements of the various nodes in the structure, some of which may be points of application of loads, and finally the deformation of the various members.

(1) It is first necessary to determine the degree of redundancy of the structure, and then to select the various quantities that will be known as the redundants $\mathbf{R}$.

(2) If the loading on the structure is not all applied through the joints it will be necessary to replace any of these loads by equivalent loads acting at the joints. These will of course be in the opposite sense to the member end reactions and fixed-end moments. The total applied loads $\mathbf{L}$ are known.

(3) The flexibility matrices for the individual members are next found and hence the flexibility matrix $\mathbf{F}_M$ is assembled for all the members in the structure.

(4) We next consider the structure with the redundants removed. Unit loads are applied in the directions of $\mathbf{L}$ and $\mathbf{R}$ and the following transformation matrices determined. $\mathbf{T}_{ML}, \mathbf{T}_{MR}, \mathbf{T}_{CL}$, and $\mathbf{T}_{CR}$.

The further transformation matrix $\mathbf{T}_{MJ}$ is also determined. Unit loads are applied in the direction of joint or node displacements that are required and the resulting member end forces are determined.

(5) The flexibility matrices $\mathbf{F}_{LL}, \mathbf{F}_{LR}, \mathbf{F}_{RL}$ and $\mathbf{F}_{RR}$ are determined from equation 8.39.

(6) The matrix $\mathbf{d}_R$ is next obtained; this may very well be a null matrix.

$$\mathbf{d}_R = \mathbf{F}_{RL}\mathbf{L} + \mathbf{F}_{RR}\mathbf{R}$$

Thus

$$\mathbf{R} = \mathbf{F}_{RR}^{-1}(\mathbf{d}_R - \mathbf{F}_{RL}\mathbf{L}) \tag{8.43}$$

If however $\mathbf{d}_R$ is zero

$$\mathbf{R} = -\mathbf{F}_{RR}^{-1}\mathbf{F}_{RL}\mathbf{L} \tag{8.44}$$

the redundants will now be known.

(7) The support reactions can now be found

$$\mathbf{P}_C = \mathbf{T}_{CL}\mathbf{L} + \mathbf{T}_{CR}\mathbf{R} \tag{8.45}$$

(8) The member end forces can be found from

$$\mathbf{P}_M = \mathbf{T}_{ML}\mathbf{L} + \mathbf{T}_{MR}\mathbf{R} \tag{8.46}$$

Some of the end forces will have to be corrected if the loads $\mathbf{L}$ were not all applied through the joints. The method of correction will become apparent in an example that follows.

(9) The displacements of joints can be found from

$$\mathbf{d}_J = \mathbf{T}_{MJ}^T\mathbf{F}_M\mathbf{P}_M \tag{8.47}$$

or in the case of a joint to which a load $\mathbf{L}$ is applied

$$\mathbf{d}_L = \mathbf{F}_{LL}\mathbf{L} + \mathbf{F}_{LR}\mathbf{R} \tag{8.48}$$

8.13 Member flexibility matrix

It is a fairly straightforward matter to derive the flexibility matrix for a member shown in figure 8.15. Once again the x axis has been made to coincide with the longitudinal axis of the member; the y and z axes are both principal ones. We require the displacements that result from applying a unit force in a particular direction at node 2 when node 1 is fixed. It is obvious for an axial force of unity in the x direction, 1, that the only displacement is in the x direction and has a value of l/EA. Similarly a unit moment about the x axis, 4, will produce a rotation of l/GJ_x.

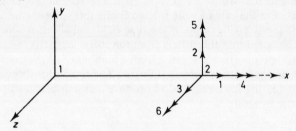

Figure 8.15

The rest of the flexibility coefficients can be determined from cantilever theory. For example a unit force in direction 2 will result in a displacement in that direction of $l^3/3EI_z$, and a rotation or slope at the end (corresponding to 6) of $l^2/2EI_z$.

The rest of the terms are left for the reader to derive.

The complete matrix can be assembled for the member and will be 6 × 6. Note that d_1 to d_3 are deflections, d_4 to d_6 rotations, P_1 to P_3 forces, and P_4 to P_6 moments. This numbering system has been chosen in preference to that used in the case of a stiffness matrix since it will be easier to describe member end displacements and actions.

$$
\begin{bmatrix} d_1 \\ d_2 \\ d_3 \\ d_4 \\ d_5 \\ d_6 \end{bmatrix}
=
\begin{bmatrix}
\dfrac{l}{EA} & 0 & 0 & 0 & 0 & 0 \\[2mm]
0 & \dfrac{l^3}{3EI_z} & 0 & 0 & 0 & \dfrac{l^2}{2EI_z} \\[2mm]
0 & 0 & \dfrac{l^3}{3EI_y} & 0 & -\dfrac{l^2}{2EI_y} & 0 \\[2mm]
0 & 0 & 0 & \dfrac{l}{GJ_x} & 0 & 0 \\[2mm]
0 & 0 & -\dfrac{l^2}{2EI_y} & 0 & \dfrac{l}{EI_y} & 0 \\[2mm]
0 & \dfrac{l^2}{2EI_z} & 0 & 0 & 0 & \dfrac{l}{EI_z}
\end{bmatrix}
\begin{bmatrix} P_1 \\ P_2 \\ P_3 \\ P_4 \\ P_5 \\ P_6 \end{bmatrix}
\tag{8.49}
$$

This matrix is of course written for a general member in space and may be reduced for a particular class of problem. We shall in fact treat two problems, the first being the pin-jointed truss that was solved using a stiffness approach in section 8.4. For a case of this type, the flexibility matrix will degenerate to one term only, namely l/EA.

If a little thought is given to the matter it will be seen that the complete flexibility matrix for the members will be formed from the individual flexibility matrices without changing any of the terms, the reason for this being that end 1 of the member has been regarded as being built in, and we are only interested in the displacements that will occur at end 2. It can be seen in general we shall have a band matrix. For the case of a pin jointed frame this will become a matrix with the leading diagonal having terms of the type l/EA and all other terms will be zero.

It should be noted that this differs from the stiffness matrix, where a particular term in the system matrix was formed from a number of terms from the member matrices transformed into system coordinates. The reason for this was that we were interested in the force required to produce a unit displacement in a particular direction in the structure.

8.14 Example of a pin-jointed redundant truss

The problem to be solved by a flexibility approach has already been solved by stiffness in section 8.4. The members all have the same cross-section and are made from the same material (figure 8.16a). The vertical and horizontal members have the same length l. It will be found convenient to letter each of the members. Every node in the frame is numbered such that a horizontal component of force or displacement can be recognised (b).

Following the various steps that we have set down in section 8.12 it is seen at once that the structure is singly redundant. It is suggested that member d is considered as the redundant member.

The uniformly distributed load is replaced by loads each of $P/2$ acting at nodes 2 and 3 (figure 18.16c).

The flexibility matrix for the structure can now be assembled. Considering the members in alphabetical order

$$
\mathbf{F}_M = \frac{l}{EA}
\begin{bmatrix}
1 & 0 & 0 & 0 & 0 \\
0 & 1 & 0 & 0 & 0 \\
0 & 0 & 1 & 0 & 0 \\
0 & 0 & 0 & \sqrt{2} & 0 \\
0 & 0 & 0 & 0 & \sqrt{2}
\end{bmatrix}
\begin{matrix}
a \\ b \\ c \\ d \\ e
\end{matrix}
$$

Unit loads are applied at nodes D, B, and C, these being the nodes at which external loads are applied. The unit loads will be along the same line as the external loads but will be applied in the positive sense according to the directions of the node displacements or forces 4, 6, and 8. The forces are found in the

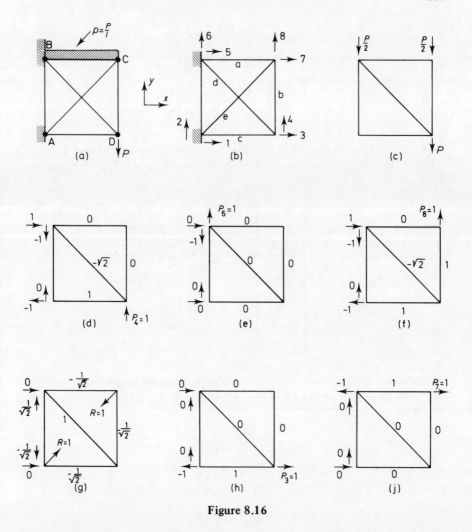

Figure 8.16

individual members as are the reactions at the various supports (figures 8.16d, e and f).

We can now write down the transformation matrices T_{ML} and T_{CL}

$$
T_{ML} =
\begin{array}{c}
\\ \\ \\ \\ \\
\end{array}
\begin{array}{ccc}
P_4 & P_6 & P_8 \\
\end{array}
\left[
\begin{array}{ccc}
0 & 0 & 0 \\
0 & 0 & 1 \\
1 & 0 & 1 \\
-\sqrt{2} & 0 & -\sqrt{2} \\
0 & 0 & 0 \\
\end{array}
\right]
\begin{array}{c}
a \\
b \\
c \\
d \\
e \\
\end{array}
$$

$$
\mathbf{T}_{CL} = \begin{array}{c} \begin{array}{ccc} P_4 & P_6 & \cdot \; P_8 \end{array} \\ \left[\begin{array}{ccc} -1 & 0 & -1 \\ 0 & 0 & 0 \\ 1 & 0 & 1 \\ -1 & -1 & -1 \end{array} \right] \begin{array}{c} 1 \\ 2 \\ 5 \\ 6 \end{array} \end{array}
$$

A unit load is next applied in the direction of the redundant member (figure 8.16g). This will enable the transformation matrices $\mathbf{T}_{MR}$ and $\mathbf{T}_{CR}$ to be written down.

$$
\mathbf{T}_{MR} = \left[\begin{array}{c} -\dfrac{1.}{\sqrt{2}} \\[2mm] -\dfrac{1}{\sqrt{2}} \\[2mm] -\dfrac{1}{\sqrt{2}} \\[2mm] 1 \\[1mm] 1 \end{array} \right] \begin{array}{c} a \\ b \\ c \\ d \\ e \end{array} \qquad \mathbf{T}_{CR} \left[\begin{array}{c} 0 \\[2mm] -\dfrac{1}{\sqrt{2}} \\[2mm] 0 \\[2mm] \dfrac{1}{\sqrt{2}} \end{array} \right] \begin{array}{c} 1 \\ 2 \\ 5 \\ 6 \end{array}
$$

If a complete solution to the problem is required it will be necessary to apply unit loads in directions 3 and 7 (figures 8.16h and j). The transformation matrix $\mathbf{T}_{MJ}$ is then obtained.

$$
\mathbf{T}_{MJ} = \begin{array}{c} \begin{array}{cc} P_3 & P_7 \end{array} \\ \left[\begin{array}{cc} 0 & 1 \\ 0 & 0 \\ 1 & 0 \\ 0 & 0 \\ 0 & 0 \end{array} \right] \begin{array}{c} a \\ b \\ c \\ d \\ e \end{array} \end{array}
$$

The only other matrix that will be required is that of the load matrix $\mathbf{L}$.

$$
\mathbf{L} = P \left[\begin{array}{c} -1 \\ -\frac{1}{2} \\ -\frac{1}{2} \end{array} \right] \begin{array}{c} 4 \\ 6 \\ 8 \end{array}
$$

We are now in a position to determine the matrices $\mathbf{F}_{LL}$, $\mathbf{F}_{LR}$, $\mathbf{F}_{RL}$, and $\mathbf{F}_{RR}$. The method used is set down in equations 8.39.

$$
\mathbf{F}_{LL} = \frac{l}{EA}
\begin{bmatrix}
0 & 0 & 1 & -\sqrt{2} & 0 \\
0 & 0 & 0 & 0 & 0 \\
0 & 1 & 1 & -\sqrt{2} & 0
\end{bmatrix}
\begin{bmatrix}
1 & 0 & 0 & 0 & 0 \\
0 & 1 & 0 & 0 & 0 \\
0 & 0 & 1 & 0 & 0 \\
0 & 0 & 0 & \sqrt{2} & 0 \\
0 & 0 & 0 & 0 & \sqrt{2}
\end{bmatrix}
\begin{bmatrix}
0 & 0 & 0 \\
0 & 0 & 1 \\
1 & 0 & 1 \\
-\sqrt{2} & 0 & -\sqrt{2} \\
0 & 0 & 0
\end{bmatrix}
$$

$$
= \frac{l}{EA}
\begin{bmatrix}
3 \cdot 83 & 0 & 3 \cdot 83 \\
0 & 0 & 0 \\
3 \cdot 83 & 0 & 4 \cdot 83
\end{bmatrix}
$$

$$
\mathbf{F}_{LR} = \frac{l}{EA}
\begin{bmatrix}
-2 \cdot 707 \\
0 \\
-3 \cdot 414
\end{bmatrix}
$$

$$
\mathbf{F}_{RL} = \frac{l}{EA} \begin{bmatrix} -2 \cdot 707 & 0 & -3 \cdot 414 \end{bmatrix}
$$

$$
\mathbf{F}_{RR} = \frac{l}{EA} \begin{bmatrix} 4 \cdot 332 \end{bmatrix}
$$

The force in the redundant member can now be found. The value of $\mathbf{d}_R$ is zero, hence

$$
\mathbf{R} = -\mathbf{F}_{RR}^{-1} \mathbf{F}_{RL} \mathbf{L}
$$
$$
\mathbf{R} = -P[4 \cdot 332]^{-1}[-2 \cdot 707 \quad 0 \quad -3 \cdot 414]
\begin{bmatrix} -1 \\ -\tfrac{1}{2} \\ -\tfrac{1}{2} \end{bmatrix}
$$

$$
R_5 = -1 \cdot 02P
$$

The support reactions are found from

$$
\mathbf{P}_C = \mathbf{T}_{CL} \mathbf{L} + \mathbf{T}_{CR} \mathbf{R}
$$

$$
\begin{bmatrix} P_{C1} \\ P_{C2} \\ P_{C5} \\ P_{C6} \end{bmatrix} = \mathbf{P}
\begin{bmatrix}
-1 & 0 & -1 \\
0 & 0 & 0 \\
1 & 0 & 1 \\
-1 & -1 & -1
\end{bmatrix}
\begin{bmatrix} -1 \\ -\tfrac{1}{2} \\ -\tfrac{1}{2} \end{bmatrix} + \mathbf{P}
\begin{bmatrix} 0 \\ -\dfrac{1}{\sqrt{2}} \\ 0 \\ \dfrac{1}{\sqrt{2}} \end{bmatrix}
\begin{bmatrix} -1 \cdot 02 \end{bmatrix}
$$

$$
P_{C1} = 1 \cdot 5P; \ P_{C2} = 0 \cdot 72P; \ P_{C3} = -1 \cdot 5P; \ P_{C4} = 1 \cdot 28P.
$$

The member end forces are found from

$$\mathbf{P}_M = \mathbf{T}_{ML}\mathbf{L} + \mathbf{T}_{MR}\mathbf{R}$$

$$
\begin{bmatrix} P_a \\ P_b \\ P_c \\ P_d \\ P_e \end{bmatrix} = \mathbf{P}
\begin{bmatrix} 0 & 0 & 0 \\ 0 & 0 & 1 \\ 1 & 0 & 1 \\ -\sqrt{2} & 0 & -\sqrt{2} \\ 0 & 0 & 0 \end{bmatrix}
\begin{bmatrix} -1 \\ -\frac{1}{2} \\ -\frac{1}{2} \end{bmatrix}
+ \mathbf{P}
\begin{bmatrix} -\dfrac{1}{\sqrt{2}} \\ -\dfrac{1}{\sqrt{2}} \\ -\dfrac{1}{\sqrt{2}} \\ 1 \\ 1 \end{bmatrix}
\begin{bmatrix} -1\cdot02 \end{bmatrix}
$$

$P_a = 0\cdot72P$; $P_b = 0\cdot22P$; $P_c = -0\cdot78P$; $P_d = 1\cdot10P$; $P_e = -1\cdot02P$.

The displacements of the positions at which external loads applied are given by

$$\mathbf{d}_L = \mathbf{F}_{LL}\mathbf{L} + \mathbf{F}_{LR}\mathbf{R}$$

$$
\begin{bmatrix} d_4 \\ d_6 \\ d_8 \end{bmatrix} = \frac{Pl}{EA}
\begin{bmatrix} 3\cdot83 & 0 & 3\cdot83 \\ 0 & 0 & 0 \\ 3\cdot83 & 0 & 4\cdot83 \end{bmatrix}
\begin{bmatrix} -1 \\ -\frac{1}{2} \\ -\frac{1}{2} \end{bmatrix}
+ \frac{Pl}{EA}
\begin{bmatrix} -2\cdot707 \\ 0 \\ -3\cdot414 \end{bmatrix}
\begin{bmatrix} -1\cdot02 \end{bmatrix}
$$

$d_4 = -2\cdot99(Pl/EA)$; $d_6 = 0$; $d_8 = 2\cdot77(Pl/EA)$.

The displacements of other joints can be found from

$$\mathbf{d}_J = \mathbf{T}_{MJ}^T \mathbf{F}_M \mathbf{P}_M$$

$$
\begin{bmatrix} d_3 \\ d_7 \end{bmatrix} = \frac{Pl}{EA}
\begin{bmatrix} 0 & 0 & 1 & 0 & 0 \\ 1 & 0 & 0 & 0 & 0 \end{bmatrix}
\begin{bmatrix} 1 & 0 & 0 & 0 & 0 \\ 0 & 1 & 0 & 0 & 0 \\ 0 & 0 & 1 & 0 & 0 \\ 0 & 0 & 0 & \sqrt{2} & 0 \\ 0 & 0 & 0 & 0 & \sqrt{2} \end{bmatrix}
\begin{bmatrix} 0\cdot72 \\ 0\cdot22 \\ -0\cdot78 \\ 1\cdot10 \\ -1\cdot02 \end{bmatrix}
$$

$d_3 = 0\cdot78(Pl/EA)$; $d_7 = 0\cdot72(Pl/EA)$.

The problem has now been completely solved giving the forces in all the members, the reactions, and the displacements of the nodes in the framework.

8.15 Example of a plane frame

As a further example of the application of the flexibility approach we shall solve the problem of a plane frame taking axial loads into account. The frame is shown in figure 8.17a and is the same frame that was solved in section 8.9 using a stiffness approach. It will be possible to reduce the member flexibility matrix to a 3 x 3 matrix.

$$
\begin{bmatrix} d_1 \\ d_2 \\ d_6 \end{bmatrix} = \frac{l}{EI} \begin{bmatrix} \dfrac{I}{A} & 0 & 0 \\ 0 & \dfrac{l^2}{3} & \dfrac{l}{2} \\ 0 & \dfrac{l}{2} & 1 \end{bmatrix} \begin{bmatrix} P_1 \\ P_2 \\ P_6 \end{bmatrix} \tag{8.50}
$$

The frame is redundant to the second degree and it is proposed to make the frame statically determinate by the removal of the horizontal and vertical reaction at C. Directions of system displacements and forces are shown numbered 1 to 9 in figure 8.17b. The frame consists of two members a and b and in figure 8.17c we have each member coordinate system together with member end load or displacement directions, which are labelled a2 b6 etc.

The fixed-end moments and loads are next calculated, and when these are reversed in direction they will become the loading system applied at the nodes (d). Loads are applied in system directions 1, 3, 4, 5, 6, 8, and 9.

The two members a and b are identical and so the flexibility matrices will be the same. For the case where $I = 5 \times 10^8$ mm^4; $A = 10^4$ mm^2; $l = 5000$ mm

$$
\mathbf{F}_{Ma} = \mathbf{F}_{Mb} = \frac{l}{EI} \begin{bmatrix} 5 \times 10^4 & 0 & 0 \\ 0 & \dfrac{25}{3} \times 10^6 & 2500 \\ 0 & 2500 & 1 \end{bmatrix}
$$

The complete member flexibility matrix can now be assembled and will again be a banded matrix.

Unit loads are now applied at the nodes in the positive sense of directions 1, 3, 4, 5, 6, 8, and 9, that is, the points at which external loads are applied. The reactions are found at the supports and the forces at ends 2 of the members. The effects are shown in figure 8.17e to 1 where two diagrams are shown for each unit load, the first showing the load applied and the reactions that result, and the second the member end actions. k for example shows the effect of a unit load being applied in direction 8. This will result in a reaction of -1 at 2 and moment of -4000 at 3. For member a there is a force of 1 at a1 and a moment of 4000 at a6; for member b there are two forces, 3/5 at b1 and 4/5 at b2.

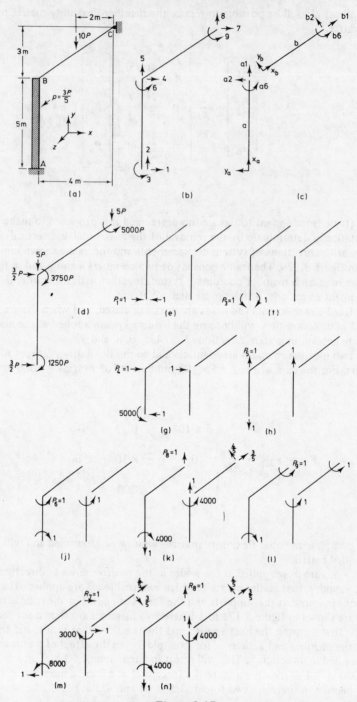

Figure 8.17

The matrices $\mathbf{T}_{ML}$ and $\mathbf{T}_{CL}$ can now be written down.

$$
\mathbf{T}_{ML} =
\begin{array}{c}
\begin{array}{ccccccc} P_1 & P_3 & P_4 & P_5 & P_6 & P_8 & P_9 \end{array} \\
\left[
\begin{array}{ccccccc}
0 & 0 & 0 & 1 & 0 & 1 & 0 \\
0 & 0 & -1 & 0 & 0 & 0 & 0 \\
0 & 0 & 0 & 0 & 1 & 4000 & 1 \\
0 & 0 & 0 & 0 & 0 & \frac{3}{5} & 0 \\
0 & 0 & 0 & 0 & 0 & \frac{4}{5} & 0 \\
0 & 0 & 0 & 0 & 0 & 0 & 1
\end{array}
\right]
\begin{array}{c} a1 \\ a2 \\ a6 \\ b1 \\ b2 \\ b6 \end{array}
\end{array}
$$

$$
\mathbf{T}_{CL} =
\begin{array}{c}
\begin{array}{ccccccc} P_1 & P_3 & P_4 & P_5 & P_6 & P_8 & P_9 \end{array} \\
\left[
\begin{array}{ccccccc}
-1 & 0 & -1 & 0 & 0 & 0 & 0 \\
0 & 0 & 0 & -1 & 0 & -1 & 0 \\
0 & -1 & 5000 & 0 & -1 & -4000 & -1
\end{array}
\right]
\begin{array}{c} 1 \\ 2 \\ 3 \end{array}
\end{array}
$$

Unit loads are now applied in the directions of the two chosen redundants, that is, 7 and 8 (figures 8.17m and n). This enables the matrices $\mathbf{T}_{MR}$ and $\mathbf{T}_{CR}$ to be found. The load matrix $\mathbf{L}$ is also written down.

$$
\mathbf{T}_{MR} =
\begin{array}{c}
\begin{array}{cc} R_7 & R_8 \end{array} \\
\left[
\begin{array}{cc}
0 & 1 \\
-1 & 0 \\
-3000 & 4000 \\
\frac{4}{5} & \frac{3}{5} \\
-\frac{3}{5} & \frac{4}{5} \\
0 & 0
\end{array}
\right]
\begin{array}{c} a1 \\ a2 \\ a6 \\ b1 \\ b2 \\ b6 \end{array}
\end{array}
\qquad
\mathbf{T}_{CR} =
\begin{array}{c}
\begin{array}{cc} R_7 & R_8 \end{array} \\
\left[
\begin{array}{cc}
-1 & 0 \\
0 & -1 \\
8000 & -4000
\end{array}
\right]
\begin{array}{c} 1 \\ 2 \\ 3 \end{array}
\end{array}
\qquad
\mathbf{L} = P
\left[
\begin{array}{c}
\frac{3}{2} \\ -1250 \\ \frac{3}{2} \\ -5 \\ -3750 \\ -5 \\ 5000
\end{array}
\right]
\begin{array}{c} 1 \\ 3 \\ 4 \\ 5 \\ 6 \\ 8 \\ 9 \end{array}
$$

The following matrices can now be found from equations 8.39.

$$
\mathbf{F}_{LL} = \frac{l}{EI}
\left[
\begin{array}{ccccccc}
0 & 0 & 0 & 0 & 0 & 0 & 0 \\
0 & 0 & 0 & 0 & 0 & 0 & 0 \\
0 & 0 & \frac{25}{3} \times 10^6 & 0 & -2500 & -1 \times 10^7 & -2500 \\
0 & 0 & 0 & 5 \times 10^4 & 0 & 5 \times 10^4 & 0 \\
0 & 0 & -2500 & 0 & 1 & 4000 & 1 \\
0 & 0 & -1 \times 10^7 & 5 \times 10^4 & 4000 & 214 \times 10^5 & 6000 \\
0 & 0 & -2500 & 0 & 1 & 6000 & 2
\end{array}
\right]
$$

$$\mathbf{F}_{LR} = \frac{l \times 10^4}{EI} \begin{bmatrix} 0 & 0 \\ 0 & 0 \\ 1583 & -1000 \\ 0 & 5 \\ -0\cdot55 & 0\cdot4 \\ -2598 & 2140 \\ -0\cdot7 & 0\cdot6 \end{bmatrix}$$

$$\mathbf{F}_{RR} = \frac{l \times 10^4}{EI} \begin{bmatrix} 3537 & -2598 \\ -2598 & 2140 \end{bmatrix}$$

$$\mathbf{F}_{RR}^{-1} = \frac{EI}{l \times 10^{10}} \begin{bmatrix} 2606 & 3163 \\ 3163 & 4307 \end{bmatrix}$$

The values of the two redundants are found from

$$\mathbf{R} = -\mathbf{F}_{RR}^{-1}\mathbf{F}_{RL}\mathbf{L}$$

where $\mathbf{F}_{RL} = \mathbf{F}_{LR}$.

Performing the matrix operation gives

$$R_7 = -2\cdot364P \quad R_8 = 2\cdot142P$$

The support reactions can be found from

$$\mathbf{P}_C = \mathbf{T}_{CL}\mathbf{L} + \mathbf{T}_{CR}\mathbf{R}$$

$$\mathbf{P}_C = P \begin{bmatrix} -1 & 0 & -1 & 0 & 0 & 0 & 0 \\ 0 & 0 & 0 & -1 & 0 & -1 & 0 \\ 0 & -1 & 5000 & 0 & -1 & -4000 & -1 \end{bmatrix} \begin{bmatrix} \frac{3}{2} \\ -1250 \\ \frac{3}{2} \\ -5 \\ -3750 \\ -5 \\ 5000 \end{bmatrix}$$

$$+ P \begin{bmatrix} -1 & 0 \\ 0 & -1 \\ 8000 & -4000 \end{bmatrix} \begin{bmatrix} -2\cdot364 \\ 2\cdot142 \end{bmatrix}$$

$$P_1 = -0\cdot636P; \ P_2 = 7\cdot858P; \ P_3 = 17P.$$

The member end forces are given by

$$P_M = T_{ML}L + T_{MR}R$$

It will however be necessary to correct the member end forces since the external loads are not applied through the nodes. The true values of the member end forces will be obtained by the addition of the fixed-end actions. These are the same as the load matrix L but with the sign changed. For end 2 of member b it is of course necessary to resolve the loads in the directions b1 and b2.

$$
\begin{bmatrix} P_{a1} \\ P_{a2} \\ P_{a6} \\ P_{b1} \\ P_{b2} \\ P_{b6} \end{bmatrix}
= P
\underbrace{\begin{bmatrix} -10 \\ -1\cdot5 \\ -18\,750 \\ -3 \\ -4 \\ 5000 \end{bmatrix}}_{T_{ML}L}
+ P
\underbrace{\begin{bmatrix} 2\cdot142 \\ 2\cdot364 \\ 15\,661 \\ -0\cdot606 \\ 3\cdot132 \\ 0 \end{bmatrix} + P \begin{bmatrix} 0 \\ 1\cdot5 \\ -1250 \\ 3P \\ 4P \\ -5000 \end{bmatrix}}_{T_{MR}R}
= P
\begin{bmatrix} -7\cdot858 \\ 2\cdot364 \\ -4339 \\ -0\cdot606 \\ 3\cdot132 \\ 0 \end{bmatrix}
$$

Finally the displacements of the ends of the members can be found from

$$d_J = T_{MJ}^T F_M P_M$$

Or in the system coordinates where there are any loaded nodes

$$d_L = F_{LL}L + F_{LR}R$$

These latter values are set out below

$$
\begin{bmatrix} d_1 \\ d_3 \\ d_4 \\ d_5 \\ d_6 \\ d_8 \\ d_9 \end{bmatrix}
= \frac{Pl}{EI}
\begin{bmatrix} 0 \\ 0 \\ 5\cdot938 \times 10^7 \\ -5 \times 10^5 \\ -22\,500 \\ -1\cdot073 \times 10^8 \\ -27\,500 \end{bmatrix}
+ \frac{Pl}{EI}
\begin{bmatrix} 0 \\ 0 \\ -5\cdot886 \times 10^7 \\ 1\cdot071 \times 10^5 \\ 21\,571 \\ 1\cdot073 \times 10^8 \\ 29\,402 \end{bmatrix}
= \frac{Pl}{EI}
\begin{bmatrix} 0 \\ 0 \\ 0\cdot052 \times 10^7 \\ -3\cdot929 \times 10^5 \\ -929 \\ 0 \\ 1902 \end{bmatrix}
$$

Substituting $l = 5000$ mm; $E = 210$ kN/mm^2; and $I = 5 \times 10^8$ mm^4: $d_1 = 0$; $d_3 = 0$; $d_4 = 0\cdot0248P$ mm; $d_5 = 0\cdot0187P$ mm; $d_6 = -4\cdot421 \times 10^{-5}P$ rad; $d_8 = 0$; $d_9 = 9\cdot058 \times 10^{-5}P$ rad.

8.16 Choice of stiffness or flexibility approach

It could perhaps be argued that the choice is simple to make and depends on comparing the degree of kinematic indeterminacy with the degree of redundancy in a particular structure. If there are more unknown displacements than redundants

the flexibility method should give the shorter solution. This may well be so when
working by hand.

However if a computer is used the stiffness approach is generally to be preferred. It is possible to write a general programme which can quite easily be
modified to take account of specific cases. In the case of flexibility a general
programme is much more difficult to write. Again a stiffness approach requires
no particular choice of redundants whereas with flexibility it is necessary to
decide at the outset what the redundants should be. A particular set of redundants
may very well give a more rapid solution than another set. This means that the
speed of the result can depend on the skill of the operator.

One topic has deliberately been omitted from the subject material covered.
This is the connection matrix, which can be used to advantage when using a computer. This and other more advanced methods will be found in books that are
devoted to the use of matrix methods in structures.

Problems

8.1 For the spring of stiffness k shown in P8. 1a write down the matrix relating
the loads P, applied to the spring, and the displacements u of the spring.

The spring is now inclined at an angle θ to the x direction, as in P8. 1b. Show
that the relationship between the forces and displacements at 1 and 2 in the x and
and y directions is

$$
\begin{Bmatrix} P_{1(x_1)} \\ P_{1(y_1)} \\ P_{2(x_2)} \\ P_{2(y_2)} \end{Bmatrix} = k \begin{bmatrix} c^2 & sc & -c^2 & -sc \\ sc & s^2 & -sc & -s^2 \\ -c^2 & -sc & c^2 & sc \\ -sc & -s^2 & sc & s^2 \end{bmatrix} \begin{Bmatrix} u_1 \\ v_1 \\ u_2 \\ v_2 \end{Bmatrix}
$$

where $s = \sin \theta$ and $c = \cos \theta$.

Making use of this result, determine the stiffness matrix relating the forces
and displacements at point **B** for the elastic pin-jointed truss shown in P8. lc. All
members are of cross-sectional area A and modulus E. Hence determine the
horizontal and vertical displacements of B.

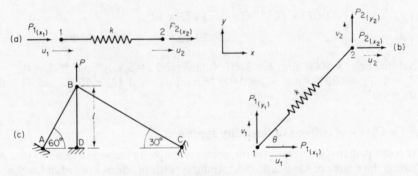

P8.1

8.2 The members of the pin-jointed steel structure P8.2 all have the same cross-section. Using a stiffness approach, determine the displacements of D and the forces in the members.

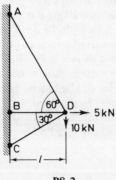

P8.2

8.3 The uniform member AB shown in P8.3a is of length l, cross-section A, and flexural rigidity EI about the z axis, which is a principal axis. Determine the stiffness matrix relating the 'forces' and 'displacements' at each end of the member in the directions indicated. A double arrowhead denotes a moment or rotation.

Determine the overall stiffness matrix for the beam problem shown in P8.3b. Axial effects may be neglected.

Discuss briefly how this matrix may be used in the determination of the various reactions.

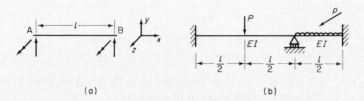

(a)   (b)

P8.3

8.4 Find the stiffness matrix for the beam shown in P8.4.

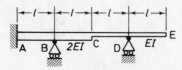

P8.4

8.5 (a) A cantilever ABC is built-in at A, C is at the free end and B is the mid-point. Describe how, from first principles, you might find: (i) the flexibility matrix, (ii) the stiffness matrix, relating vertical loads and deflections at points B and C. The values of individual terms are *not* required.

(b) Derive the stiffness matrix relating the 'forces' and 'displacements' in the directions shown at B for the cantilever in P8.5a. The single arrowhead denotes a force, the double arrowhead the axis of a moment or torque.

(c) Part of a grid structure is shown in P8.5b. The two members DE and FE are of the same section, they are each built-in at one end and are rigidly joined at E. They intersect at right angles and lie in a horizontal plane. Axial and shear effects can be neglected and it may be assumed that $GJ = EI$.

Derive the stiffness matrix relating the externally applied 'forces' and the corresponding 'displacements' indicated in P8.5b.

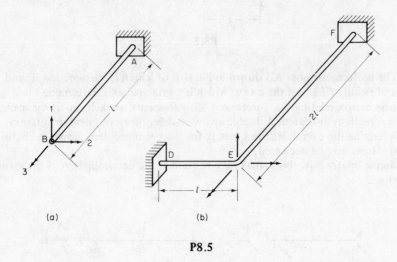

(a)   (b)

P8.5

8.6 Find the stiffness matrix for the frame shown in P8.3. All members have a cross-section of A and a second moment of area of I.

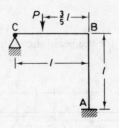

P8.6

If axial displacements can be neglected determine the rotation of the joint at B.

8.7 The load P is applied at the centre of the beam of the portal frame in P8.7. Determine the displacement of B using a stiffness approach. It may be assumed that B and C do not deflect horizontally.

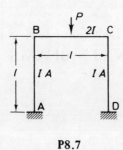

P8.7

8.8 Solve problem 8.2 using a flexibility approach.

8.9 Solve problem 8.7 using a flexibility approach.

9 INFLUENCE LINES FOR STATICALLY INDETERMINATE BEAMS

9.1 Beams with two spans

In chapters 1 and 2 the virtual-work or virtual-displacement method was found to be the most convenient approach when sketching the influence lines for various functions in statically determinate structures. A similar method can be employed when the system is statically indeterminate. The application of virtual work for determining the influence lines both for statically determinate and indeterminate cases is known as the Müller–Breslau principle. Basically this states that the influence line for a function will be represented by the deflection of the load line of the structure when a displacement is applied corresponding to the function. For the case of a statically determinate structure the influence line will consist of a series of straight lines, whereas that for the indeterminate structure will be curved.

A two-span continuous beam is shown in figure 9.1a. We shall first discuss how the influence line for the reaction at A might be found.

If end A is given a unit vertical displacement and the beam is restrained to stay on the supports at B and C, it will take up a shape as shown in b. The actual displacement curve can be found by beam deflection theory.

Making use of virtual work

$$V_A \times 1 - Pd = 0$$

If P is now set equal to unity then $V_A = d$. This means that the deflection curve represents the influence line for the vertical reaction at A.

It is possible to obtain the influence line for shear force or bending moment in a similar manner. For the shear force at E a parallel mechanism is inserted as in section 1.20. Equal and opposite forces Q are then applied. The resulting displacement curve which will represent the influence line for shear force is shown in c. Note that at E the tangents to the two parts of the curve must have the same slope. It would be very easy to sketch the incorrect diagram for the left-hand portion of the curve. If Q is imagined to act alone this would appear to be straight. There is however a moment in addition to the shear force and the curvature shown in the diagram will result.

For bending moment a pin is inserted and equal and opposite moments applied to either side of the pin, which will cause the beam to deflect as shown in d. The

300

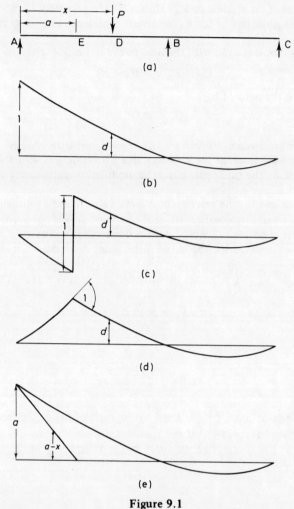

Figure 9.1

values of the applied moments are increased until the relative displacement of the two portions of the beam is equal to one radian.

By virtual work

$$M\theta - Pd = 0$$

Put $\theta = 1$ and $P = 1$; then $M = d$.

Instead of obtaining the influence line for the shear force and bending moment at E from first principles, it is possible to modify the influence line for V_A.

Considering the shear force first, with a unit load applied—when the load is to the left of E the shear force at E has a value of $(V_A - 1)$, and when the load is to the right of E the shear is equal to V_A. Thus it is only necessary to modify the first part of the influence line for V_A, to give the influence line for Q shown in figure 9.1c.

The bending moment at E, with the load to the left of E is given by

$$M_E = V_A \, a - 1(a - x)$$

With the load to the right of E

$$M_E = V_A a$$

Thus if a similar diagram to figure 9.1b is drawn, but with a height a instead of unity at A, this will represent the influence line for bending moment when the load is to the right of E. The other part has to be modified by subtracting $a - x$. This is shown in e.

The influence line for the reaction at A can be also found by a slightly different approach using flexibility coefficients. If the support at A is removed, when the applied load P is in position A would deflect downwards a distance $f_{AD}P$. With the external load equal to V_A applied at A the point A would deflect upwards an amount $f_{AA}V_A$.

Thus $-f_{AD}P + f_{AA}V_A = 0$ since the true deflection at A is zero. Therefore

$$V_A = \frac{f_{AD}}{f_{AA}} P$$

Now $f_{AD} = f_{DA}$ and putting $P = 1$

$$V_A = \frac{f_{DA}}{f_{AA}}$$

so that the influence line for V_A is given by the ratio of the deflection at D to that at A when a unit load is applied at A.

The influence lines for both shear force and bending moment can be found by a similar approach.

9.2 Betti's reciprocal theorem

This is an extension of the Maxwell reciprocal relationship that was derived in chapter 6.

Two different loading systems are applied to the same linear elastic structure (figure 9.2). With the loading system in 9.2a the deflections d_{3a} and d_{4a} are measured, corresponding to the directions of the loads that are applied in system b. Similarly for this system where the loads P_3 and P_4 are applied, the deflections d_{1b} and d_{2b} are measured in the directions of the loads that are applied to system a. Now

$$d_{3a} = f_{31}P_1 + f_{32}P_2$$
$$d_{4a} = f_{41}P_1 + f_{42}P_2$$

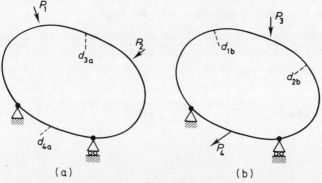

Figure 9.2

and

$$d_{1b} = f_{13}P_3 + f_{14}P_4$$
$$d_{2b} = f_{23}P_3 + f_{24}P_4$$

Multiply the loads on system a by the corresponding displacements on system b

$$P_1(f_{13}P_3 + f_{14}P_4) + P_2(f_{23}P_3 + f_{24}P_4) \tag{9.1}$$

Next multiply the loads on system b by the corresponding displacements on system a

$$P_3(f_{31}P_1 + f_{32}P_2) + P_4(f_{41}P_1 + f_{42}P_2) \tag{9.2}$$

Now we know that from the Maxwell reciprocal relationship that $f_{13} = f_{31}$, etc. If the two expressions are multiplied out and compared term by term they will be found to be identical.

Betti's theorem states, for a linear elastic system with two different sets of loads applied, that if the first set of loads is multiplied by the corresponding displacements produced by the second set of loads, the result equals that obtained by multiplying the second set of loads by the corresponding displacements produced by the first set of loads.

This theorem can also be demonstrated by the use of simple matrix algebra

$$\mathbf{d}_a = \mathbf{F}\mathbf{P}_a$$
$$\mathbf{d}_b = \mathbf{F}\mathbf{P}_b$$
$$\mathbf{P}_a^T\mathbf{d}_b = \mathbf{P}_a^T\mathbf{F}\mathbf{P}_b = (\mathbf{P}_a^T\mathbf{F}\mathbf{P}_b)^T$$

since $\mathbf{P}_a^T\mathbf{d}_b = \mathbf{d}_b^T\mathbf{P}_a$, or

$$\mathbf{P}_a^T\mathbf{d}_b = \mathbf{P}_b^T\mathbf{F}\mathbf{P}_a = \mathbf{P}_b^T\mathbf{d}_a$$

since $\mathbf{F} = \mathbf{F}^T$

9.3 Applications of Betti's theorem to influence lines

We shall consider the problem that has already been discussed, that of finding the influence line for the bending moment at E (figure 9.3a).

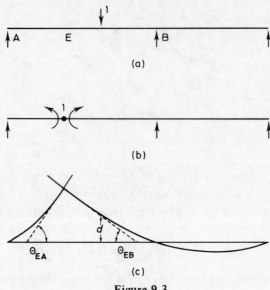

(a)

(b)

(c)

Figure 9.3

The same beam with a pin inserted at E and unit couples applied at either side of the pin is shown at b and in the displaced position at c. θ_{EA} is the slope of member AE at E and θ_{EB} is the slope of BE at E. d is the deflection corresponding to the unit load. θ_E is the actual slope at E in the original problem and M_E is the bending moment corresponding to this.

Applying Betti's theorem

$$M_E \theta_{EA} + M_E \theta_{EB} - 1 \times d = 1 \times \theta_E - 1 \times \theta_E$$

therefore

$$M_E = \frac{d}{\theta_{EA} + \theta_{EB}}$$

$(\theta_{EA} + \theta_{EB})$ is the rotation of the two parts of the beam at E, and if this is set equal to unity the same expression results as that obtained by the application of virtual work.

The influence line for bending moment at B is required for the beam in figure 9.4a. Note that there is a different cross-section for each span. The deflected form of the beam with a pin inserted at B and unit moments applied is shown at b, and the resulting bending-moment diagram at c.

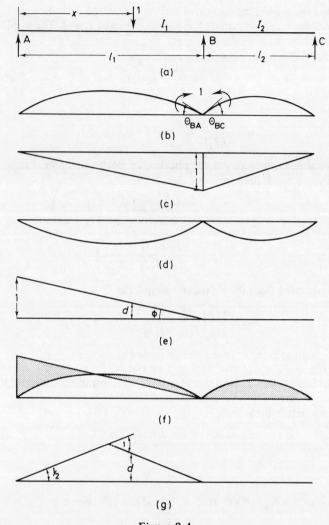

Figure 9.4

Applying the second moment area theorem about A for span AB, and about C for span BC, the slopes at B can be found. Then the first theorem can be applied to find the slopes at A and C.

$$\theta_{BA} = \frac{l_1}{3EI_1} \quad \theta_{BC} = \frac{l_2}{3EI_2}$$

$$\theta_{AB} = \frac{l_1}{6EI_1} \quad \theta_{CB} = \frac{l_2}{6EI_2}$$

Applying the second moment area theorem again the deflection under the unit load at a distance x from the left-hand support can be found. The displacement in the y-direction is

$$v = \frac{xl_1}{6EI_1} - \frac{x^3}{6EI_1l_1}$$

therefore

$$v = \frac{x}{6EI_1l_1}(l_1^2 - x^2) \quad 0 < x < l_1$$

We have already shown by the application of Betti's theorem that $M_B = v/(\theta_{BA} + \theta_{BC})$, therefore

$$M_B = \frac{x(l_1^2 - x^2)}{6EI_1l_1} \bigg/ \frac{1}{3E}\left(\frac{l_1}{I_1} + \frac{l_2}{I_2}\right)$$

$$= \frac{xl_2}{2l_1}\left(\frac{l_1^2 - x^2}{l_1 I_2 + l_2 I_1}\right)$$

If x is measured from the right-hand support at C

$$M_B = \frac{xl_1}{2l_2}\left(\frac{l_2^2 - x^2}{l_1 I_2 + l_2 I_2}\right) \quad 0 < x < l_2$$

The moment at B is negative and is sketched at d.

The reaction V_A could now be found by statics. An alternative approach however would be as follows: the hinge inserted at B is left in position and the end A of the beam is given a unit vertical displacement (e).

Applying Betti's theorem

$$V_A \times 1 - 1 \times d + M_B \times \phi = 0$$

therefore

$$V_A = d - M_B\phi$$

The value of d is $(l_1 - x)/l_1$ for $x < l_1$ and $\phi = 1/l_1$. For $x > l_1$, $d = 0$.

$$V_A = \frac{l_1 - x}{l_1} - \frac{xl_2}{2l_1^2}\left(\frac{l_1^2 - x^2}{l_1 I_2 + l_2 I_1}\right);$$

$0 < x < l_1$, x measured from A.

$$V_A = \frac{-xl_1}{2l_1 l_2}\left(\frac{l_2^2 - x^2}{l_1 I_2 + l_2 I_1}\right)$$

$0 < x < l_2$, x measured from C.

A sketch of the influence line is shown at f.

The influence line for bending moment at any other point can easily be determined. Since a pin has been inserted at B the statically determinate bending-moment diagram can be drawn; this has then to be corrected for continuity at B. Consider the bending moment at the mid-span of AB. A pin is inserted at the point and moments are applied until the two parts of the beam have a unit rotation (figure 9.4g).

$$M \times 1 - 1 \times d + M_B \times \tfrac{1}{2} = 0$$

$$M = d - \frac{M_B}{2}$$

The influence line can now be drawn.

In the last two cases the system has been made statically determinate, and the influence line for the required quantity drawn and then corrected. This method of approach can be used for beams with a higher degree of redundancy.

9.4 Multi-span beams

The influence line is required for the reaction at A (figure 9.5). If the supports at B and C are removed (b) the influence line can easily be drawn. This will need to be corrected since the values of V_A should be zero when the unit load is at B and C.

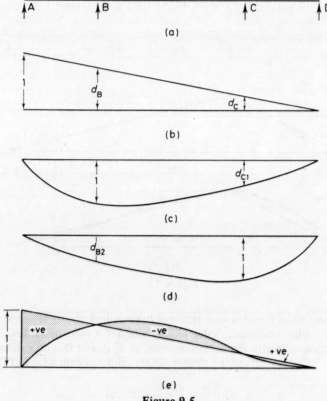

Figure 9.5

The influence line for the reaction at B is next determined when the beam is simply supported at A, B, and D. This will be the deflection curve for the beam simply supported at A and D with a load applied at B such that the deflection at B is unity (c). In a similar way the influence line for the reaction at C when the beam is simply supported at A, C and D can be drawn (d).

The original influence line can now be corrected. The apparent reaction d_B can be reduced to zero by adding the correct proportion of the reactions given by c and d. Thus

$$1 \times K_1 + d_{B2} \times K_2 + d_B = 0$$

and

$$d_{C1} \times K_1 + 1 \times K_2 + d_C = 0$$

From these two equations we can determine the value of the two constants K_1 and K_2. The correct influence line for V_A can now be drawn. It will consist of figure b plus K_1 times figure c plus K_2 times figure d. A sketch of this is shown at e.

The influence line for bending moment at a point on a beam can be found by a similar approach. As a numerical example we shall find the influence line for bending moment at B for the beam with three equal spans (figure 9.6).

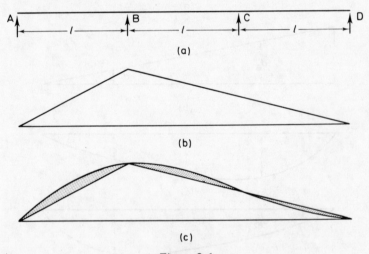

(a)

(b)

(c)

Figure 9.6

If the supports at B and C were removed the required influence line would be as shown at b with a maximum value of $2l/3$ at B and a value of $l/3$ at C.

The reaction at B with the beam supported at A, B, and D is now found. We require the deflection curve for a simply supported beam with a load at B.

$$EI \frac{d^2v}{dx^2} = \frac{2Px}{3} - P[x - l]$$

x measured from A.

$$EI\frac{dv}{dx} = \frac{2Px^2}{6} - P\frac{[x-l]^2}{2} + A$$

$$EIv = \frac{2Px^3}{18} - P\frac{[x-l]^3}{6} + Ax + B$$

At $x = 0$, $v = 0$ therefore $B = 0$. At $x = 3l$, $v = 0$ therefore $A = -5Pl^2/9$.

We shall confine out attention to a few salient points. These are set out in table 9.1 together with the corresponding deflection if the value at B is taken as unity. $k = Pl^3/EI$.

Table 9.1

x	v	Relative value
$\dfrac{l}{2}$	$-\dfrac{19}{72}k$	0·594
l	$-\dfrac{32}{72}k$	1·0
$\dfrac{3l}{2}$	$-\dfrac{69}{144}k$	1·078
$2l$	$-\dfrac{28}{72}k$	0·875
$\dfrac{5l}{2}$	$-\dfrac{31}{144}k$	0·484

We have considered the beam supported at A, B, and D and have the influence line for the reaction at B. Due to the reaction at B alone the reactions at A and C will be some proportion of R_B and the bending moment at any point will be proportional to a length times the value of R_B.

Since the problem is symmetrical the values in the table may be used to find the influence line for the reaction at C when the beam is supported at A, C and D. The influence line for the bending moment at B will be a constant times the value of this influence line.

It should now be possible to combine the three influence lines to find the value of the constants, as in the original problem the bending moment is zero when the load is at B or C. At B

$$\frac{2l}{3} + K_1 + 0.875K_2 = 0$$

At C

$$\frac{l}{3} + 0 \cdot 875K_1 + K_2 = 0$$

whence $K_1 = -1 \cdot 63l$ and $K_2 = 1 \cdot 09l$.

The influence line for bending moment at B can now be found; the values are set out below at $l/2$, $3l/2$ and $5l/2$.

	$\dfrac{l}{2}$	$\dfrac{3l}{2}$	$\dfrac{5l}{2}$
B.M. stat. det.	$0 \cdot 333l$	$0 \cdot 5l$	$0 \cdot 167l$
Correction for V_B	$-0 \cdot 968l$	$-1 \cdot 755l$	$-0 \cdot 79l$
Correction for V_C	$0 \cdot 528l$	$1 \cdot 172l$	$0 \cdot 646l$
Σ	$-0 \cdot 107l$	$-0 \cdot 083l$	$0 \cdot 023l$

A sketch of the final bending moment diagram is shown in figure 9.6c.

Problems

9.1 A uniform beam is of length l and uniform stiffness EI. The left-hand end is simply supported and the right-hand end is fixed. Calculate the deflection curve when a rotation θ is induced at the left-hand end by applying a moment to it. Hence find the influence line for the moment induced at the left-hand end by a unit vertical load acting on the beam when both ends are fixed.

9.2 Starting from first principles prove the Müller–Breslau theorem. Obtain the deflection curve for a simply supported uniform beam of flexural rigidity EI and length l when a moment is applied at one of the supports. Use this result to determine the influence line for the support moment at the centre B of a two-span uniform beam of length $2l$, given that the supports A, B, and C are level and unyielding. Deduce the influence line for the moment at a point D at the mid-span of AB.

9.3 A cantilever of length l has a prop placed at the free end. Find the influence line for the force in the prop.

9.4 A simply supported beam of span $3l$ has an intermediate support at l from one end. The flexural rigidity of the beam is EI. Determine the influence line for the reaction at the intermediate support.

9.5 A symmetrical parabolic arch of span l and rise h is pinned at the abutments. The slope of the arch at any point is ϕ and the flexural rigidity is given by $EI \cos \phi$, where EI is the value at the crown. Determine the influence line for the horizontal thrust.

10 STABILITY OF COLUMNS

10.1 Introduction

Let us consider the case of a perfectly straight long slender rod of uniform cross-section and made from a linear elastic material. The rod is first tested in tension. An axial load is steadily applied and increased until the yield stress is approached. If a small lateral disturbing force were applied during the test the rod would bend slightly, but would return to the original straight state as soon as the lateral load is removed.

If the same rod is next tested in axial compression it will remain in a state of equilibrium until a certain critical load is reached. Thus it initially behaves in a similar manner to the rod in tension. At this critical load the rod will remain bent and will not return to its original straight state after the application and removal of a small lateral load. The average compressive stress would be found to be very much less than the yield stress of the material in compression.

There is then a very big difference in the behaviour of such a member when it is loaded in tension or compression, and we shall seek to examine the compression behaviour in more detail.

In practice it is not possible to obtain a perfectly straight uniform member, and the material will not be perfectly homogeneous. Any deviation from perfection is equivalent to having a slight eccentricity of loading, producing bending in addition to the axial thrust. The bending action is equivalent to the small lateral load that was applied when the axial load reached a certain value at which instability occurred. We shall refer to the load at which instability occurs as the *Euler critical load*.

10.2 The use of potential energy

In order to gain further insight into the question of stability we shall examine in detail what would appear at first sight to be a trivial problem. In the introduction we could refer to the member in compression as being in stable equilibrium if it returned to its original position when a small lateral disturbing force was applied. An alternative approach would be to make use of the potential energy, V, of the system.

311

It is shown in many books on statics that if a body with potential energy V is given a virtual displacement in the x direction, then the body is in equilibrium if dV/dx is zero. It can further be shown that the equilibrium is stable if $d^2 V/dx^2$ is positive, neutral if $d^2 V/dx^2$ is zero and unstable if $d^2 V/dx^2$ is negative. In the case of a structure the potential energy would consist of the sum of the strain energy in the members and the potential energy of the external forces.

Figure 10.1

The problem that we shall consider is shown in figure 10.1, and consists of a rigid rod OG of length l that will be considered massless; the end O is pinned to a rigid foundation but is restrained by a torsional spring of stiffness k, the spring having zero torque in it when the rod is horizontal. A horizontal force P acts at end G; the force is assumed to remain horizontal when the rod displaces through an angle θ as shown.

The strain energy stored in the spring is $\frac{1}{2}k\theta^2$, and the loss of potential energy of the load P can be written $Pl(1 - \cos\theta)$, so that

$$V = \tfrac{1}{2}k\theta^2 - Pl(1 - \cos\theta)$$

$$\frac{dV}{d\theta} = k\theta - Pl\sin\theta$$

$$\frac{d^2 V}{d\theta^2} = k - Pl\cos\theta$$

An equilibrium situation exists when $dV/d\theta = 0$; there are in fact two conditions

$$\theta = 0 \text{ and } \frac{\theta}{\sin\theta} = \frac{Pl}{k}$$

For the case $\theta = 0$, $d^2 V/d\theta^2 = k - Pl$, so that if this is to be a stable position

$$k > Pl \text{ or } P < \frac{k}{l}$$

Obviously if $P > k/l$ the rod is in a position of unstable equilibrium. We shall put $P_c = k/l$, the value that divides the stable from the unstable situation.

The graph shown in figure 10.2 gives a plot of P/P_c against θ. The region discussed so far is the line AB which is stable and the line BC which is unstable, the division occuring at B where $P/P_c = 1$.

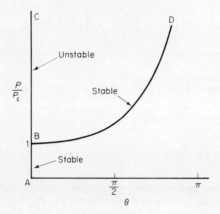

Figure 10.2

Next consider the other possible solution

$$\frac{\theta}{\sin \theta} = \frac{Pl}{k} = \frac{P}{P_c}$$

A plot of this results in the curve **BD**. Now

$$\frac{d^2 V}{d\theta^2} = k - Pl \cos \theta$$

and substituting for P

$$\frac{d^2 V}{d\theta^2} = k(1 - \theta \cot \theta)$$

Since $(1 - \theta \cot \theta)$ is positive for all values of θ between 0 and π, $d^2 V/d\theta^2$ is positive and we have a stable situation.

So far we have only considered the mathematics of the problem and it is enlightening to look at the configuration when $\theta = 90°$ and $180°$. At the $90°$ position the force P will be at right angles to the rod, thus bending the rod. At $180°$ the rod will be in tension.

When the load is steadily increased from zero, the P-θ relation initially follows the line AB, that is $\theta = 0$. At B the critical load P_c is reached, and there is a choice of two equilibrium paths, BC or BD, the former being unstable and the latter stable. The division is referred to as the bifurcation of the equilibrium path. In general the structure is more likely to follow the stable path. It should also be noted that the equilibrium is stable when $P = P_c$.

We can approach the solution to the problem in a slightly different way. All the theory that we have considered so far in elasticity has referred to small deformations, in which case it would be reasonable to consider a small change in the value of θ from the position $\theta = 0$. This can easily be done by substituting an expansion for $\sin \theta$ and $\cos \theta$, that is, $\sin \theta = \theta$ and $\cos \theta = 1 - \theta^2/2$. In which case

$$V = \tfrac{1}{2}k\theta^2 - Pl\left(1 - 1 + \frac{\theta^2}{2}\right) = \tfrac{1}{2}(k - Pl)\,\theta^2$$

giving

$$\frac{\mathrm{d}V}{\mathrm{d}\theta} = (k - Pl)\,\theta$$

and

$$\frac{\mathrm{d}^2 V}{\mathrm{d}\theta^2} = (k - Pl)$$

So that $\mathrm{d}V/\mathrm{d}\theta = 0$ when $\theta = 0$, the rod is straight, and from the second differential the situation is stable so long as $P < k/l$ or $P < P_c$. This again correstonds to the line AB. There is a further solution when $\mathrm{d}V/\mathrm{d}\theta = 0$ given by $P = k/l = P_c$; note that if this is so then any value of θ will satisfy, indicating that the equilibrium is neutral. This would be represented by a horizontal line through B. Remember, however, that the value of θ is small in the approximation, and the result would not be far wrong for the initial portion of the line BD. It follows that small displacement theory gives the correct value of the load at which bifurcation occurs, but states that the equilibrium is neutral rather than stable for small values of θ.

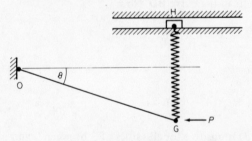

Figure 10.3

Next consider a slight variation in the problem, figure 10.3. The torsion spring at the pinned end has been replaced by a tension spring GH which is pinned to the free end of the rod at G and fixed in a suitable manner at H so that it is perfectly free to move in a horizontal direction, such that the spring always remains vertical as the rod rotates. The load P is again assumed to act in a horizontal direction, and the force in the spring is zero when the rod is horizontal.

In this case

$$V = \tfrac{1}{2}k\,(l\sin\theta)^2 - Pl\,(1 - \cos\theta)$$

$$\frac{\mathrm{d}V}{\mathrm{d}\theta} = kl^2\sin\theta\cos\theta - Pl\sin\theta$$

$\mathrm{d}V/\mathrm{d}\theta$ is zero when $\theta = 0$ or π or when $\cos\theta = P/kl = P/P_c$ where $kl = P_c$. Thus there are three equilibrium conditions.

$$\frac{\mathrm{d}^2 V}{\mathrm{d}\theta^2} = kl^2 \cos 2\theta - Pl \cos \theta$$

When $\theta = 0$, $\mathrm{d}^2 V/\mathrm{d}\theta^2 = kl^2 - Pl$. This will be positive so long as $P < P_c$. When $\theta = \pi$, $\mathrm{d}^2 V/\mathrm{d}\theta^2 = kl^2 + Pl$. This will always be positive for positive values of P. It is of course a slightly artificial case because the rod has turned through $180°$ and the original compressive force has now become tensile. If the sign of P were changed the condition for stability would be that $P/P_c > -1$.

The last equilibrium condition was $\cos \theta = P/kl$, and substituting into the second differential will give

$$\frac{\mathrm{d}^2 V}{\mathrm{d}\theta^2} = -kl^2 \sin^2 \theta$$

It can be seen that the equilibrium is neutral if $\theta = 0$ or π and unstable for $0 < \theta < \pi$.

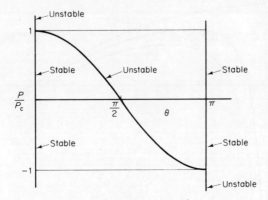

Figure 10.4

The stable and unstable regions are shown in figure 10.4. Note that there is again a bifurcation at $P/P_c = 1$, but this time we go from a stable to two unstable regimes. This means that as soon as the load reaches P_c, instability occurs and the rod will rotate through $180°$ until a stable situation is reached with the rod in pure tension. This of course differs from the previous case where the situation was stable when P exceeded the value P_c. The difference lies in that for the first case the restoring torque continues to increase with θ in the torsion spring; in the linear spring the force increases until $\theta = 90°$ but the moment arm is decreasing in length and is zero at $\theta = 90°$.

It is important to consider the effect of what might be called an initial imperfection. In the problem that we have just discussed, it was assumed that the rod OG was horizontal when the spring force was zero. Let us now investigate the case when the rod is inclined at a small angle ϕ below the horizontal when the spring force is zero.

The expression for the potential energy will be modified as follows. θ is again measured from the horizontal

$$V = \tfrac{1}{2}kl^2 \,(\sin\theta - \sin\phi)^2 - Pl\,(\cos\phi - \cos\theta)$$

$$\frac{dV}{d\theta} = kl^2\,(\sin\theta - \sin\phi)\cos\theta - Pl\sin\theta$$

Putting $dV/d\theta = 0$ and substituting $kl = P_c$

$$P_c\,(\sin\theta - \sin\phi)\cos\theta = P\sin\theta$$

$$\frac{P}{P_c} = (\sin\theta - \sin\phi)\cot\theta \tag{10.1}$$

When ϕ is zero equation 10.1, as would be expected, gives the same result as was obtained previously with the rod horizontal.

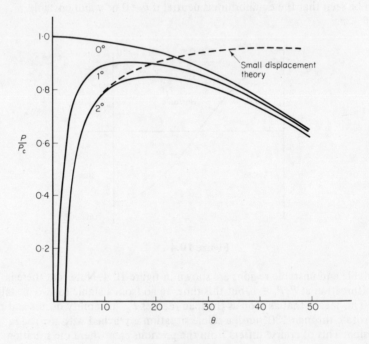

Figure 10.5

A plot of P/P_c against θ is shown in figure 10.5 for values of $\phi = 0°$, $1°$, and $2°$. It is important to note that a small imperfection can cause a considerable reduction in the maximum load that can be applied before the system becomes unstable. The value of θ when the maximum load is reached, or when instability occurs can be found by differentiating the expression for P/P_c with respect to θ in equation 10.1.

It will be found that

$$\sin\theta = (\sin\phi)^{\tfrac{1}{3}}$$

giving

$$\frac{P}{P_c} = \left\{1 - (\sin \phi)^{\frac{2}{3}}\right\}^{\frac{3}{2}}$$

If small displacement theory is used, that is, substituting θ for $\sin \theta$ and $1 - \theta^2/2$ for $\cos \theta$ in equation 10.1 it will be found that

$$\frac{P}{P_c} = \frac{(\theta - \phi)}{\theta}$$

The broken line in figure 10.5 is drawn for a value of $\phi = 2°$. The agreement with large displacement theory is good up to a value of P/P_c of about 0·7. After this there is a divergence and the graph indicates that the value of P steadily increases with θ giving no hint of a maximum value and an unstable situation. It is of course unreasonable to use small displacement theory for such large values of θ. However, there is a lesson to be learnt in that it is quite possible to draw the wrong conclusions on the question of stability when using small displacement approximations.

As a conclusion to this section, let us examine the stability if the linear spring is replaced with a non-linear one with $\phi = 0°$. A fairly simple form for the spring would be such that the relation between the load P and the displacement d would be given by

$$P = kd + k_1 d^2$$

The strain energy stored in the spring can easily be found by integrating this expression, hence

$$U = \tfrac{1}{2}kd^2 + \tfrac{1}{3}k_1 d^3$$

substituting $d = l \sin \theta$ will give the following expression for the total potential energy

$$V = \frac{kl^2 \sin^2 \theta}{2} + \frac{k_1 l^3 \sin^3 \theta}{3} - Pl(1 - \cos \theta)$$

$$\frac{dV}{d\theta} = kl^2 \sin \theta \cos \theta + k_1 l^3 \sin^2 \theta \cos \theta - Pl \sin \theta = 0$$

for stability, whence

$$P = l \cos \theta \, (k + k_1 l \sin \theta)$$

substituting $P_c = kl$

$$\frac{P}{P_c} = \cos \theta \, (1 + \frac{k_1 l}{k} \sin \theta)$$

$$\frac{d(P/P_c)}{d\theta} = -\sin \theta + \frac{k_1 l}{k} \cos 2\theta$$

this last expression is of course equivalent to finding $d^2 V/d\theta^2$ and we have a stable situation if

$$\frac{k_1 l}{k} \cos 2\theta > \sin \theta$$

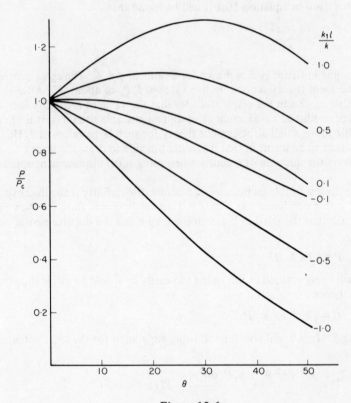

Figure 10.6

A series of graphs of P/P_c against θ is shown in figure 10.6 for six different values of $k_1 l/k$; the value of P_c has again been taken as kl. Three of the graphs are for positive values of the stiffness ratio and three for negative values. The first point to note is that all the graphs have a common value of $P/P_c = 1$ when $\theta = 0$. For positive values of stiffness ratio there will always be an initial region which is stable. Instability sets in as soon as $(k_1 l/k) \cos 2\theta = \sin \theta$. The value of θ at which this occurs increases with increasing value of k_1. For negative values of the stiffness ratio the system is always unstable; this can be seen at once by inspecting the three graphs in figure 10.6.

10.3 Euler critical loads

Relations between the vertical load, shear force and bending moment were derived for a beam element in section 1.9. We shall now consider the slightly more complicated case shown in figure 10.7, where a beam element of length d*x* has a

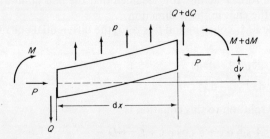

Figure 10.7

vertical load of p per unit length applied and in addition a compressive force P is applied at the ends. A relative displacement dv occurs at the ends. Equilibrium of the beam is maintained by the addition of shear forces and bending moments.

Resolving vertically gives

$$dQ + p\, dx = 0$$

or

$$\frac{dQ}{dx} = -p$$

Taking moments and omitting second-order terms

$$dM + Q\, dx + P\, dv = 0$$

$$\frac{dM}{dx} + Q + P\frac{dv}{dx} = 0$$

differentiating with respect to x

$$\frac{d^2M}{dx^2} + \frac{dQ}{dx} + \frac{d}{dx}\left(P\frac{dv}{dx}\right) = 0$$

now $M = EI d^2v/dx^2$ and $dQ/dx = -p$; thus

$$\frac{d^2}{dx^2}\left(EI\frac{d^2v}{dx^2}\right) + \frac{d}{dx}\left(P\frac{dv}{dx}\right) = p \qquad (10.2)$$

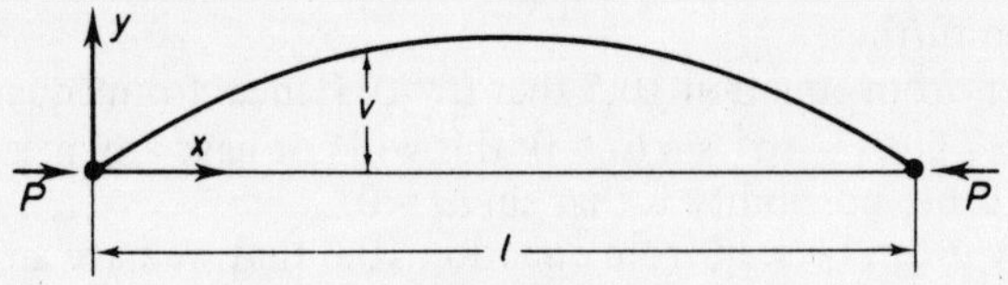

Figure 10.8

Let us apply the differential equation to the problem of a thin elastic rod of constant cross-section shown in figure 10.8. The rod is pinned at each end, where there is an axial force P acting. It is assumed that the pins do not restrict relative movements at the ends in the x direction.

For this case EI is constant and p is zero; the differential equation will simplify to

$$EI \frac{d^4 v}{dx^4} + P \frac{d^2 v}{dx^2} = 0 \qquad (10.3)$$

The general solution to this equation is

$$v = A \sin \alpha x + B \cos \alpha x + Cx + D$$

where $\alpha^2 = P/EI$; thus

$$\frac{d^2 v}{dx^2} = -A \alpha^2 \sin \alpha x - B \alpha^2 \cos \alpha x$$

the end conditions are such that both the deflection and the bending moment are zero at each end of the beam.

$x = 0, v = 0$

$$0 + B + 0 + D = 0 \qquad (10.4)$$

$x = 0, M = 0$

$$0 - B\alpha^2 + 0 + 0 = 0 \qquad (10.5)$$

$x = l, v = 0$

$$A \sin \alpha l + B \cos \alpha l + Cl + D = 0 \qquad (10.6)$$

$x = l, M = 0$

$$-A\alpha^2 \sin \alpha l - B\alpha^2 \cos \alpha l + 0 + 0 = 0 \qquad (10.7)$$

This can be written in matrix form

$$\begin{bmatrix} 0 & 1 & 0 & 1 \\ 0 & -\alpha^2 & 0 & 0 \\ \sin \alpha l & \cos \alpha l & l & 1 \\ -\alpha^2 \sin \alpha l & -\alpha^2 \cos \alpha l & 0 & 0 \end{bmatrix} \begin{bmatrix} A \\ B \\ C \\ D \end{bmatrix} = 0 \qquad (10.8)$$

There is one trivial solution to this set of equations: $A = B = C = D = 0$; this will not be discussed further.

It can be seen from equation 10.5 that $B = 0$. Hence, from equation 10.4, $D = 0$.

Equation 10.7 gives $-A\alpha^2 \sin \alpha l = 0$; this will not have any meaning if A is zero. The only other possibility is that $\sin \alpha l = 0$.

Alternatively it can be seen from equation 10.8 that we have an eigenvalue problem, in which case, for an non-trivial solution the determinant of the matrix must be zero, so that

$$-\alpha^4 l \sin \alpha l = 0$$

If $\alpha = 0$ then the value of P is zero, so that there is no compressive force acting. Thus we are again left with the only possibility that

$$\sin \alpha l = 0$$

This implies that $\alpha l = 0$ or $n\pi$ where $n = 1, 2, 3$, etc. Now $\alpha^2 = P/EI$, so that

$$P = \frac{n^2 \pi^2 EI}{l^2} \tag{10.9}$$

The smallest value of P for an equilibrium state will be when $n = 1$. In which case $P = \pi^2 EI/l^2$. We shall refer to this value as being the Euler critical load for a pin-ended strut, and denote it by P_E. Higher values of n correspond to different buckling modes. For example the case for $n = 2$ could be obtained if the centre of the strut were constrained such that it could not move laterally.

An attempt will next be made to determine the deflection curve for the strut when the critical load is applied.

It has been shown from equations 10.4 to 10.7 that B, D, and $A \sin \alpha l$ are all zero; so that from equation 10.6 C must be zero

$$v = A \sin \alpha x = A \sin \sqrt{\left(\frac{P_E}{EI}\right)}\, x = A \sin \frac{\pi}{l} x \tag{10.10}$$

The value of A is indeterminate. All that can be said is that a pin-ended strut will bend into the shape of a sine curve, the amplitude of which is indeterminate.

In the first part of this chapter potential energy was applied to determine the stability of certain structural problems. It should be possible to examine the stability of a pin-ended strut by this means. It will of course be necessary to try to calculate the potential energy, including the strain energy due to bending.

The strain energy due to bending will be given by the expression.

$$U = \int_0^l \frac{M^2 \, dx}{2EI}$$

Writing $M = EI \, d^2 v/dx^2$

$$U = \int_0^l \frac{EI}{2} \left(\frac{d^2 v}{dx^2}\right)^2 \, dx$$

The energy due to axial and shear effects will be omitted as small compared with the bending effect.

The applied compressive force P will move through a small distance due to the axial compression of the rod; the external work done will be omitted as was the energy due to axial compression. It will be assumed that the strut remains straight until the critical load is reached and then suddenly bows out. The force P will move through a distance that can be calculated as follows.

$$ds^2 = dx^2 + dv^2$$

or

$$ds = dx \left\{ 1 + (\frac{dv}{dx})^2 \right\}^{\frac{1}{2}}$$

expanding

$$ds \approx dx \left\{ 1 + \frac{1}{2} (\frac{dv}{dx})^2 \right\}$$

Assuming that the length of the strut does not change, if the original length was l the bowed length is also l. Let the distance moved through by the load be δ, then

$$l = \int_\delta^l ds = \int_\delta^l dx + \frac{1}{2}(\frac{dv}{dx})^2 \ dx = l - \delta + \int_\delta^l \frac{1}{2}(\frac{dv}{dx})^2 \ dx$$

so that

$$\delta = \int_\delta^l \frac{1}{2}(\frac{dv}{dx})^2 \ dx$$

As δ is a small quantity there will only be a very small error in the integral if the limits are taken as 0 to 1. Hence

$$\delta = \int_0^l \frac{1}{2} (\frac{dv}{dx})^2 \ dx$$

It is now possible to write down an expression for the potential energy

$$V = \int_0^l \frac{EI}{2} (\frac{d^2 v}{dx^2})^2 \ dx - \frac{P}{2} \int_0^l (\frac{dv}{dx})^2 \ dx \qquad (10.11)$$

For the case of the pin-ended strut we have already determined that $v = A \sin \pi x/l$, so that

$$\frac{dv}{dx} = \frac{A\pi}{l} \cos \frac{\pi x}{l} \text{ and } \frac{d^2 v}{dx^2} = - \frac{A\pi^2}{l^2} \sin \frac{\pi x}{l}$$

$$V = \int_0^l \frac{EI}{2} \frac{A^2 \pi^4}{l^4} \sin^2 \frac{\pi x}{l} \ dx - \frac{P}{2} \int_0^l \frac{A^2 \pi^2}{l^2} \cos^2 \frac{\pi x}{l} \ dx$$

$$= \frac{\pi^4 A^2 EI}{4l^3} - \frac{\pi^2 A^2 P}{4l}$$

For equilibrium $dV/dA = 0$ or

$$\frac{\pi^2 A}{2l} (\frac{\pi^2 EI}{l^2} - P) = 0$$

There are two possible equilibrium situations; the first is when $A = 0$. This implies that the strut remains straight. The alternative is when $P = \pi^2 EI/l^2$, the result that was found previously.

$$\frac{d^2 V}{dA^2} = \frac{\pi^2}{2l} (\frac{\pi^2 EI}{l^2} - P)$$

and there is a stable condition if $P < \pi^2 EI/l^2$, neutral if $P = \pi^2 EI/l^2$ and unstable if $P > \pi^2 EI/l^2$.

It was established that the differential equation 10.2 was a fourth-order equation. If it is assumed right from the start of the derivation that there is no vertical load p applied, then the equation of equilibrium can be written as a second-order equation.

$$M = EI \frac{d^2 v}{dx^2} \tag{10.12}$$

where the value of M for the pin-ended strut would be given by $-Pv$, thus

$$EI \frac{d^2 v}{dx^2} + Pv = 0$$

and

$$v = A \sin \alpha x + B \cos \alpha x$$

In general it will be found that the majority of simple strut problems can be solved by writing down an equation of the form given by equation 10.12.

The particular case that we explored was for a strut with pinned ends. Suppose that the ends had been built in; this would mean that in addition to the axial load P there would be a fixing moment M_1. The differential equation would be modified to

$$EI \frac{d^2 v}{dx^2} = -Pv + M_1 \tag{10.13}$$

The solution of the complementary function would be identical to the pinned-end case, but there would be a particular integral—a constant—to consider as well.

It can be seen that Euler critical loads can be found for other end conditions by a similar approach. It is perhaps easier to make use of the solution that we have already obtained for the pin-ended strut. For a number of other cases the solution of the differential equation will always be part of the sine curve $v = A \sin \alpha x$. So long as suitable axes are chosen the load P_C which gives the correct displacement curve will be the Euler critical load for the particular set of end conditions.

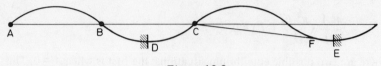

Figure 10.9

Figure 10.9 shows a continuous sine curve, and it can be seen that for the case of a pin-ended strut the first Euler critical load $\pi^2 EI/l^2$ would correspond to the length AB. AC would correspond to the second critical load. Denoting the distance between two points of contraflexure as L, that is, AB = L

$$P_E = \frac{\pi^2 EI}{L^2}$$

If we express L as a fraction of the actual length of the strut we can obtain the values of the critical loads corresponding to the higher modes. For AC, $2L = l$ therefore $P_C = 4\pi^2 EI/l^2$, the second Euler critical load.

We shall next examine other possible end conditions.

CD represents a strut with end D built in and end C free from any restraint. When l = CD, $L = 2l$.

$$P_C = \frac{\pi^2 EI}{4l^2} = \frac{P_E}{4}$$

DE represents a strut with both ends built in. When l = DE, $L = l/2$ and

$$P_C = \frac{4\pi^2 EI}{l^2} = 4P_E$$

It is also possible to derive the case for a strut that is built in at one end while the other end is constrained such that it can only move along the original axis of the strut. This is represented by CF. C is a point of contraflexure and CF is a tangent to the sine curve. The approximate length of the equivalent strut will be given by the horizontal component of CF.

$$v = A \sin \frac{\pi x}{L}$$

$$\frac{dv}{dx} = \frac{A\pi}{L} \cos \frac{\pi x}{L}$$

Now for line CF, $v = mx$ where m is the slope. At F

$$m = \left(\frac{dv}{dx}\right)_F = \frac{A\pi}{L} \cos \frac{\pi x_F}{L}$$

and

$$m = \frac{v_F}{x_F} = \frac{A}{x_F} \sin \frac{\pi x_F}{L}$$

therefore

$$\frac{A}{x_F} \sin \frac{\pi x_F}{L} = \frac{A\pi}{L} \cos \frac{\pi x_F}{L}$$

The solution of this equation gives an approximate value of x_F/L as $1\cdot43$; or $l = 1\cdot43L$. Thus

$$P_C \approx \frac{1\cdot43^2\pi^2 EI}{l^2} = 2\cdot05 P_E$$

It is suggested that the reader should attempt to obtain this result by the conventional method of writing down the differential equation. A very common error in setting this equation down is to omit the restraining force that is required at the pinned end normal to the applied load P.

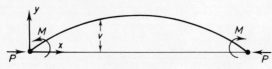

Figure 10.10

There is one further case that we shall consider with particular end conditions, and that is the problem of a strut with equal and opposite end moments applied in addition to the axial load (figure 10.10). The ends are pinned.

The differential equation is

$$EI \frac{d^2v}{dx^2} = -Pv - M$$

or

$$\frac{d^2v}{dx^2} + \alpha^2 v = -\frac{M}{P}\alpha^2$$

whence

$$v = A \sin \alpha x + B \cos \alpha x - \frac{M}{P}$$

when $x = 0$, $v = 0$, therefore $B = M/P$. When $x = l/2$, $dv/dx = 0$ therefore

$$A = \frac{M}{P} \tan \frac{\alpha l}{2}$$

thus

$$v = \frac{M}{P}\left(\tan \frac{\alpha l}{2} \sin \alpha x + \cos \alpha x - 1\right)$$

Note that a relation has been obtained between x and v in this case.
The maximum deflection occurs when $x = l/2$ and

$$v = \frac{M}{P}\left(\sec\frac{\alpha l}{2} - 1\right) \tag{10.14}$$

It is of interest to find the stiffness of the strut, that is, the moment for unit rotation.

The slope at the ends will be required, and this has a value of

$$\frac{M\alpha}{P}\tan\frac{\alpha l}{2}$$

Thus

$$s = \frac{M}{\theta} = \frac{P}{\alpha\tan\alpha l/2}$$

Ths stiffness will become zero if $\tan\alpha l/2 = \infty$, that is

$$\frac{\alpha l}{2} = \frac{\pi}{2}$$

or

$$P = \frac{\pi^2 EI}{l^2} = P_E$$

The Euler critical load is reached at zero stiffness.

10.4 Strut with initial deformation

It has already been mentioned that it is impossible to obtain a perfectly straight structural member. The initial deviation from perfect straightness could be written down in the form $v_0 = f(x)$.

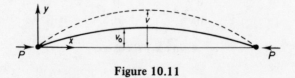

Figure 10.11

Figure 10.11 shows an imperfect column of uniform section with pinned ends. When the load P is applied there will be a resulting change of curvature that can be represented by

$$\frac{\mathrm{d}^2 v}{\mathrm{d}x^2} - \frac{\mathrm{d}^2 v_0}{\mathrm{d}x^2}$$

Hence

$$EI\left(\frac{d^2v}{dx^2} - \frac{d^2v_0}{dx^2}\right) = -Pv$$

or

$$EI\frac{d^2v}{dx^2} + Pv = EI\frac{d^2v_0}{dx^2} \tag{10.15}$$

It is worth noting at this stage that the term $EI(d^2v_0/dx^2)$ could arise from a different problem—that of a strut with a lateral load—where it would represent the bending moment at a point on the strut when the lateral load is considered alone.

We shall make the assumption that the initial deformation can be represented as a Fourier series.

$$v_0 = \sum_{n=1}^{n=\infty} a_n \sin\frac{n\pi x}{l}$$

thus

$$\frac{d^2v}{dx^2} + \alpha^2 v = -\sum_{n=1}^{n=\infty} \frac{a_n n^2 \pi^2}{l^2} \sin\frac{n\pi x}{l}$$

$$v = A\sin\alpha x + B\cos\alpha x + \sum_{n=1}^{n=\infty} \frac{a_n n^2 \pi^2}{l^2} \sin\frac{n\pi x}{l}\left/\frac{n^2\pi^2}{l^2} - \alpha^2\right.$$

For pinned ends $v = 0$ when $x = 0$ or 1. Thus $A = B = 0$.

$$v = \sum_{n=1}^{n=\infty} \frac{a_n \sin(n\pi x/l)}{1 - (P/n^2 P_E)} \tag{10.16}$$

If we take the ratio of v/v_0 for the nth term, that is,

$$1\left/1 - \frac{P}{n^2 P_E}\right.$$

we have what is termed an 'amplification factor'—the factor by which the initial component of deflection would be multiplied when an axial load P is applied to the strut.

It was mentioned that equation 10.15 also applies for the case of a laterally loaded strut. If it is possible to represent the bending moment due to the lateral loading in the form of a Fourier series, the resulting deflection equation will be identical with equation 10.16.

10.5 Struts made from ideal elasto-plastic materials

So far it has been assumed that the material of the strut behaves in an elastic manner and that the yield point of the material has not been exceeded. The Euler critical load was found to be

$$P_E = \frac{\pi^2 EI}{l^2}$$

The second moment of area I can be represented in the form Ak^2 where A is the cross-sectional area and k the radius of gyration. Thus

$$\sigma_E = \frac{\pi^2 Ek^2}{l^2} \tag{10.17}$$

This is the stress that would arise at the Euler critical load. The ratio of l/k is referred to as the 'slenderness ratio' of the strut.

Figure 10.12 shows a plot of σ_E against l/k for a steel strut. This graph will of course be incorrect once the value of the yield stress has been reached. Assuming a mild steel with a yield stress of 240 N/mm^2, the corresponding value of l/k is about 93. This means that the theory that we have derived will only apply for the case of struts with a slenderness ratio that is greater than this value.

If an ideal elasto-plastic stress–strain relation is assumed, when the slenderness ratio is less than 93 the column will not fail by elastic buckling. The value of the stress would reach σ_y and the column could be said to squash. Hence the load applied is often referred to as the 'squash load' P_S.

The above discussion has been based on the assumption that the strut was initially straight and remained straight. We shall now describe what may take place with imperfections in the strut.

Consider first the case of an ideal pin-ended strut of linear elastic material. The Euler critical load would be reached before any deflection occurred. A graph of

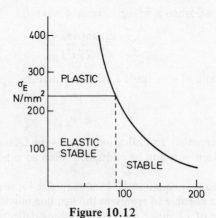

Figure 10.12

load P against central deflection would be a horizontal line A of value P_E (figure 10.13). If the strut was imperfect and there was initial eccentricity present the central displacement would be given by equation 10.16, and a curve similar to B would result assuming the material remains elastic.

Again consider a perfect strut of elasto-plastic material. The Euler load will be reached before any central deflection occurs and the stress is uniformly distributed over the section. For a small displacement, bending stresses will occur and these

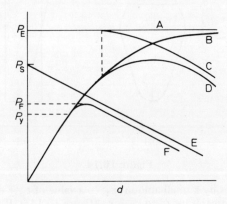

Figure 10.13

will reduce the stresses on the convex side and increase those on the concave side. At some stage the stress on the concave side will reach σ_y. Further displacement may mean that the stress on the convex side will become tensile and the load carrying capacity is reduced; there will be a deviation from the P_E line and a curve similar to C will result. If initial imperfections are present the curve would initially follow B until yield is reached and then lie below the C curve at D. For the case when the squash load P_S is reached before the Euler load, if a small displacement is imposed, the load would have to be reduced to maintain equilibrium and a curve such as E would result. If initial eccentricity is present the curve would be similar to F.

Most practical struts would have a load–deflection characteristic similar to that shown at F.

The reader should compare the curves from figure 10.13 with the results obtained in section 10.2 for initial imperfections and non-linear springs.

10.6 Double-modulus theory

This is also referred to as the reduced-modulus theory. The stress–strain curve for the material of the strut is assumed to be similar to that shown in figure 10.14a. The elastic limit is denoted by σ_y and the load applied to the column is such that $(P/A) > \sigma_y$. When the load is applied a small lateral displacement is allowed to take place near the centre of the strut; this will tend to reduce the stress on the convex side and increase the stress on the concave side.

It will be assumed that the strut has an axis of symmetry and that the displacement is applied in the direction of this axis; and that plane sections remain plane.

The cross-section of the strut is shown in figure 10.14b. Before the lateral deformation took place the stress was σ_p and the corresponding strain ϵ_p. With the displacement applied, the upper surface of the strut becomes concave and the value

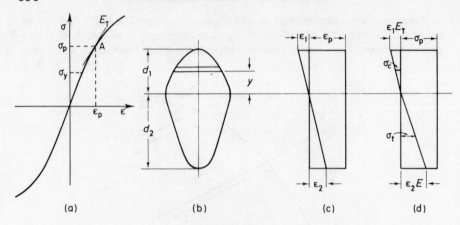

Figure 10.14

of the strain increases by a small amount ϵ_1, to a value of $\epsilon_p + \epsilon_1$, while that on the lower surface decreases by ϵ_2 to $\epsilon_p - \epsilon_2$ (figure 10.14c). If the tangent modulus on the stress–strain curve at A in figure 10.14a is E_T, the increase in stress corresponding to the strain ϵ_1, is approximately $\epsilon_1 E_T$. At the bottom surface the strain has been reduced and the unloading process will occur elastically and the stress change will be $\epsilon_2 E$. The final stress distribution is shown in figure 10.14d.

The total load applied to the cross-section has remained unchanged. Thus

$$\int_0^{d_1} \sigma_c\, b\; \mathrm{d}y - \int_0^{d_2} \sigma_t b\; \mathrm{d}y = 0$$

Now

$$\sigma_c = \epsilon_1 E_T \frac{y}{d_1} \qquad \sigma_t = \epsilon_2 E \frac{y}{d_2}$$

If the radius of curvature is R, $\epsilon_1 = d_1/R$, $\epsilon_2 = d_2/R$.

$$\frac{E_T}{R} \int_0^{d_1} by\; \mathrm{d}y - \frac{E}{R} \int_0^{d_2} by\; \mathrm{d}y = 0$$

therefore

$$E_T A_1 \bar{y}_1 - E A_2 \bar{y}_2 = 0 \qquad (10.18)$$

where $A_1\bar{y}_1$ and $A_2\bar{y}_2$ represent the first moments of area about the horizontal axis.

Equation 10.18 together with the relationship $d = (d_1 + d_2)$ enables the position of the neutral axis to be found.

Taking moments about the neutral axis

$$\int_0^{d_1} \sigma_c\, by\; \mathrm{d}y + \int_0^{d_2} \sigma_t by\; \mathrm{d}y = M$$

By making the substitutions for σ and ϵ

$$\frac{1}{R}(E_T I_1 + E I_2) = M \tag{10.19}$$

where I_1 and I_2 represent second moments of area about the neutral axis for each part.

This means that the flexural rigidity has been modified to $E_T I_1 + E I_2$.

If I is the second moment of area of the cross-section about the centroid, we could write

$$E_D I = E_T I_1 + E I_2$$

or

$$E_D = \frac{E_T I_1}{I} + \frac{E I_2}{I} \tag{10.20}$$

and this represents a reduced modulus of elasticity or double modulus.

For the case of a rectangular beam of depth d and breadth b, equation 10.18 becomes

$$E_T \frac{b d_1^2}{2} - \frac{E b d_2^2}{2} = 0$$

therefore

$$\frac{d_1}{d_2} = \left(\frac{E}{E_T}\right)^{1/2}$$

$d_1 + d_2 = d$ therefore

$$d_1 = \frac{d\sqrt{E}}{\sqrt{E_T} + \sqrt{E}} \qquad d_2 = \frac{d\sqrt{E_T}}{\sqrt{E_T} + \sqrt{E}}$$

$$I_1 = \frac{b d^3 E^{3/2}}{3(E_T^{1/2} + E^{1/2})^3} \qquad I_2 = \frac{b d^3 E_T^{3/2}}{3(E_T^{1/2} + E^{1/2})^3}$$

Now $I = b d^3 / 12$, therefore

$$E_D = \frac{4 E_T E}{(E_T^{1/2} + E^{1/2})^2} \tag{10.21}$$

The critical load P_D will of course be given by

$$P_D = \frac{\pi^2 E_D I}{l^2} \tag{10.22}$$

10.7 Tangent-modulus theory

The double-modulus theory was based on the assumption that the axial load remained constant while a small lateral displacement was applied. If however the load continued to increase the stress and strain distributions over the cross-section would differ from those shown in figure 10.14. It is quite possible that the strain could increase at both of the edges of the section, the rate of increase being greater at the top than at the bottom. If it is assumed that E_T remains constant for fairly small changes of strain, the stress increase at the top could be written $\epsilon_1 E_T$, and that at the bottom $\epsilon_2 E_T$, with a linear distribution over the cross-section.

For this particular case equation 10.19 would become

$$\frac{E_T I}{R} = M \tag{10.23}$$

where I is the second moment of area about the centroid of the cross-section.

The critical load P_T of the strut using the tangent modulus can be written down straight away as

$$P_T = \frac{\pi^2 E_T I}{l^2} \tag{10.24}$$

The value of the tangent-modulus critical load will be less than the value obtained using double-modulus theory, and this in turn will be smaller than the Euler critical load.

A number of tests have been carried out on practical struts and it is usually found that the actual critical load lies somewhere between the values given by the tangent-modulus and double-modulus methods; the tangent modulus generally gives the closer result.

10.8 Practical strut formulae

In section 10.4 we derived an expression for the lateral displacement of a strut that initially was not straight, and for which the deviations could be written down in the form of a Fourier series.

If equation 10.16 is examined it will be seen that the first term will predominate when the applied load approaches the value of the Euler critical load. If all the other terms are omitted

$$v = \frac{a \sin (\pi x/l)}{1 - (P/P_E)}$$

The maximum deflection occurs when $x = l/2$, that is

$$v_{max} = \frac{a P_E}{P_E - P}$$

thus

$$M_{max} = Pv_{max} = \frac{aP_E P}{P_E - P}$$

The maximum stress in the strut will occur when the value of the bending moment is a maximum, and this will be at the surface of the strut on the concave side.

When the maximum stress reaches the value σ_y

$$\sigma_y = \frac{P}{A} + \frac{aP_E P}{P_E - P}\frac{d_1}{Ak^2}$$

where d_1 is the distance from the neutral axis to the outer fibre and Ak^2 is the value of the second moment of area about the neutral axis.

Put $\sigma_E = P_E/A$, $\sigma = P/A$ and $\eta = ad_1/k^2$

$$\sigma_y = \sigma + \frac{\eta\sigma_E\sigma}{\sigma_E - \sigma} \tag{10.25}$$

or

$$\sigma^2 - [\sigma_y + \sigma_E(1 + \eta)]\sigma + \sigma_y\sigma_E = 0$$

Solving this quadratic equation gives

$$\sigma = \tfrac{1}{2}\{[\sigma_y + \sigma_E(1 + \eta)] - \{[\sigma_y + \sigma_E(1 + \eta)]^2 - 4\sigma_y\sigma_E\}^{1/2}\} \tag{10.26}$$

Hence the value of P_y (figure 10.13), is known, the value of the load at which yield first occurs in the strut. This particular expression was first obtained by Perry. It is not possible to solve equation 10.26 unless the value of η is known for a particular strut.

The value of P_y will be less than the actual failure load P_F of a strut. Robertson conducted a large number of tests on struts and equated the failure load to P_y; he then solved equation 10.25 for values of η. The values of η would be less than the true values for η for particular struts as $P_y < P_F$. An average value of η obtained by this method was $0 \cdot 003$ l/k. If this value is used, equation 10.26 may next be solved to obtain a value for the collapse load. This method is referred to as the Perry–Robertson formula.

In practice the actual failure load of a strut is lower than either the critical load P_E or the squash load P_S.

We shall consider the ratios P_F/P_E and P_F/P_S. For struts with a high value of the slenderness ratio $(P_F/P_E) \to 1$ and for a low value of the slenderness ratio $P_F/P_S \to 1$.

If a linear relationship is assumed between P_F/P_E and P_F/P_S such that $P_F/P_E = 1$ when $P_F/P_S = 0$, and $P_F/P_S = 1$ when $P_F/P_E = 0$, the equation of such a line may be written

$$\frac{P_F}{P_S} = -\frac{P_F}{P_E} + 1$$

or

$$\frac{1}{P_F} = \frac{1}{P_E} + \frac{1}{P_S} \qquad (10.27)$$

If $P_E = \pi^2 EI/l^2$ and $P_S = A\sigma_y$

$$P_F = A\sigma_y \bigg/ 1 + \frac{\sigma_y}{\pi^2 E}\left(\frac{l}{k}\right)^2 \qquad (10.28)$$

This is a very similar expression to that suggested by Rankine apart from the fact that $\sigma_y/\pi^2 E$ was replaced by an arbitrary constant.

10.9 The Southwell method

This is an experimental method that can be used to obtain the critical load of a strut. The strut is loaded in an axial manner and it is possible from certain observations to predict the value of the critical load without the necessity of applying a force of sufficient magnitude to cause failure of the strut.

We shall use the expression from equation 10.16 once again. If the loading P applied to the strut is somewhere near the value of P_E, the first term in the series will predominate and the deflection can be written as

$$v = \frac{a_1 \sin(\pi x/l)}{1 - (P/P_E)}$$

The fundamental component of v_0 was given by $a_1 \sin(\pi x/l)$ so that the displacement from the initial state is

$$a_1 \sin\frac{\pi x}{l}\left(\frac{P}{P_E - P}\right)$$

The maximum deflection will occur at $x = l/2$, and has a value

$$\Delta = \frac{a_1 P}{P_E - P} \qquad (10.29)$$

Equation 10.16 was derived for the case of a strut with both ends pinned, but it is possible to apply the above analysis to struts with other end conditions. Instead of assuming a Fourier series it may be assumed that the critical modes v_{C1}, v_{C2}, etc. are known for the strut and that it is possible to express the initial deflection of the strut in terms of these modes. Hence

$$v_0 = \sum a_n v_{Cn}$$

It can also be shown that when a load P is applied to the strut

$$v = \sum_{1}^{\infty} \frac{a_n v_{Cn}}{1 - (P/P_{Cn})} \qquad (10.30)$$

This equation will apply for any end conditions. Thus

$$v - v_0 = \sum_1^\infty a_n v_{Cn} \bigg/ \left(\frac{P_{Cn}}{P} - 1\right)$$

Once again the first term will predominate when the applied load is somewhere near P_{C1}, therefore

$$\Delta = \frac{a_1 P}{P_C - P} \tag{10.31}$$

where Δ is the deflection measured at a suitable point on the strut. This expression is identical to that given in equation 10.29 except that P_E is replaced by P_C. For the following discussion P_C will be used since it represents a more general case.

Equation 10.31 can be written in the following form

$$\frac{\Delta}{P} = \frac{\Delta}{P_C} + \frac{a_1}{P_C} \tag{10.32}$$

This gives a straight line relation between Δ/P and Δ. The slope of the line is $1/P_C$ and the intercept on the Δ-axis gives the value of a_1 (figure 10.15a). In a practical test however there will be considerable deviation from the straight line at lower values of P; the reason for this is that only the fundamental term was considered, but other terms can have a considerable effect when P is small when compared with P_C.

Equation 10.32 can be written in a different form

$$\frac{P}{\Delta} = -\frac{P}{a_1} + \frac{P_C}{a_1} \tag{10.33}$$

This equation gives a straight-line relationship between P/Δ and P. The intercept on the P-axis gives the value of the critical load, and the reciprocal of the slope gives the value of a_1 (figure 10.15b). Once again in a practical test there will be a deviation from the straight line when P is small compared with P_C.

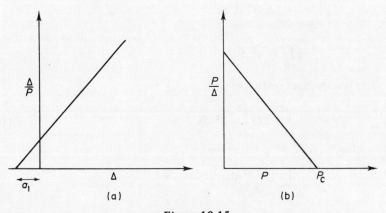

Figure 10.15

The results from an actual experiment on a strut are incorporated in an example at the end of the chapter.

It is possible to extend the use of a Southwell plot to a complete structure and hence obtain an estimate of the critical loading that would cause collapse.

10.10 Energy methods

When the case of a pin-jointed strut was treated in section 10.3, we considered the stability of the system by the application of potential energy. A general expression for the potential energy for this type of problem was obtained in equation 10.11, repeated below.

$$V = \int_0^l \frac{EI}{2} \left(\frac{d^2 v}{dx^2}\right)^2 dx - \frac{P}{2} \int_0^l \left(\frac{dv}{dx}\right)^2 dx$$

If the deflection curve for the strut were known it should be possible to make use of this general expression to determine the critical load. Unfortunately the shape of this curve is only known for a very few cases. However, let us assume for the time being that we can write down a suitable expression, and we represent the deflection curve by

$$v = af(x)$$

where a represents the amplitude. The expression for the potential energy can now be written

$$V = a^2 \int_0^l \frac{EI}{2} (f'')^2 dx - \frac{a^2 P}{2} \int_0^l (f')^2 dx$$

For stability $dV/da = 0$ so that

$$P = \frac{\int_0^l EI(f'')^2 \ dx}{\int_0^l (f')^2 \ dx}$$

This is sometimes written as

$$P_R = \frac{\int_0^l EI \left(\frac{d^2 v}{dx^2}\right)^2 \ dx}{\int_0^l \left(\frac{dv}{dx}\right)^2 \ dx} \qquad \cdot \text{(10.34)}$$

This expression is the Rayleigh formula for the critical load of a strut. If the deflected form of a particular strut is known or assumed the expression can be evaluated. If the true deflected form is known the result obtained will be equal to the true critical load of the strut. This however is very unlikely to be known and in general a guess will have to be made for the deflection curve. The resulting value of P_R will always be higher than the true critical load. This point will be discussed after the Timoshenko method has been derived.

For the case of struts with both ends simply supported or with one end built in and the other end free, the equilibrium equation can be written in the form

$$EI \frac{d^2 v}{dx^2} = -Pv \tag{10.35}$$

We have already shown that the energy stored in the strut due to bending action is

$$\int_0^l \frac{EI}{2} (\frac{d^2 v}{dx^2})^2 \ dx$$

Making use of equation 10.28 this may be expressed as

$$\int \frac{P^2 v^2 \ dx}{2EI}$$

substituting this into the expression for V and differentiating will give

$$P_T = \frac{\int_0^l (\frac{dv}{dx})^2 \ dx}{\int_0^l \frac{v^2 \ dx}{EI}} \tag{10.36}$$

Equation 10.36 is known as the Timoshenko method for the critical load of a strut. It is often more difficult to evaluate than the Rayleigh expression since the term involving I appears in the denominator, and for a variable value of the second moment of area it may not be possible to perform an exact integral. This often means that a numerical method of integration must be used.

It can be shown mathematically that the Timoshenko method gives a result that is nearer to the true critical load, and that both methods will give values that are higher or at best equal to the true critical load. However an exact proof of this is rather difficult and we shall confine ourselves to the following discussion. If the true deflection curve be chosen then either method will give a result that is equal to the true critical load. For any other chosen deflection curve however, the strut has to be forced into this mode by the addition of external restraints such as moments and forces. The effect of these restraints is to make the strut more rigid and hence the critical load is raised above the true value.

Example

A tapered column of length l is built in at the bottom end. The second moment of area at any cross-section is given by

$$I_x = I\left(\frac{x}{a} + 1\right)^2$$

I is the second moment of area at the top of the column, and x is measured from the top of the column. Find the load at which the column will buckle when $a = (1 + \sqrt{2})l$.

Assume $v = v_0 \sin \pi x/2l$

$$\frac{dv}{dx} = \frac{\pi}{2l} v_0 \frac{\cos \pi x}{2l}$$

$$\frac{d^2v}{dx^2} = -\frac{\pi^2}{4l^2} v_0 \frac{\sin \pi x}{2l}$$

$$\left(\frac{dv}{dx}\right)^2 = \frac{\pi^2}{4l^2} \frac{v_0^2}{2}\left(1 + \cos\frac{\pi x}{l}\right)$$

$$\left(\frac{d^2v}{dx^2}\right)^2 = \frac{\pi^4}{16l^4} \frac{v_0^2}{2}\left(1 - \cos\frac{\pi x}{l}\right)$$

$$P_R = \frac{EI}{a^2} \frac{\pi^2}{4l^2} \int_0^l (x^2 + 2ax + a^2)\left(1 - \cos\frac{\pi x}{l}\right) dx \Bigg/ \int_0^l \left(1 - \cos\frac{\pi x}{l}\right) dx$$

$$= \frac{EI}{a^2} \frac{\pi^2}{4l^2}\left[\frac{l^3}{3} + al^2 + a^2l + \frac{2l^3}{\pi^2} + \frac{4al^2}{\pi^2}\right]\Bigg/l$$

Now $a = 2.41l$ and $a^2 = 5.8l^2$, therefore

$$P_R = \frac{4.12EI}{l^2}$$

$$P_T = \frac{\pi^2EI}{4la^2}\Bigg/\int_0^l \frac{1 - \cos(\pi x/l)}{x^2 + 4.82\,lx + 5.8l^2}$$

The denominator has to be evaluated by a numerical method. Dividing the column into ten equal parts and making use of Simpson's rule, gives a value for the denominator of $0.104/l$ and $P_T = 4.07EI/l^2$.

It is in fact possible to obtain a theoretical result for this particular column. The value of the Euler critical load being $P_C = 4.046EI/l^2$.

We can see that both Rayleigh and Timoshenko give results that are too high, but Timoshenko lies closer to the true value.

In this chapter, the strut has been discussed in isolation. It is of course perfectly possible for members of a structure to collapse through instability. A pin-jointed structure presents no difficulty as the force in the members is axial. A rigid jointed structure on the other hand is a much more difficult problem since the members will be subject to end moments in addition to axial loads.

Other forms of instability can occur in structural members. In addition to collapse by strut action certain types of cross-section even though subject to an axial load may fail due to a rotation about a longitudinal axis; this is referred to as *torsional instability*. A further form of buckling can occur in a loaded beam; here it is possible for a sideways deflection to occur together with a rotation—termed *lateral buckling*. It is felt that these topics are beyond the scope of this present work and it is suggested that the reader should consult works of a more specialised nature.

Problems

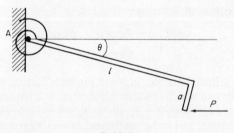

P.10.1

10.1 The rigid rod of length l in P10.1 is pinned at end A and restrained by a torsion spring of stiffness k. The torque in the spring is zero when the rod is horizontal. A load P, which may be assumed to remain horizontal, is applied to the free end of the rod; this acts as shown at the end of an arm of length a. Determine the condition for equilibrium and the requirement for stability. Show that if both a and θ are small then the critical load will be given by k/l.

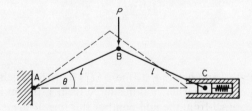

10.2 Two light rigid bars AB and BC each of length l, are pin-jointed together at B. The end A is pin-jointed to a rigid foundation, and the end C to a slider which moves in a horizontal guide, the line of which passes through A, and is restrained by a spring of stiffness k (P10.2). When $\cos \theta = 0.9$ a vertical load P is gradually applied at B. Show that an equilibrium position is reached when $\sin \theta \approx 0.26$. Determine the value of the critical load.

10.3 A strut of length l is clamped at one end and pin-jointed at the other. For length $l/2$ from the clamped end the flexural rigidity is EI, and for the remainder of the length the strut can be treated as having infinite flexural rigidity. Find an equation from which the Euler critical load can be obtained.

10.4 A slender uniform column is vertical and built in at the foot. It supports an axial load at the top, which is free to move. In order to raise the value of the critical load it is proposed to stiffen the upper half of the column. Show that the critical load cannot be increased by more than about 20 per cent by this means. The equation $\theta = \cot \theta$ is satisfied by $\theta = 0{\cdot}86$ approximately.

10.5 A strut initially straight, of length l and flexural rigidity EI, is built in at one end. An axial load P and a moment M are applied at the other end, but lateral movement there is prevented. Show that the slope of the strut at this end is given by the following expression, where $\alpha^2 = P/EI$

$$\theta = \frac{\alpha M}{P} \left(\frac{2 - 2 \cos \alpha l - \alpha l \sin \alpha l}{\sin \alpha l - \alpha l \cos \alpha l} \right)$$

10.6 Show that an initially straight, uniform pin-ended strut, which is not allowed to extend at its ends and which is uniformly heated to a temperature T above the surroundings, will buckle when $T = T_c$ where $T_c = \pi^2 k^2 / \alpha L^2$, k is the least radius of gyration, L the length, and α the coefficient of linear expansion.

A uniform boiler tube of length $2L$ is built in to heavy diaphragms and passes through a supporting diaphragm, which constrains it in position, but not in direction, at its mid-point. Show that it will buckle when the temperature is uniformly raised by approximately $2{\cdot}04\pi^2 k^2 / \alpha l^2$ provided the ends are immovable.

10.7 A strut of length l is encastré at one end; the other end is supported in such a way that if a transverse displacement d occurs there, it is resisted by a transverse restraining force kd. Show that the buckling load P is given by the solution of the equation

$$\frac{\tan \alpha l}{\alpha} = l - \frac{P}{k}$$

where $\alpha^2 = P/EI$.

10.8 A tie of length l and bending stiffness EI is imperfect, being slightly curved such that $v = v_0 \cos (\pi x/l)$ where v is measured from a straight line between the ends and x is measured from the mid-point of this line. If the ends of the tie are pinned, find the tensile force P required to reduce the initial eccentricity v_0 at the centre of the tie by half. A lateral force F is now applied to the centre of the tie in addition to the tension. Find the magnitude of F required to reduce the central eccentricity to zero.

10.9 A pin-ended strut of length l has an initial bow given by $v_0 = A \sin (\pi x/l)$ where A is a length small compared with l, and the axes x and y are at one end of the strut. Show that the lateral deflection d at the centre of the strut due to an end load P is given by $d = A/(P_E/P - 1)$, where P_E is the Euler critical load.

10.10 A pin-ended strut has length l and flexural rigidity EI. It is laterally supported by an elastic medium. The medium resists the lateral movement d of each point on the strut by applying a force kd per unit length at that point. Taking the buckling mode to be a half sine-wave, find the critical value of the axial force P either by solving the differential equation or by an energy method.

10.11 In a test on a pin-ended strut the applied axial compressive force P, and the resulting lateral central deflection v, were measured as follows

v(mm)	0·23	0·38	0·55	0·75	0·96	1·27	1·63	2·04
P(kN)	6·85	8·90	9·80	10·54	11·20	11·75	12·10	12·50

Using a Southwell plot, determine the critical load.

11 PLASTIC ANALYSIS OF BEAMS AND FRAMES

11.1 Introduction

When elastic theory is used to design a statically indeterminate structure in mild steel, the stresses that arise when the complete external loading is applied must not exceed a certain value, termed the working stress. The ratio of the yield stress of the material to the working stress is called the safety factor. In a structure made from a linearly elastic material the safety factor is also the ratio of the load required to produce this yield stress to the working load.

At the design stage it is not possible to predict all the applied loads exactly. It is quite possible for an overload to be applied sometime in the life of a structure. Also there may be defects in the materials used or there may have been poor workmanship during the course of construction. Extra stresses can arise due to differential settlement of supports and from a variety of other causes. It can be seen that if a structure were to be designed to a stress very close to the elastic limit, it is quite possible that the yield stress might be exceeded sometime in the life of the structure. This would not necessarily mean that the structure would collapse since the yield stress is well below the ultimate stress. However if a suitable safety factor is introduced the design stress will be well below the yield point. As an example a typical value for mild steel would be about 1·5.

In some cases a design based on elastic theory can be extremely conservative and wasteful of material. As an illustration we shall consider the case of a beam with a constant cross-section with both ends built in, and carrying a uniformly distributed load. From elastic theory we can easily show that the maximum bending moment occurs at the ends of the beam and that the value there is twice the value at the mid-point of the beam. The ends of the beam are therefore the critical sections and the maximum load that is allowed to be carried by the beam is governed by the working stress not being exceeded at the ends.

Let us suppose that the beam is loaded until the yield stress is reached at the supports. Further load can still be applied without the beam collapsing. As the load is increased *plastic hinges* will form at the ends of the beam. Even when these are fully developed the bending moment at the centre of the beam will not be sufficiently great for the yield stress to be reached there. Loading could in fact be continued until a plastic hinge occurs at the centre of the beam. At this stage collapse is said to have taken place. This example has been discussed to show that there is still a considerable reserve of strength in a redundant structure even though the yield stress may have been reached at some point.

The plastic design method is based on calculating the load required to produce sufficient plastic hinges in the structure to turn it, or at least part of it, into a mechanism. This load would then be divided by the *load factor* as opposed to the safety factor and the value of the working load determined. In practice of course the problem would be presented the other way round. The working loads would be known approximately and for a particular load factor the sections of the various members could be determined.

One word of warning should be introduced at this stage. If there is the possibility of instability occurring or if the deflections have to be kept to a minimum it may not always be possible to use a plastic method of design.

11.2 Collapse of redundant beams

In chapter 5 the moment–curvature relationship was derived for a rectangular cross-section made from an ideal elasto-plastic material. It was also discussed for a universal beam. In this chapter we shall idealise this relationship as shown in figure 11.1. This will mean that once M_p is reached at a particular point on a beam, large changes of curvature can occur for no increase in bending moment. It is as though a hinge had been inserted in the beam, hence the term *plastic hinge*.

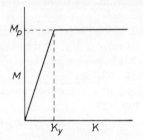

Figure 11.1

The beam in figure 11.2a is built in at both ends. It carries a uniformly distributed load that increases in value until collapse of the beam occurs. The bending-moment diagram when the beam is completely elastic is shown at b. This can be considered as two separate diagrams, the free bending-moment diagram, a parabola of maximum height $pl^2/8$, and the reactant diagram, a rectangle of height $pl^2/12$. It can be seen that the maximum bending moment occurs at the ends of the beam and the value at the centre is only half this value.

If a moment–curvature relationship similar to that shown in figure 11.1 is assumed, the value of the load when the bending moment has reached the fully plastic value of M_p at the supports is given by

$$p_y = \frac{12M_p}{l^2}$$

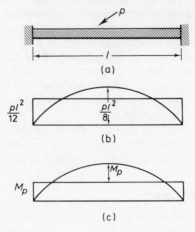

Figure 11.2

The bending moment cannot increase beyond this value but an increase in curvature can take place and plastic hinges form at each end of the beam. The beam from now on will behave as though it is simply supported at the ends. The value of the bending moment at the centre of the beam at this stage is $M_p/2$ or $p_y l^2/24$.

As the uniformly distributed load is further increased the central bending moment can increase until its value is M_p. At this stage a further hinge will occur at the centre and the beam has become a mechanism. Any further increase of load would cause the beam to collapse.

The incremental load p_i required to produce the further increase in bending moment of $M_p/2$ at the centre is given by

$$\frac{p_i l^2}{8} = \frac{M_p}{2}$$

therefore

$$p_i = \frac{4M_p}{l^2}$$

We can now find the total value of the load that causes collapse

$$p_c = (12 + 4)\frac{M_p}{l^2} = \frac{16M_p}{l^2}$$

Note that the collapse load is only dependent on the value of M_p and does not depend in any way on the shape of the moment–curvature relationship. It has however been useful in our analysis to use a particular relation since we have been able to predict the load that caused the first plastic hinges to form at the ends of the beam.

We shall next endeavour to find out something about the central deflection of the beam.

It has been assumed that the beam behaves in an elastic manner until the load p_y is reached, so that the first part of the deflection curve will be linear and the value of the deflection is given by

$$d = \frac{p_y l^4}{384EI} = \frac{M_p l^2}{32EI}$$

Once the two end-hinges have formed the beam behaves as though it is simply supported, and the incremental load-deflection relationship will again be linear with a maximum incremental deflection

$$d_i = \frac{5}{384} \frac{p_i l^4}{EI} = \frac{5M_p l^2}{96EI}$$

The total deflection at the centre of the beam when all hinges have formed and collapse is imminent is

$$d = \frac{M_p l^2}{EI} \left(\tfrac{1}{32} + \tfrac{5}{96}\right) = \frac{M_p l^2}{12EI}$$

The load–deflection curve consists of distinct linear phases as shown in the sketch (figure 11.3). Point X corresponds to the formation of hinges at the ends of the beam and point Y to the formation of the central hinge. It must be emphasised that the shape of the curve is dependent upon the moment–curvature relationship that is assumed. In actual practice the sharp kinks in the diagram would be rounded as the formation of a hinge would not take place instantaneously

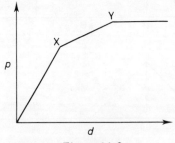

Figure 11.3

One interesting point is that the value of the collapse load is independent of any displacements that may occur at the supports as the beam is loaded, or indeed of any initial imperfections. It is required only that the supports should be capable of developing the fully plastic bending moment.

To demonstrate this fact we shall consider the extreme case of a rotation at the ends of the beam, such that before the load is applied there is an initial bending moment at each end of the beam of $-M_p/2$, that is, hogging. The supports at the ends of the beam are now clamped in this position and the load applied. Plastic hinges will form at the ends when the bending moment there is $-M_p$, an increase in value of $-M_p/2$. The load required to produce this state of affairs will be precisely half that of the original case.

The value of the bending moment at the centre with no load applied was $-M_p/2$ and the change due to the application of p_y is $M_p/4$. Thus the value at the centre when hinges form at the ends is

$$-\frac{M_p}{2} + \frac{M_p}{4} = -\frac{M_p}{4}$$

To form a plastic hinge at the centre requires a final value of bending moment M_p, an increase of $5M_p/4$ therefore

$$\frac{p_i l^2}{8} = \frac{5M_p}{4}$$

$$p_i = \frac{10M_p}{l^2}$$

So that the total load needed to produce a complete collapse is $16M_p/l^2$, the same value as that obtained with no initial rotation at the ends. What in fact we are saying is that the final bending moment (figure 11.2c) is independent of the deflections and rotations of the supports.

Next consider a continuous beam of two spans, each of length l. The applied loading is shown in figure 11.4a. The value of M_p is required such that collapse just occurs under the loading system.

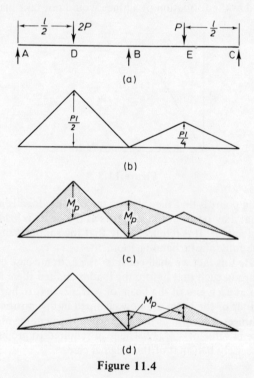

(a)

(b)

(c)

(d)

Figure 11.4

Using the same basic method of approach the statically determinate bending-moment diagram can be drawn. It is.possible to make the problem statically determinate in several different ways, but probably the simplest is to put a cut through the beam at B. This will give rise to triangular bending-moment diagrams of heights $Pl/2$ and $Pl/4$ in spans AB and BC respectively (figure 11.4b). It is next necessary to add the reactant bending-moment diagram. A bending-moment diagram has to be introduced at B to restore continuity. The reactant bending-moment diagram will be triangular with a maximum height at B equal to the value of the final bending moment at B. The combined bending-moment diagram is shown at c. There are three possibilities for a maximum bending moment, at B, D, or E.

We shall first assume that there is a maximum bending moment at B and D, and that plastic hinges have formed at these points. This means that span ADB has collapsed and

$$\tfrac{3}{2} M_p = \frac{Pl}{2}$$

or $M_p = Pl/3$. So a beam must be selected capable of carrying a fully plastic moment of $Pl/3$.

So far we have satisfied two requirements. Part of the beam has become a mechanism, and the system of applied forces and the moments are in equilibrium. It is quite possible that the solution obtained might give a bending moment greater than M_p at some point in span BC. If this is so the solution is an unsafe one. It is an easy matter to check. The maximum bending moment in span BC will occur at E and the value there is

$$\frac{Pl}{4} - \frac{M_p}{2} = \frac{Pl}{12}$$

a value well below the derived value of M_p. This implies that the solution we have obtained is the correct one and that $M \leqslant M_p$ at all points. The latter inequality satisfies the yield criterion. So that three conditions have in fact been satisfied, equilibrium, mechanism, and yield.

Let us suppose by some mischance that originally the incorrect failure mechanism was chosen and it was assumed that hinges would form at B and E (figure 11.4d). Thus

$$\tfrac{3}{2} M_p = \frac{Pl}{4}$$

therefore

$$M_p = \frac{Pl}{6}$$

Equilibrium and mechanism have both been satisfied, but a check on the value of the bending moment at D shows that the value there is $5Pl/12$, very much in excess of the suggested value of $M_p = Pl/6$. In other words the yield criterion is not satisfied. We do know that from this piece of analysis the beam chosen must have a fully plastic bending moment greater than $Pl/6$ and we have found a *lower bound* to the problem. It is easy to obtain an *upper bound* from the same analysis. If a beam was chosen with $M_p = 5Pl/12$ failure would not occur. So that we may state

$$\frac{Pl}{6} \leqslant M_p \leqslant \frac{5Pl}{12}$$

The correct value of M_p obtained from the first solution must of course lie between these bounds.

11.3 Load factor

The term load factor was mentioned in the introduction to this chapter. If a particular statically determinate structure has a set of working loads P_w applied, the bending moment at any point will be some function of P_w. If the loads are all increased by the same factor λ, since the structure is statically determinate, the bending moments will all increase by λ. When M_p is reached at any point on the structure, a hinge forms and the structure becomes a mechanism and collapse occurs. The value of λ_c to cause collapse is called the *collapse load-factor*.

If σ_w is the working stress in the material, the working bending moment M_w would be given by $\sigma_w Z_e$ where Z_e is the elastic section modulus. Now $\lambda_c M_w$ must be equal to the fully plastic bending moment M_p. In chapter 3 it was stated that $M_p = \sigma_y Z_p$ where Z_p was the plastic section modulus.

$$\lambda_c M_w = \sigma_y Z_p$$

$$M_w = \sigma_w Z_e$$

therefore

$$\lambda_c = \frac{\sigma_y Z_p}{\sigma_w Z_e} \tag{11.1}$$

A reasonable value of σ_y/σ_w for mild steel is $1 \cdot 5$, so that

$$\lambda = 1 \cdot 5 \frac{Z_p}{Z_e} = 1 \cdot 5\alpha$$

For a rectangular section $\alpha = 1 \cdot 5$ $\lambda_c = 2 \cdot 25$
For a universal beam section $\alpha = 1 \cdot 15$ $\lambda_c = 1 \cdot 725$

Using these load factors would mean that the working load for a statically determinate case would be the same whether elastic or plastic design is used. In a redundant problem using elastic design the bending moment at any point is a function of the load applied, but using plastic design the bending moment would be a different function. In general the allowable working load is greater using plastic design.

11.4 Basic theorems

Certain conditions apply when a structure is on the point of collapse.

Equilibrium

The system of bending moments must be in equilibrium with the applied loads.

Yield

The bending moment at any point on the structure must not exceed M_p.

Mechanism

The bending moment must be equal to M_p at a sufficient number of points such that the structure, or at least part of it, becomes a mechanism.

There are three basic theorems, which we shall not prove here, concerned with the three conditions set out above. A set of working loads is postulated and a load factor λ is determined with certain conditions applicable.

Uniqueness theorem

If all the conditions are satisfied at the same time then the value of the load factor is unique.

Lower-bound or safe theorem

If a load factor is found such that equilibrium and yield are satisfied then a safe solution has been obtained. This means that M_p has not been exceeded anywhere, but it also is possible that the structure has not yet collapsed, since there may not be sufficient plastic hinges formed to produce a mechanism. In fact further load would have to be applied, so $\lambda \leqslant \lambda_c$.

Upper-bound or unsafe theorem

If a load factor is found so that a mechanism is formed then possibly an unsafe solution has been obtained. It may well be that for the chosen mechanism the bending moment at some other point in the structure is greater than M_p. The yield condition has been forgotten. If this is so, then an incorrect mechanism has been chosen and a different mechanism would have formed at a lower load, $\lambda \geqslant \lambda_c$.

It can be seen that it should be possible to work from a lower-bound solution towards the unique solution or from an upper-bound to a unique solution.

11.5 Graphical analysis

This method has already been applied to simple beam problems. Basically it consists of drawing a statically determinate bending-moment diagram, and then superimposing the reactant bending-moment diagram. A decision has then to be made as to where the plastic hinges are likely to occur. As an illustration of this method we shall first discuss the solution of a portal frame with pinned feet and then proceed to the more complicated case of a portal with a pitched roof and built-in feet.

The portal frame in figure 11.5a has pinned feet and has a load P_1 applied at the centre of BC, and a side load P_2 at B. The members all have the same cross-section.

To make the frame statically determinate the horizontal force at D is assumed to be zero (b), thus the vertical reaction at D is $(P_1/2) + (P_2h/l)$. The bending-moment diagram for this loading system is shown at d. The relative values of the bending moments at B and E are dependent on the values of P_1 and P_2 and also on the dimensions of the frame, so that it would be perfectly possible for the

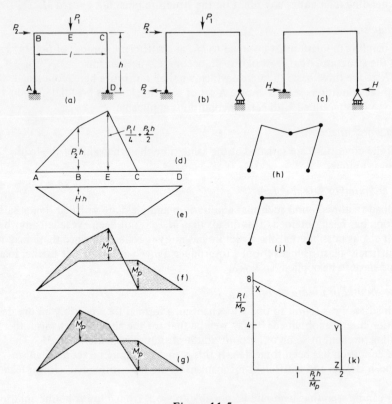

Figure 11.5

bending moment at B to exceed that at E. The reactant bending-moment diagram due to the loading at c is drawn at e. The two bending-moment diagrams are superimposed and the resultant shaded in f or g.

Next it is necessary to decide how many plastic hinges are required for collapse and where these are located. It is fairly obvious in this case that two hinges are required—these would turn the portal into a four-bar chain. For the bending-moment diagram f the hinges will occur at E and C since the resultant bending moment at E is greater than at B. For g the hinges occur at B and C.

From the bending-moment diagrams we can see that either

$$2M_p = \frac{P_1 l}{4} + \frac{P_2 h}{2} \tag{11.2}$$

or

$$2M_p = P_2 h \tag{11.3}$$

The two different collapse modes corresponding to these two equations are shown at h and j. It is possible to plot the equations in a particular way. This is known as an interaction diagram, sometimes referred to as a yield-surface diagram, shown at k. Equation 11.2 is represented by the line XY and equation 11.3 by the line YZ. So that for any point on the line XYZ the portal is just collapsing. Any point inside the lines represents a safe combination of P_1 and P_2.

A point outside the line represents an impossible combination of loads since collapse will already have taken place. At Y both modes of collapse occur simultaneously and three hinges form.

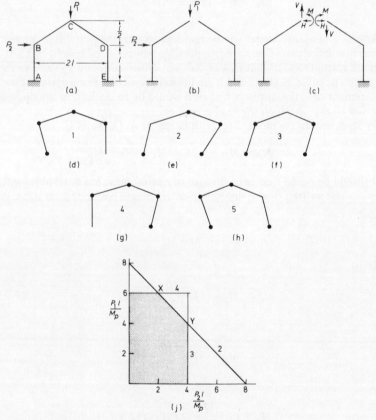

Figure 11.6

The solution for the pitched-roof portal with built-in feet (figure 11.6a) will be more complicated. The portal is made statically determinate by removing the connection at C (b). The reactant forces are shown at c. We shall assume that a positive bending moment produces tension on the inside edges of the members. The bending moments can now be written down for possible hinge points.

$$M_A = M + Vl + \frac{3Hl}{2} - P_1 l - P_2 l$$

$$M_B = M + Vl + \frac{Hl}{2} - P_1 l$$

$$M_C = M$$

$$M_D = M - Vl + \frac{Hl}{2}$$

$$M_E = M - Vl + \frac{3Hl}{2}$$

A total of 4 plastic hinges will be required to turn the portal into a mechanism, again a four-bar chain. There are five possible modes of collapse (d to h). If the loads are restricted such that P_1 can only act downwards and P_2 only to the right, two of the possible modes d and h are unlikely to occur. For a vertical downward displacement of C the displacement of B would be to the left, in the opposite direction to that in which the load is applied.

We shall investigate mode 2 (e) with hinges at A, C, D and E.

$$M_A = M_D = -M_p \quad M_C = M_E = M_p$$

It should be noted that once the sign of one moment has been obtained, the others can be written down, since consecutive hinges have opposite signs.

At C

$$M = M_p$$

At D

$$M_p - Vl + \frac{Hl}{2} = -M_p$$

At E

$$M_p - Vl + \frac{3Hl}{2} = M_p$$

At A

$$M_p + Vl + \frac{3Hl}{2} - P_1 l - P_2 l = -M_p$$

From M_D and M_E, $Hl = 2M_p$. From M_E and M_A

$$Vl = \frac{P_1 l}{2} + \frac{P_2 l}{2} - M_p$$

Substituting in M_D gives

$$P_1 l + P_2 l = 8M_p$$

For mode 4

$$M_B = M_D = -M_p \quad M_C = M_E = M_p$$

At C

$$M = M_p$$

At D

$$M_p - Vl + \frac{Hl}{2} = -M_p$$

At E

$$M_p - Vl + \frac{3Hl}{2} = M_p$$

At B

$$M_p + Vl + \frac{Hl}{2} - P_1 l = -M_p$$

From M_D and M_E, $Hl = 2M_p$. From M_D and M_B, $Vl = P_1 l/2$. Substituting in M_D gives

$$P_1 l = 6M_p$$

It is possible to investigate all modes in this way and the following equations will result

Mode 1 $P_1 l - P_2 l = 6M_p$

Mode 2 $P_1 l + P_2 l = 8M_p$

Mode 3 $P_2 l = 4M_p$

Mode 4 $P_1 l = 6M_p$

Mode 5 $P_1 l - 2P_2 l = 8M_p$

The interaction diagram is drawn in figure 11.6j. It can be seen at once that modes 1 and 5 will not be of interest for positive values of P_1 and P_2—this was seen earlier. At X modes 2 and 4 occur at the same time and at Y modes 2 and 3, there being five hinges in each case.

The analysis by this method is somewhat tedious, particularly if the framework is at all complicated. The method of virtual work together with combined mechanisms is generally to be preferred. This will be discussed in the following sections.

11.6 Virtual-work approach

An equilibrium equation using virtual work could be written down relating M_A, M_B, and M_C to the load P, for the beam problem in figure 11.7a. It would be necessary to use a compatible set of deformations, and that shown at b would be suitable.

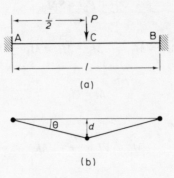

(a)

(b)

Figure 11.7

The moment sign-convention is the same as that used in the previous section. A joint rotation will be considered positive if the joint appears to open if viewed from the underside of a beam or the inside of a frame.

For the deformations suggested

$$\theta_A = -\theta \quad \theta_C = 2\theta \quad \theta_B = -\theta \quad d = l\theta/2$$

The virtual-work equation can be written

$$M_A(-\theta) + M_C(2\theta) + M_B(-\theta) = \frac{Pl}{2}\theta$$

or

$$-M_A + 2M_C - M_B = \frac{Pl}{2} \tag{11.4}$$

If the beam is allowed to reach collapse the values of the bending moments at A, B, and C are known, $M_A = M_B = -M_p$; $M_C = M_p$.

So equation 11.4 simplifies to $M_p = Pl/8$ and this will be the collapse equation for the beam.

It should be noted that when the virtual-work equation is formed with values of M_p substituted, all terms on the left-hand side become positive, that is a positive bending moment is associated with a positive joint rotation.

The same collapse equation could be written down using real work, as long as all elastic deformations are disregarded as small compared with plastic deformations.

So far we have not considered any loading apart from concentrated loads. If a uniformly distributed load of value p per unit length is applied in figure 11.7 instead of P, the hinges which form at the points of maximum bending moment would occur in identical positions. It is only necessary to determine the work done by the external load: this is equivalent to taking the total load on the beam through a distance equal to one half the maximum displacement. The right-hand side of the virtual-work equation would be

$$pl \times \frac{1}{2}\left(\frac{l}{2}\,\theta\right)$$

and at collapse

$$M_\mathrm{p} = \frac{pl^2}{16}$$

The case shown in figure 11.8—that of a cantilever with a prop—will cause a little more trouble. Two hinges will form at collapse. It is obvious that one will be at A, the built-in end. Presumably the other will occur at the point of maximum

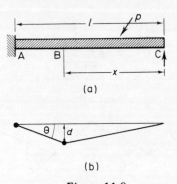

(a)

(b)

Figure 11.8

bending moment. This is unfortunately unknown. We shall assume it to be at a distance x from C. If the hinge at A rotates through θ, that at B will rotate through $l\theta/x$.

By virtual work

$$M_\mathrm{p}\theta + M_\mathrm{p}\,\frac{l}{x}\,\theta = pl \times \tfrac{1}{2}(l - x)\theta$$

Note that all terms on the left-hand side of the equation have been written down as positive.

$$M_\mathrm{p} = \frac{plx(l - x)}{2(l + x)} \tag{11.5}$$

The value of M_p that we are looking for is the maximum. Knowing this we could determine the value of Z_p and hence the size of section that is capable of carrying this maximum bending moment. This means that we have a safe design.

$$\frac{\mathrm{d}M_\mathrm{p}}{\mathrm{d}x} = 0$$

which gives

$$x^2 + 2xl - l^2 = 0$$

or

$$x = (\sqrt{2} - 1)l$$

Substituting this value into equation 11.5 will give the value of M_p

$$M_p = 0.086pl^2$$

As a further example of the virtual-work approach we shall endeavour to find the collapse load-factor λ for the beam in figure 11.9a, where the fully plastic moment is M_p. This implies that the loads applied are working loads and that we must scale then up by λ to investigate collapse.

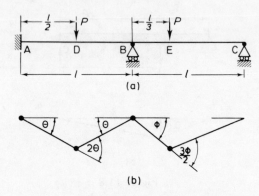

Figure 11.9

It is possible for the beam to collapse in one of two ways—failure in span AB with three hinges at A, B, and D; or failure in span BC with two hinges at B and E. The possible failure mechanisms are shown in figure 11.9b.

For collapse of AB

$$M_p\theta + 2M_p\theta + M_p\theta = \lambda_1 \frac{Pl}{2}\theta \quad \lambda_1 = \frac{8M_p}{Pl}$$

For collapse of BC

$$M_p\phi + M_p \times \frac{3\phi}{2} = \lambda_2 \frac{Pl}{3}\theta \quad \lambda_2 = 7\tfrac{1}{2}\frac{M_p}{Pl}$$

Hence span BC will collapse before span AB and λ_2 is the correct load factor.

The portal frame in figure 11.10a can be solved very easily by virtual work.

Possible hinge points are at A, B, C, D, and under the load P_1 at E. Four hinges would be required for the sidesway mechanism of figure 11.10b. Also four are

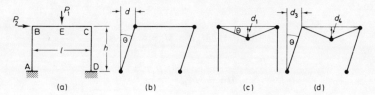

Figure 11.10

required for the 'combined' mechanism (d). Partial collapse can occur by failure of the beam with three hinges (c).

For the sidesway mode

$$M_A \theta_A + M_B \theta_B + M_C \theta_C + M_D \theta_D = P_2 d$$

The values of the hinge rotations can all be set equal to $\pm \theta$ and the moments to $\pm M_p$.

$$4M_p \theta = P_2 h \theta \quad 4M_p = P_2 h$$

For beam collapse

$$4M_p \theta = P_1 \frac{l}{2} \theta \quad 8M_p = P_1 l$$

For the combined mode

$$M_A \theta_A + M_E \theta_E + M_C \theta_C + M_D \theta_D = P_1 d_4 + P_2 d_3$$

If $\theta_A = -\theta$ it will be found that $\theta_E = 2\theta; \theta_C = -2\theta; \theta_D = \theta; d_3 = h\theta; d_4 = l\theta/2$, therefore

$$6M_p \theta = \frac{P_1 l}{2} \theta + P_2 h \theta$$

or

$$6M_p = \frac{P_1 l}{2} + P_2 h$$

To make sure that the correct mode of collapse has been chosen for a problem it is necessary to show that the value of M_p has not been exceeded at any point, in other words the yield condition is satisfied. For the combined mode it would only be necessary to show that $M_B \leqslant M_p$. Probably the easiest way of determining the bending moment at B is by virtual work. A suitable set of displacements would be those in figure 11.10b or c. Choosing the beam mechanism

$$\theta_B = \theta_C = -\theta \quad \theta_E = 2\theta$$
$$M_E = M_p \quad M_C = -M_p$$

therefore

$$-M_B\theta + 2M_p\theta + M_p\theta = \frac{P_1 l}{2}\theta$$

$$M_B = 3M_p - \frac{P_1 l}{2}$$

Substituting for M_p from $6M_p = (P_1 l/2) + P_2 h$ gives

$$M_B = \frac{P_2 h}{2} - \frac{P_1 l}{4}$$

The reader should satisfy himself that the sidesway displacement diagram gives an identical result.

11.7 Combination of mechanisms

For a particular framework and loading system it would be perfectly possible to sketch every mechanism. Using virtual-work principles a load factor could then be found for each mechanism. The lowest value of λ would then be the true collapse load-factor for the system. This approach could become somewhat tedious and it is always possible to overlook a particular mechanism which might be the vital one. The combination of mechanisms is a method that is more systematic in its approach.

If we consider the portal frame again (figure 11.10a) and introduce load factors
for beam collapse

$$\lambda_1 P_1 \frac{l}{2}\theta = 4M_p\theta \quad \lambda_1 = \frac{8M_p}{P_1 l}$$

for sidesway collapse

$$\lambda_2 P_2 h\theta = 4M_p\theta \quad \lambda_2 = \frac{4M_p}{P_2 h}$$

Now it is possible to combine the mechanisms of figures 11.10b and c into a further mechanism (d). This is formed by the elimination of the hinge at B which has different directions of rotation for the two basic mechanisms. The total external work equals the sum of the values of the external work for each of the basic mechanisms; and the total internal work equals the sum of the values of internal work. The value, however, will be too large since it contains the work for both a positive and a negative rotation of joint B, which have cancelled each other out. This means that we should have to subtract an amount of work $2M_p\theta$ from the total.

Adding the two equations

$$\lambda_3 P_1 \frac{l}{2}\theta + \lambda_3 P_2 h\theta = 8M_p\theta$$

$$-2M_p\theta$$

$$\lambda_3 P_1 \frac{l}{2}\theta + \lambda_3 P_2 h\theta = 6M_p\theta$$

$$\lambda_3 = \frac{6M_p}{P_1(l/2) + P_2 h}$$

To make the example more specific the following values have been assumed for the variables

$$l = 4 \text{ m}; \quad h = 3 \text{ m}; \quad P_1 = 40 \text{ kN}; \quad P_2 = 20 \text{ kN}; \quad M_p = 40 \text{ kNm}$$

$$\lambda_1 = \frac{8 \times 40}{40 \times 4} = 2 \quad \lambda_2 = \frac{4 \times 40}{20 \times 3} = 2\cdot67$$

So beam collapse would occur before sidesway collapse. For the combined mechanism

$$\lambda_3 = \frac{6 \times 40}{20 \times 7} = 1\cdot714$$

This is lower than either of the other load factors so the combined mode will be the one to occur. Before we finally accept this result we ought to make quite sure that the yield criterion is not invalidated. This would involve finding the bending moment at B. An expression for M_B was found at the end of the last section.

$$M_B = \left(\frac{P_2 h}{2} - \frac{P_1 l}{4} \right)\lambda_3 = (30 - 40)1\cdot714 = -17\cdot14 \text{ kNm}$$

The final bending-moment diagram would consist of a series of straight lines and, as the value of M_B is very much less than the value of M_p, the correct mode of collapse has been assumed.

It can be seen that if the method of combined mechanisms is to be used to advantage, then it is first necessary to determine the number of independent mechanisms. Let n be the number of bending moments that are required to specify completely the distribution of bending moments throughout a structure. If the structure is statically determinate the values of all the bending moments can be found. One method of doing this would be by using virtual work to form equations of equilibrium. Suitable mechanisms would have to be chosen. The number of mechanisms m would be equal to the number of equations of equilibrium which in turn would equal n. If the structure is redundant and d is the degree of statical indeterminacy of the structure, we could make use of the m mechanisms but the structure would not be determinate unless d bending moments were specified in some way or another. Thus we may write

$$n = m + d$$

This means that the numberof basic mechanisms m will be given by $(n - d)$.

We may easily check this for the portal. The value of n is 5. The complete bending-moment distribution would be known given the values of the bending moments at A, B, C, D, and E. The frame is three times statically indeterminate. Thus $n - d = 2$, giving two basic mechanisms.

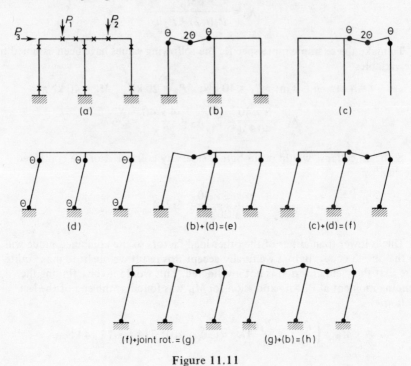

Figure 11.11

For the two-bay portal frame in figure 11.11a the complete distribution of bending moments would be known if the bending moments could be determined at all the points marked with a cross. This means that n is equal to 10. Note that we must know the three bending moments at the junction of the three beams. We are not allowed to state that if two of the bending moments are known then the third can be found by summing all the bending moments to zero. If this is done an equation of equilibrium has been used. The framework will be found to be indeterminate to the sixth degree: so there should be four independent mechanisms. Three of these are obvious, collapse of the left-hand beam, collapse of the right-hand beam, and sideway of the whole framework.

Each of these mechanisms corresponds to an equilibrium equation. The equilibrium equation for the tee junction of the three members is that the sum of the bending moments should be zero. This corresponds to a rotation of the joint. This is in fact the fourth mechanism. Taken by itself it would not mean a great deal but it can be combined with certain basic mechanisms to form other mechanisms. For the sake of simplicity it will be assumed that the value of the fully plastic moments for the beams and the columns have the same value.

Three of the basic mechanisms are shown in figures 11.11b, c and d. The combination of sidesway and beam collapse requires no further comment. These are shown at e and f. f can be combined with a joint rotation of the tee joint. This will mean that the two joints, one in a beam and the other in a column, can be replaced by a single joint in the other beam, resulting in a lower value of the load factor. A load factor would be obtained for every mechanism and the lowest value selected. All that would remain would be to check that the yield criterion had not been violated.

As a numerical example we shall consider the slightly more difficult case of a two-storey portal-type structure. The problem is presented in a slightly different manner. The dimensions of the members are given, and so also are the relative values of Z_p or M_p for each member. These are shown in circles on the line diagram in figure 11.12a. The structure has to be designed for a load factor of 2. This will mean that we would have to find the value of M_p for each member, from which the value of Z_p would be determined and a suitable cross-section could be selected for the members.

To specify the complete distribution of bending moments in the structure it would be necessary to know the bending moments at twelve points (b). The frame is statically indeterminate to the sixth degree so there should be six basic mechanisms. The basic frame is shown again at c with the loads increased by the value of the load factor.

Using virtual work the value of M_p has been calculated for the four basic mechanisms d to g; the working is shown under each diagram. We shall of course be looking for the highest value of M_p, instead of the smallest value of the load factor. The worst case so far is the collapse of the lower beam. Note that for collapse of the top beam the hinges form in the columns rather than at the ends of the beam.

e and g are combined together at h. The easiest way of finding the value of M_p is to sum the two virtual-work equations. This leads to a value of M_p of 25·7. We can, however, modify this collapse mechanism if joint B is allowed to rotate forming the mode at h'. This would reduce the internal work done by an amount $5\,M_p$. A value of 31·3 is obtained for M_p, the highest so far. The rest of the diagrams are self-explanatory, and the worst case is at k giving a value of M_p = 33·1.

It is quite possible to form other mechanisms, for example by combining d and f, but since the value of M_p for these mechanisms is on the low side it is unlikely that the combination will approach the maximum found.

It is essential to carry out a check to find out if the yield condition is satisfied before accepting the value of M_p.

Assuming a value of M_p = 33·1 kNm and attaching the correct signs M_{AB} = $-99·3$; M_{FE} = 99·3; M_G = 99·3; M_H = 132·4; M_{DE} = $-66·2$; M_{EB} = $-132·4$.

The original structure was six times redundant. We now have plastic hinges at six points, or six equations of condition. Unfortunately in spite of this

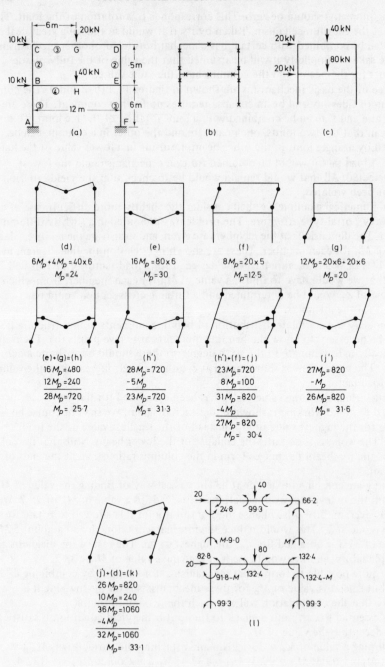

Figure 11.12

portal is still singly redundant. However, it should be possible to find bending moments in terms of one unknown. We shall proceed using virtual work.

For a displacement mode corresponding to collapse of beam CD

$$2M_G - M_{CB} - M_{DE} = \tfrac{1}{2} \times 40 \times 12$$
$$198 \cdot 6 - M_{CB} + 66 \cdot 2 = 240$$
$$M_{CB} = 24 \cdot 8$$

For collapse of beam BE

$$2M_H - M_{BE} - M_{EB} = 480$$
$$M_{BE} = -82 \cdot 8$$

All the bending moments so far determined have been entered on the line diagram in figure 11.12l where the frame is split into two parts. If we now assume that the bending moment M_{ED} is M, then M_{EF} can be written down.

Using virtual work with a sidesway displacement system for the top storey

$$M_{BC}(-\theta) + 24 \cdot 8\theta - 66 \cdot 2(-\theta) + M\theta = 100$$
$$M_{BC} = M - 9 \cdot 0$$

M_{BA} follows at once from equilibrium of joint B.

The limits are known for the value of the bending moment at any point. For example $-66 \cdot 2 \leqslant M_{ED} \leqslant 66 \cdot 2$.

The following inequalitites may be written down

$$-66 \cdot 2 \leqslant M - 9 \cdot 0 \leqslant 66 \cdot 2 \quad \text{or} \quad -57 \cdot 2 \leqslant M \leqslant 75 \cdot 2$$
$$-99 \cdot 3 \leqslant M - 91 \cdot 8 \leqslant 99 \cdot 3 \quad \text{or} \quad 7 \cdot 5 \leqslant M \leqslant 191 \cdot 1$$
$$-99 \cdot 3 \leqslant M - 132 \cdot 4 \leqslant 99 \cdot 3 \quad \text{or} \quad 33 \cdot 1 \leqslant M \leqslant 231 \cdot 7$$

Also

$$-66 \cdot 2 \leqslant M \leqslant 66 \cdot 2$$

The limits of M can now be written down

$$33 \cdot 1 \leqslant M \leqslant 66 \cdot 2 \tag{11.6}$$

This means that as long as M lies anywhere between these limits the value of M_p will not be exceeded at any point on the structure.

If a system of bending moments can be found that is in equilibrium with the applied loading and the yield condition is satisfied, and if a mechanism has also formed, then the solution is unique. Most certainly we have equilibrium as is shown by the set of moments in figure 11.12l, and we also have a mechanism. Any value of M can be chosen between 66·2 and 33·1 kNm and we have a unique solution. This means that the value of M_p is the correct one. If the incorrect mechanism had been chosen it would be found impossible to obtain a value of M to satisfy the equivalent expression to equation 11.6.

It is not always possible to write down the geometric relationships between the hinge angles in all cases. The pitched-roof portal in figure 11.6a is somewhat more complicated. Usually a sketch of the displacement diagram for a small displacement will provide the relationships quite easily. If the reader has a knowledge of instantaneous centres from applied mechanics, this can often provide a rapid solution.

We shall conclude with the case of a portal with a uniformly distributed load on the top member, together with a side load (figure 11.13). Values of M_p are shown in circles.

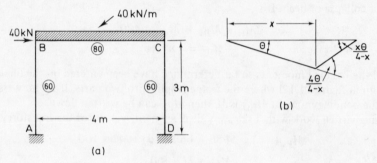

Figure 11.13

For a pure sidesway collapse

$$\lambda_1 \times 40 \times 3 = 4 \times 60$$

$$\lambda_1 = 2$$

For collapse of the beam with a hinge distant x from B (figure 11.13b) the hinges will form in the top of the column rather than at the ends of the beam.

$$\lambda_2 \times 40 \times \frac{4x\theta}{2} = 60\theta + \frac{60x\theta}{4-x} + \frac{320\theta}{4-x}$$

For a pure beam collapse the hinge would occur at the centre, that is $x = 2$ m

$$\lambda_2 = 1 \cdot 75$$

For a combined collapse it is not known where the hinge will form in the beam. Adding the solutions for sidesway and for a hinge distant x from the left-hand side and correcting for the positive and negative rotation at B

$$\lambda_3(120 + 80x) = 240 + 60 + \frac{60x}{4-x} + \frac{320}{4-x} - 2 \times 60$$

$$\lambda_3 = \frac{26 - 3x}{12 + 5x - 2x^2}$$

Setting $d\lambda_3/dx = 0$

$$3x^2 - 52x + 83 = 0$$

$$x = 1 \cdot 8 \text{ m}$$

The value of λ_3 is $1{\cdot}42$. The hinge forms to the left of the centre of BC and as this value of λ is lower than the other two, failure will occur by a combined mechanism. The error introduced by assuming a hinge in the middle of BC is not very large, and $\lambda = 1{\cdot}43$.

It will be noted that in all the problems covered in this chapter the loading has been of a proportional type, the loads in effect being steadily applied such that they all reach their maximum values just as the structure collapses. It should be pointed out that this does not occur in practice; it may well be that loads can vary between certain limits and may well be applied in a random fashion to a structure. The methods that we have been using for the determination of a load factor will not cater for this type of loading and the reader who is interested will need to read further in this subject.

Problems

11.1 A beam of uniform cross-section with plastic moment M_p and length $2l$ rests on simple supports at its ends and on a central prop. Equal concentrated loads are applied at the centre of each span. Find the value of the collapse load. Find also the collapse load if the ends had been built in.

11.2 A continuous beam of length l is of uniform section; it is simply supported at each end and at two intermediate points such that the end spans are of the same length. A uniformly distributed load of p per unit length is applied over the entire length. For the most economic plastic design locate the position of the intermediate supports.

11.3 A continuous beam ABC is freely supported at the ends A and C, and also at B, where $AB = L$ and $BC = 2L$. The maximum plastic moment in AB is M and in BC $1{\cdot}5M$. The beam carries a uniformly distributed load p over its entire length. Find using plastic theory the maximum value of p.

11.4 The beam shown at P 11.4 is to be designed according to the principles of plastic design. The length of the beam and the positions of the built-in end and the applied forces are prescribed, but the uniform plastic moment of resistance M_p

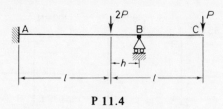

P 11.4

and the position of the simple support defined by the length h, are at the designer's disposal. Determine the values of M_p and h such that M_p shall be as small as possible, consistent with the safety of the structure.

11.5 A portal frame made from members of equal plastic moment of resistance $Pl/2$ is loaded as shown in P 11.5. Determine the greatest value of λ for which the portal is just safe and find the bending moments at B, Ċ, and E.

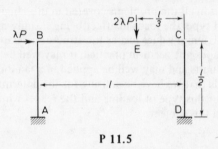

P 11.5

11.6 The symmetrical frame in P 11.6 has each leg with moment of resistance M_p and a beam of value $2M_p$. Determine how the collapse load and the collapse mechanism vary with k.

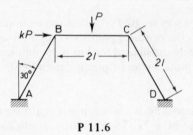

P 11.6

11.7 All the members of the frame P 11.7 are of the same cross-section, and the vertical loads are applied at the mid-points of the beams. If $M_p = 150$ kNm determine the collapse load-factor.

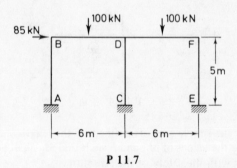

P 11.7

11.8 For the framework in P 11.8 the relative values of M_p for the members are shown in circles. Show that suitable values for the plastic moment of resistance of the various members are given by (2 or 3) x 1320/111.

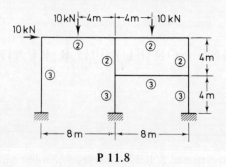

P 11.8

SUGGESTIONS FOR FURTHER READING

Chapter 6   J. H. Argyris and S. Kelsey. *Energy Theorems and Structural Analysis*, Butterworth, London, 1968

Chapter 7   E. Lightfoot. *Moment Distribution*, Spon, London, 1961.
J. Rygol. *Structural Analysis by Direct Moment Distribution*, Crosby Lockwood, St Albans, 1968

Chapter 8   R. K. Livesley. *Matrix Methods of Structural Analysis*, Pergamon, Oxford, 1964

Chapter 9   T. M. Charlton. *Model Analysis of Plane Structures*, Pergamon, Oxford, 1966
W. J. Larnach. *Influence Lines for Statically Indeterminate Plane Structures*, Macmillan, London, 1964

Chapter 10   F. Bleich. *Buckling Strength of Metal Structures*, McGraw-Hill, New York, 1952
M. R. Horne and W. Merchant. *The Stability of Frames*, Pergamon, Oxford, 1965
S. P. Timoshenko and J. M. Gere. *Theory of Elastic Stability*, McGraw-Hill, New York, 1961

Chapter 11   J. Baker and J. Heyman. *Plastic Design of Frames Vol. 1*, Cambridge University Press, 1969
J. Heyman. *Plastic Design of Frames Vol. 2*, Cambridge University Press, 1971
M. R. Horne. *Plastic Theory of Structures*, Nelson, London, 1971.

ANSWERS TO PROBLEMS

Chapter 1

1.1 $V_A = 19.5$ kN; $V_B = 25.5$ kN; $Q = -3.5$ and -6 kN; $M = 46$ and 0 kNm

1.2 $\frac{2}{3}p$; $\dfrac{l}{\sqrt{12}}$; $\dfrac{pl^2}{36\sqrt{3}}$

1.4 $14\frac{1}{8}$ kN; $22\frac{7}{8}$ kN; $30\frac{7}{8}$ kN; $22\frac{1}{8}$ kN

1.6 $Pl/8$

1.7 x $l/8$ $l/4$ $3l/4$ $7l/8$
y $0.208l$ $0.379l$ $0.472l$ $0.273l$

1.9 13.5 kNm; 4.5 kN

1.10 8.46 m; 192.5 kNm

1.11 22.5 kN; 186.7 kNm; 35.8 kN; 210.1 kNm

Chapter 2

2.1 AC: $-5/6$ kN; AD: $5\sqrt{2}/6$ kN; CD: $35/3$ kN

2.2 DE: 12.5 kN; AB: -12.5 kN; CA: $35/12$ kN; CD: $-35/12$ kN;
CB: $-35\sqrt{13}/24$; CE: $35\sqrt{13}/24$

2.3 $-P/4$

2.4 DG: $2.12P$; DE: $-0.72P$; AF: $-0.6P$

2.5 $V_D = \frac{7}{6}P$; $V_E = \frac{13}{12}P$; $V_F = \frac{7}{4}P$; $H_F = 2P$

2.6 AD = DF = FB = -40 kN; DC = 40 kN; CB = $20\sqrt{3}$ kN

2.7 41.2 kN

Chapter 3

3.1 $\dfrac{P}{\sigma}\exp\left(\dfrac{Px}{\sigma}\right)$; $\dfrac{\sigma l}{E}$

3.2 $\sigma_B = 30.8$ N/mm²; $\sigma_S = 64.7$ N/mm²; $\sigma_B = \sigma_S = 43.3$ N/mm²;
$\sigma_B = 12.5$ N/mm²; $\sigma_S = 108$ N/mm²

3.3 $\sigma_l = \dfrac{Pr}{2t}$; $\sigma_\theta = \dfrac{Pr}{t}$; $\sigma = \dfrac{Pr}{t}$; $\dfrac{4}{3}\pi r^3 \left[T(3\alpha - \beta) + P\left\{ \dfrac{r}{2tE}(1 - \nu) + \dfrac{1}{K} \right\} \right]$;

$$P + \frac{(3\alpha - \beta)TK\,2tE}{2tE + 3rK(1 - \nu)}$$

β is the coefficient of cubical expansion of water.

3.4 $169, -79$ N/mm^2; $(53{\cdot}3, 81{\cdot}5, -23{\cdot}5) \times 10^{-5}$

3.5 240° gauge

3.6 $5{\cdot}2$ kN/m

3.7 $\dfrac{E_2^{1/2} d}{E_1^{1/2} + E_2^{1/2}}$

3.8 $1{\cdot}65\,td^3$; $0{\cdot}56\,td^3$; $-\dfrac{1{\cdot}34Pl^3}{Etd^3}$

3.9 $\frac{5}{24}a$; $P = \dfrac{15\sqrt{3}R^3 \sigma_\mathrm{m}}{10R + 48z}$

3.10 $\frac{7}{5}$; $\frac{5}{3}\sigma_y\,bd^2$; $\dfrac{\sigma_y}{3}$; $\frac{5}{6}\sigma_y bd^2$

Chapter 4

4.1 $r = 56{\cdot}8$ mm; $0{\cdot}024$ rad/m

4.2 $\dfrac{\pi G^2 R^3}{10A}$; $\dfrac{lG}{20AR}$

4.4 $2{\cdot}05 d^3 \tau_y$

4.5 $0{\cdot}598T$; $0{\cdot}96r$

4.6 $7{\cdot}81$ N/mm^2

4.8 $2r$ from centre

4.10 $16/\pi^2$

4.11 $Ga^4/8$

4.12 $28{\cdot}4GtR^3$

Chapter 5

5.1 $-\dfrac{l^3}{48EI}(6pl + 5P)$; $\dfrac{l}{24}(6pl + 5P)$; $\dfrac{l^2}{12EI}(pl + P)$

5.2 $-\dfrac{pl^4}{1536EI}$; $\dfrac{pl^3}{256EI}$; centre $\dfrac{pl}{32}$

5.3 $\dfrac{23}{324}\dfrac{Pl^3}{EI}$

5.4 $\frac{1}{4}$ points; $-\frac{3pl^2}{64EI}(x^2)$, $x < \frac{l}{4}$; $-\frac{3pl^2}{64EI}\left(xl - x^2 - \frac{l^2}{8}\right)$, $\frac{l}{4} < x < \frac{l}{2}$

5.5 $-\frac{6pl^4}{Ebh_0^3}$

5.6 $8M/l^2$; $11\cdot6M/l^2$

5.7 12·7 per cent; 6 per cent; 3·5 per cent

Chapter 6
6.1 0

6.2 0·19 mm; 0·024 mm

6.3 0·324P

6.4 $20°\,18'$.

6.6 0·62P

6.7 $-\frac{3}{8}pd^2$

6.8 23·3 kN; 10·6 mm

6.9 4P/21; 10P/21

6.10 $\dfrac{Tl}{(EI\cos^2\alpha + GJ\sin^2\alpha)}$

6.11 $H = \dfrac{P}{6}$

6.12 0·46P; 0·11PR; 0·124$\dfrac{PR^3}{EI}$

Chapter 7
7.1 $M_{AB} = -27$ kNm; $M_{BC} = -68\cdot4$ kNm; $V_A = 29\cdot1$ kN; $V_B = 93\cdot7$ kN; $V_C = 77\cdot2$ kN

7.2 40 mm

7.3 $M_{BE} = -84$ kNm

7.4 13·3 N

7.5 $M_{AE} = -0\cdot012Pl$; $M_{BF} = -1\cdot069Pl$; $M_{CG} = -1\cdot962Pl$

7.6 $M_{AB} = 19$ kNm; $M_{BA} = 47\cdot5$ kNm; $M_{CD} = -41\cdot2$ kNm; $M_{DC} = -25\cdot3$ kNm

7.7 0·301P; 0·463P; 0·236P

7.8 $\dfrac{5}{72}\dfrac{Ml}{EI}$

7.9 $M_{AB} = 200$ kNm; $M_{BA} = 133$ kNm; $M_{CD} = M_{DC} = 67$ kNm

7.10 $M_{AB} = -11\cdot5$ kNm; $M_{BC} = -20\cdot8$ kNm; $M_{CB} = -10\cdot4$ kNm

Chapter 8

8.1 $0.577 \dfrac{Pl}{AE}$ up; $0.268 \dfrac{Pl}{AE}$ to left

8.2 $4.43 \dfrac{l}{AE}$ to right; $18.1 \dfrac{l}{AE}$ down; $F_{BD} = 4.45$ kN

8.6 $0.048 \dfrac{Pl^2}{EI}$

8.7 $v = \dfrac{Pl}{2AE}$; $\theta = \dfrac{Pl^2}{64EI}$

Chapter 9

9.1 $\dfrac{1}{l^2}(2x^2 l - x^3 - l^2 x)$

9.2 $\dfrac{\alpha l}{4}(1 - \alpha^2)$; $\alpha = \dfrac{x}{l}, \alpha \not> 1$; $M_{D\,max} = \frac{5}{64}l$

9.3 $\dfrac{Px^2(3l - x)}{2l^3}$

9.4   See Table 9.1

9.5 $\dfrac{5l}{8h}(\alpha - 2\alpha^3 + \alpha^4)$, $\alpha = x/l$

Chapter 10

10.2 $P = 0.0706kl$

10.3 $\tan\dfrac{\alpha l}{2} = \dfrac{\alpha l}{2 + \alpha^2 l^2}$

10.8 $P = \dfrac{\pi^2 EI}{l^2}$; $F = \dfrac{2\pi^2 EI v_0}{l^3[(2/\pi)\tanh(\pi/2) - 1]}$

10.10 $\dfrac{\pi^2 EI}{l^2} + \dfrac{kl^2}{\pi^2}$

10.11 13.9 kN

Chapter 11

11.2   End span $0.315l$

11.3 $P = \dfrac{3.94 M_p}{l^2}$

11.4 $h = \dfrac{\sqrt{5} - 1}{2}l$; $M_p = \dfrac{(3 - \sqrt{5})}{2}Pl$

11.5 $\lambda = 2$; $M_B = Pl/6$

11.6 Beam $P = \dfrac{6M_p}{l} \left(k < \dfrac{1}{3\sqrt{3}} \right)$; Sidesway $P = \dfrac{2\sqrt{3}M_p}{kl} \left(k > \dfrac{\sqrt{3}}{2} \right)$;

Combined $P = \dfrac{14M_p}{(2 + k\sqrt{3})l} \left(\dfrac{1}{3\sqrt{3}} < k < \dfrac{\sqrt{3}}{2} \right)$

11.7 1·865

INDEX